# THE ARAL SEA BASIN

This book offers the first multidisciplinary overview of water resources issues and management in the Aral Sea Basin, covering both the Amu Darya and Syr Darya River Basins.

The two main rivers of Amu Darya and Syr Darya and their tributaries comprise the Aral Sea Basin area and are the lifeline for about 70 million inhabitants in Central Asia. Written by regional and international experts, this book critically examines the current state, trends and future of water resources management and development in this major part of the Central Asia region. It brings together insights on the history of water management in the region, surface and groundwater assessment, issues of transboundary water management and environmental degradation and restoration, and an overview of the importance of water for the key economic sectors and overall socio-economic development of Central Asian countries, as well as of hydro politics in the region. The book also focusses on the future of water sector development in the Basin, including a review of local and international actors, as well as an analysis of the current status and progress towards the Sustainable Development Goals by Basin countries.

The book will be essential reading for those interested in sea basin management, environmental policy in Central Asia and water resource management more widely. It will also act as a reference source for decision-makers in state agencies, as well as a background source of information for NGOs.

**Stefanos Xenarios** is Associate Professor in Nazarbayev University (NU), Graduate School of Public Policy, based in Nur-Sultan (Astana), Kazakhstan.

**Dietrich Schmidt-Vogt** is Honorary Professor at the Chair of Silviculture, Faculty of Environment and Natural Resources, Freiburg University, Germany.

**Manzoor Qadir** is Assistant Director in United Nations University Institute for Water, Environment and Health (UNU-INWEH), based in Hamilton, Ontario, Canada.

**Barbara Janusz-Pawletta** is UNESCO Chair Holder in Water Management at the Central Asia Kazakh-German University based in Almaty, Kazakhstan.

**Iskandar Abdullaev** is the former Executive Director of the Regional Environmental Centre Central Asia (CAREC), based in Almaty, Kazakhstan, and the current Deputy Director of the CAREC Institute, Urumchi, China.

# EARTHSCAN SERIES ON MAJOR RIVER BASINS OF THE WORLD

*Series Editor: Vladimir Smakhtin*

Large river basins are dynamic and complex entities. Defined by hydrological boundaries, they are nearly always shared by more than one country. Encompassing a diverse range of landscapes with often huge temporal and spatial variability of resources, they are put to different and often conflicting uses, and managed by a range of institutions and organisations. While an intrinsic part of Nature, many have been extensively engineered and used by people, often with adverse consequences. Each major river basin has its own development trajectory and often fascinating history. Bringing together multidisciplinary teams of experts, this series explores these complex issues, identifies knowledge gaps and examines potential development pathways towards greater sustainability.

**The Nile River Basin**
Water, Agriculture, Governance and Livelihoods
*Edited by Seleshi Bekele Awulachew, Vladimir Smakhtin, David Molden, Don Peden*

**The Volta River Basin**
Water for Food, Economic Growth and Environment
*Edited by Timothy O. Williams, Marloes Mul, Charles Biney and Vladimir Smakhtin*

**The Ganges River Basin**
Status and Challenges in Water, Environment and Livelihoods
*Edited by Luna Bharati, Bharat R. Sharma and Vladimir Smakhtin*

**The Zambezi River Basin**
Pathways for Sustainable Development
*Edited by Jonathan Lautze, Xueliang Cai, Everisto Mapedza, Marcus Wishart, Zebediah Phiri*

**The Aral Sea Basin**
Water for Sustainable Development in Central Asia
*Edited by Stefanos Xenarios, Dietrich Schmidt-Vogt, Manzoor Qadir, Barbara Janusz-Pawletta and Iskandar Abdullaev*

https://www.routledge.com/series/ECMRBW

# THE ARAL SEA BASIN

## Water for Sustainable Development in Central Asia

*Edited by*
*Stefanos Xenarios, Dietrich Schmidt-Vogt,*
*Manzoor Qadir, Barbara Janusz-Pawletta*
*and Iskandar Abdullaev*

LONDON AND NEW YORK

First published 2020
by Routledge
2 Park Square, Milton Park, Abingdon, Oxon OX14 4RN

and by Routledge
605 Third Avenue, New York, NY 10017

First issued in paperback 2021

*Routledge is an imprint of the Taylor & Francis Group, an informa business*

*British Library Cataloguing-in-Publication Data*
A catalogue record for this book is available from the British Library

*Library of Congress Cataloging-in-Publication Data*
A catalog record has been requested for this book

Typeset in Bembo
by Swales & Willis Ltd, Exeter, Devon, UK

*The views expressed herein are those of the author(s) and do not necessarily reflect the views of the United Nations University*

ISBN 13: 978-0-367-77702-9 (pbk)
ISBN 13: 978-1-138-34888-2 (hbk)

# CONTENTS

Contents

# CHAPTER PHOTOS

1 Uzbek woman in the outskirts of Samarkand, Uzbekistan
Photo: MehmetO/Shutterstock.com

2 Sunset on Syr Darya River, Kazakhstan
Photo: Maxim Petrichuk/Shutterstock.com

3 Khisorak Reservoir
Photo: Shovkat Khodjaev, IWMI-Central Asia

4 Mountainous landscape in Kyrgyzstan
Photo: Abror Gafurov

5 Koksaray reservoirs on the Syr Darya River, Kazakhstan
Photo: Schankz/Shutterstock.com

6 Rustic boats at a ship graveyard in a desert around Moynaq in the Aral Sea, Uzbekistan
Photo: Milosz Maslanka/Shutterstock.com

7 Irrigated cotton plantations near Jalal Abad, Kyrgyzstan
Photo: Steffen Entenmann

8 Mountainous landscape in Afghanistan
Photo: Maximum Exposure PR/Shutterstock.com

9 The Syr Darya River in the deserts of Kazakhstan
Photo: Karakol/Shutterstock.com

10 The Naryn River in the Tien Shan Mountains in Kyrgyzstan
Photo: Chubykin Arkady/Shutterstock.com

11 The Amu Darya River before the confluence of the Aral Sea
Photo: Boris Rezvantsev/Shutterstock.com

12 The Vaksh River in Tajikistan
Photo: Dinara R. Ziganshina

13 A girl with national costume in Namangan, Uzbekistan
Photo: Vladimir Goncharenko/Shutterstock.com

# NOTES ON AUTHORS

**Iskandar Abdullaev** was Executive Director of the Regional Environmental Centre Central Asia (CAREC), based in Almaty, Kazakhstan, from 2013 and is now Deputy Director of the CAREC Institute, Urumchi, China. Prior to joining CAREC, Iskandar Abdullaev worked as a Regional Advisor for the German Society for International Cooperation (GIZ) Program 'Transboundary Water Management in Central Asia' from 2009. From 2007 to 2009, he worked as a Senior Researcher at the Centre for Development Research (ZEF) of the University of Bonn, Germany, and conducted research in Central Asia and Afghanistan. From 2001 to 2007, he worked as a regional researcher at the headquarters of the International Water Management Institute (IWMI) in Sri Lanka. The regional scope of his research includes Central Asia, Iran, Pakistan, Thailand and Sri Lanka. He started his professional carrier in 1991 at the Tashkent Institute of Irrigation and Melioration, Uzbekistan, and worked there until 1999 as a lecturer, head of department and assistant professor.

**Mohamed Ahmed** is Assistant Professor at Texas A&M University, Corpus Christi, Texas. In his research, he applies integrated (geophysics, remote sensing, hydrogeology, modelling, GIS) approaches to investigate a wide range of geophysical, geological, hydrological and environmental problems. Before attaining a Ph.D. in Geosciences at Western Michigan University in 2012, he received his M.Sc. (2009) and B.Sc. (2004) degrees in Applied Geophysics from Suez Canal University, Egypt. He has more than ten years of research and teaching experience, and a notable publication record. He has collaborated in research projects funded, among others, by NASA, NATO and the US State Department. He has presented papers and served as a discussant on several national and international panels, at conferences and symposia.

**Aigul Akylbekova** is the Academic Secretary of the Ahmedsafin Institute of Hydrogeology and Environmental Geoscience, Almaty, Kazakhstan. She participated as a research manager in developing the atlas of potential renewable energy sources of the Republic of Kazakhstan, and also in programs and projects implementing the national strategy of water resources for ensuring the water security of Kazakhstan. Her previous positions include President of the Association of Women of Central Asia and Afghanistan on Water Resources, chief engineer in a project on biodiversity conservation in protected natural territories of Kazakhstan and GIS specialist in the demarcation and delimitation of the borders of Kazakhstan. She currently studies changes in the chemical composition of groundwater in the context of climate change and their impact on human health. Aigul is a geodetic engineer, having graduated from the Kazakh Academy of Architecture and Civil Engineering based in Almaty, Kazakhstan.

**Iroda Amirova** is a doctoral researcher at the Leibniz Institute of Agricultural Development in Transition Economies (IAMO), Halle (Saale), Germany. The focus of her research is on institutional analysis of irrigation water management in Kazakhstan and Uzbekistan. She obtained her M.Sc. in Agricultural and Food Economics from the University of Bonn, Germany, where she majored in Agricultural Policy and Development. Before joining IAMO she worked as a consultant at the Central Asia office of the International Water Management Institute (IWMI). In her undergraduate years she studied Economics at the Westminster International University in Tashkent, Uzbekistan.

**Oyture Anarbekov** is a water resources management specialist with an academic background in water governance, economics and institutional development. He is Researcher at the IWMI Central Asia Office in Tashkent, Uzbekistan, and has more than 13 years of experience in water resources management and conservation, agricultural production and food security issues. Oyture has contributed to the Institute's research on smallholder agricultural water management, on irrigation service pricing issues at plot level, the full-cost recovery principle at the main canal level and on institutional aspects of water users' associations. He is currently leading a research project on river basins management as well as water governance in selected pilot rivers in Central Asia. He has previously worked for the Tashkent State University of Economics as Deputy Dean on Information Technologies and Science. He holds an M.Sc. in Agricultural Economics, an MBA from the Cyprus International Institute of Management and a CAS from Bern University of Applied Science, and is currently a Ph.D. candidate at the Centre of Development and Environment (CDE), University of Bern, Switzerland.

**Jamal Annagylyjova** is Program Officer at the Secretariat of the United Nations Convention to Combat Desertification, where she is responsible for Central and Eastern European countries, as well as for the Central Asian region. Previously, she worked as a monitoring and evaluation expert for the Central Asian Countries Initiative for Land Management (CACILM), the largest land management programme for Central Asia. In her early career, she worked as a researcher at the National Institute of Deserts, Flora and Fauna of the Ministry of Nature Protection of Turkmenistan. She holds a Ph.D. in Geography from the University of Cologne, Germany.

**Dagmar Balla** is Senior Scientist at the Leibniz Centre for Agricultural Landscape Research (ZALF e.V.), Müncheberg, Germany. Her research is focused on hydrology, especially on the water quality of small water bodies including wetlands, buffer zones, natural ponds and running water systems. She also has long-term experience in the restoration of peatlands, which are an essential component of the agricultural landscape of Western Europe. Her research interests are focussed on water quality aspects, the assessment of ecosystem services and, in particular, purification ability with respect to both nutrients and contaminants. Currently, Dagmar Balla is manager of the BioWat project, which tests sustainable solutions for water desalination and purification under semi-arid and arid conditions in Central Asia. She has also taught the Water Management module of the Integrated Natural Resources Management master's course at the Humboldt University, Berlin, Germany.

**Martina Barandun** is a glaciologist and Senior Researcher at the Department of Geosciences, University of Fribourg, Switzerland, alongside work as postdoc at the Paul Scherrer Institute (PSI). She studied geography and obtained her Ph.D. in Geography at the University of Fribourg. Her research on glacier mass changes in Central Asia includes field observations and modelling as well as the monitoring of different cryosphere components using remote sensing techniques. In her new position at the PSI she works on the spectral characteristics and chemical composition of light-absorbing impurities on glacier surfaces and their effect on glacier mass changes. In addition, Martina has long-standing experience in organising and leading field measurements, as well as in conducting practical and theoretical training courses for young scientists and students of glaciology in Central Asia.

**Aziza Baubekova** is Teaching Fellow in Biology at the Center for Preparatory Studies, Nazarbayev University, Nur-Sultan, Kazakhstan. Aziza holds an M.Phil. degree in Environmental Sciences and Policy from the Central European University (CEU) and a B.Sc. in Environmental Sciences from the University of Nottingham. She is a water expert focussing on the sustainable use of transboundary watercourses in Central Asia. Her research is interdisciplinary and focusses on water security, the water–food–energy nexus, integrated modelling and the Sustainable Developments Goals (SDGs) implementation, with a particular focus on SDG 6. Aziza currently conducts research in the Ili-Balkhash Basin shared between China and Kazakhstan. Her project experience ranges from hydrological modelling with application of remote sensing and GIS techniques to integrated assessment of several transboundary basins in Central Asia.

**Jelle Beekma** is a water resources specialist and consultant with more than 30 years of experience. Beekma coordinated and led various water resources management projects in Asia, Africa and South America. Beekma has successfully supported institutional strengthening in the water sector in various countries, including in Central Asia. He has led basin management, irrigation modernisation, water resources development and river restoration projects. He has also successfully developed and implemented methods to forecast river flows and flood risks based on satellite observations of snow cover. He has recently become actively involved in source-to-sea approaches and in the reduction of (plastic) pollution of oceans. He has authored various scientific papers, technical reports, proceedings and lay-audience publications on water issues and has presented his water management findings to governments, communities and peer-scientists worldwide.

**Maksud Bekchanov** is Senior Researcher in Agricultural Economy at the Centre for Development Research (ZEF), Bonn University, Germany. After successful completion at ZEF of his Ph.D. thesis on efficient water allocation options in Central Asia, he has been working for several international development organisations, including the International Food Policy Research Institute (IFPRI), the International Water Management Institute (IWMI), the International Centre for Bio-saline Agriculture (ICBA) and ZEF. As a post-doctoral researcher, he has extensively studied various environmental and resource management issues, including the prevention of deforestation, water–energy–food interlinkages, sustainable energy options, conservation agriculture, waste recycling and water quality management. He has over 15 peer-reviewed papers in international journals and over 25 publications as book chapters or discussion papers.

**Tobias Bolch** is a lecturer and leader of the Mountain Cryosphere research group at the School of Geography and Sustainable Development, University of St. Andrews, Scotland. His main research focuses on investigating the impact of climate change on glaciers and glacial hazards in various mountain ranges of the world, with a regional focus on Central Asia and the Himalaya, where he has been working for more than 15 years. He uses remote sensing and photogrammetry to study glacier characteristics and their changes in area, volume and velocity, with special emphasis on debris-covered glaciers. He also studies other components of the cryosphere, like snow cover, permafrost and permafrost bodies. His remote sensing work is complemented by regular field investigations where he and his group also measure climate parameters or perform geophysical measurements. Through his various activities and visits, Tobias Bolch has excellent contacts to international and local experts and has become a well-recognised researcher of the cryosphere in High Asia.

**Manon Cassara** is a water resources management consultant, with a specialisation in international waters and a focus on institutional, basin planning and water information management issues as well as on stakeholders' engagement strategies. Building on her diverse international experience with projects in Egypt, the Caucasus, Vietnam and Senegal, she has contributed to several institutional assessments, water reform processes and inter-state dialogue initiatives. Manon is a graduate in international law, specialising in international water issues. Since 2013, she has been working for the World Bank in the Central Asia region, supporting the implementation of the Central Asia Energy and Water Development Program (CAEWDP) and diverse water and environment activities in Kazakhstan, Kyrgyzstan, Tajikistan, Turkmenistan and Uzbekistan. She is also supporting the development of the Swiss Development Cooperation (SDC) water resources regional program for the Central Asia region.

**Dietrich Darr** has been Professor of Agribusiness at Rhine-Waal University of Applied Sciences, Kleve, Germany since 2012. He holds an M.Sc. degree in Tropical Forestry and received his doctorate from TU Dresden with a dissertation on the diffusion of agroforestry innovations in East Africa in 2008. Before his appointment at Rhine-Waal University he had been working for a leading international management consultancy. His primary research interests include the analysis of smallholders' land use-related decision-making in the Global South; the role of underutilised wild fruits and other non-timber forest resources in local livelihoods and value chains; as well as upgrading and innovation management in the processing industry. Dietrich has been leading and (co-)supervising a number of research, Ph.D. and M.Sc. thesis projects in Central Asia since 2014. In addition to his teaching and research activities, he provides consulting services to companies and organisations in the agro-food sector and to other clients.

**Kakhramon Djumaboev** is Researcher at the International Water Management Institute (IWMI). He has 15 years of experience in water resources management, specifically with issues of water resources management in a transboundary context, and of the water–energy–food nexus in lift-irrigated areas, where irrigation water use is linked with high consumption of energy. He has contributed to more than 30 research and action projects in Central Asia, and has authored and co-authored more than 20 research papers in international and national journals. Currently, he is leading a project on mitigating the competition for water in the Amu Darya River Basin by improving water and energy use efficiency.

**Andrei Dörre** is a geographer, researcher and lecturer. He is currently Visiting Professor at the Department of Development Studies, University of Vienna, Austria, and is affiliated with the Centre for Development Studies of the Freie Universitaet Berlin (Germany). His research interest focuses on social change and human–environmental interactions in Central Asia. He has dealt in particular with different aspects of change in post-Soviet transformation societies, including research on pastoral practices in Kyrgyzstan, resource management, irrigation agriculture, food security and development in the Pamirs of Tajikistan, as well as the interrelationship between international intervention, security promotion and development activities in Afghanistan.

**Joel Fiddes** is a geoscientist with over ten years of interdisciplinary experience working across academic institutions, international organisations, multilateral donors and UN bodies. His main research interests are high-mountain processes related to natural hazards, water resources and resulting impacts upon communities. He uses and develops geoscientific models and other data science tools in combination with field campaigns to understand these processes and improve mapping and forecasting methods. Fiddes has worked in numerous countries of high-mountain Asia, including Tajikistan, Kyrgyzstan, Afghanistan, Nepal and India. He obtained his Ph.D. from the Glaciology group at the University of Zurich, Switzerland, and currently works as a research scientist in Davos at the Swiss Federal Institute for Snow and Avalanche Research.

**Abror Gafurov** is a hydrologist with a focus on water cycles in Central Asia. His recent research is in the fields of surface and groundwater interaction, hydrological forecasting, the impact of climate change on the water cycle and climate change adaptation in Central Asia. Moreover, he has strong experience in using remotely sensed snow cover data for hydrological studies in Central Asia. Abror Gafurov obtained his M.Sc. in water resources engineering and management as well as his Ph.D. in Hydrology at the University of Stuttgart, Germany. In his professional career he has worked for the Asian Development Bank (ADB)-funded Amu-Zang Irrigation Rehabilitation Project at the Ministry of Agriculture and Water Resources (MAWR) in Uzbekistan. Since 2009, he has been a researcher at the German Research Centre for Geosciences and has authored more than 50 scientific papers, technical reports and proceedings on water issues. Since 2017, he has been coordinator of the Central Asian Water project and Lecturer at the Humboldt University, Berlin.

**Zafar Gafurov** is a specialist on remote sensing and GIS, as well as on climate change and resource management in Central Asia. He has obtained certificates on climate change in international training courses and a degree in Photogrammetry and Geoinformatics at Stuttgart University, Germany. His professional interests include the analysis of remote sensing datasets, mapping and natural resources management, as well as climate change adaptation strategies for the sustainable management of land and water resources in Central Asia. Zafar is currently working at the International Water Management Institute (IWMI) Central Asia office in Tashkent as a national researcher on RS/GIS and leads a research project on transboundary water resource management under climate change in the region.

**Sara Giska** is a Fulbright Student Researcher from the USA, at the UNESCO Chair for Water Management in Central Asia at Kazakh-German University (DKU), Almaty, Kazakhstan, where she is focusing on the Ile and Irtysh transboundary watercourses. Sara is pursuing a master's degree in Security Studies at Georgetown University; she intends to complete a thesis on the natural resource management–international security nexus in the post-Soviet space. Sara received her B.A. in International Relations and Applied Linguistics from Miami University (2016) and is a native of Columbus, Ohio. Sara is a former Boren Scholar to Brazil (2014–2015), where she conducted research on comparative regional economic development.

**Ali Torabi Haghighi** is Lecturer and leader of the Water-Land-Energy resources development research group at the Water, Energy and Environmental Research unit at the University of Oulu, Finland. Ali holds a Ph.D. on Water Resources and Geo-environmental Engineering from the University of Oulu. His previous positions include Visiting Researcher at the University of California, Irvine, USA, as manager of several dam and irrigation projects in Fars Regional Water Authority, Shiraz, Iran (1998–2009), and as Lecturer at the Islamic Azad University, Estahban, Iran (2002–2009). His research interests focus on water resources and more specifically on the study of lake and river flow regimes, large-scale hydrology, cold-climate hydrology, environmental flow assessment, the political and socio-economic aspects of water resource systems, sustainable water resources development, extreme hydrological events, river engineering, remote sensing in water resources, dam and hydropower operation and geotechnical engineering.

**Ahmad Hamidov** is Post-Doctoral Researcher at the Leibniz Centre for Agricultural Landscape Research (ZALF) in Müncheberg, Germany. He obtained his degree on the economics of water resources at the Tashkent Institute of Irrigation and Agricultural Mechanization Engineers, Uzbekistan, in 2004. He further acquired his Ph.D. on agricultural economics from the Humboldt University, Berlin in 2015. His research interests include the sustainability of land and water resources, the water–energy–food nexus, community-based natural resource management, irrigation governance, institutions, climate change and social–ecological–technical systems analysis. Ahmad is currently coordinating the BioWat project Resources Management in the Salinized and Drought Stress-Endangered Irrigated Areas of Central Asia for Adapting to Climate Change, funded by the German Federal Ministry of Education and Research (BMBF). He also teaches the master's course Advanced Empirical Methodology for Social-Ecological Systems Analysis at the Humboldt University, Berlin, Germany.

**Bunyod Holmatov** is a Ph.D. candidate at the University of Twente in the Netherlands. His current research is focused on quantifying the land–water–carbon footprints of different sources of energy. Prior to starting his Ph.D. programme, he worked as a consultant at the IWMI Central Asia office, and before that as a research fellow at Yale University. During his studies he also conducted internships at UNDP Uzbekistan, the Latvian NGO *homo ecos* and the Alliance for Global Water Adaptation. In these capacities, he was involved in a number of research projects focused on transboundary water management, water security, water governance, water quality and climate change. Bunyod has authored and contributed to a number of peer-reviewed articles and serves as a reviewer for water sector journals. He is multilingual and has lived and travelled extensively in Asia, Africa, Europe and North America. Bunyod holds a Master of Environmental Management degree from the Yale School of Forestry and Environmental Studies. He is originally from Kokand, Uzbekistan.

**Martin Hoelzle** is Professor of Physical Geography at the University of Fribourg, Switzerland. He is a member of the Swiss Advisory Body on Climate Change (OcCC) of the Federal Department of the Environment, Transport, Energy and Communications (DETEC), scientific advisor at the World Glacier Monitoring Service (WGMS) and former President of the Swiss Snow, Ice and Permafrost Society, as well as Vice-President of the Swiss Cryospheric Commission. He has worked for 30 years in the development of modern, interdisciplinary cryospheric monitoring strategies and has a broad knowledge about complex process chains related to interactions between climate and the mountain environment.

**Barbara Janusz-Pawletta** holds a Ph.D. from the Free University in Berlin, Germany. In 2011, she established a regional master's programme on Integrated Water Resources Management in Central Asia at the German-Kazakh University, Almaty, Kazakhstan. Since 2016 she has been the UNESCO Chairholder for water management in Central Asia and implements capacity-building activities in the region. Barbara is also the Co-Editor-in-Chief of the *Central Asian Journal of Water Research*, which is an open-access, peer-reviewed e-journal dedicated to all aspects of water management in Central Asia. She is a member of both the Technical Committee of the Global Water Partnership and the Committee on the Role of International Law in Sustainable Natural Resource Management for Development of the International Law Association. Today, Janusz-Pawletta holds a leading position at the Kazakh-German University, as Vice-Rector on International Cooperation.

**Saghit Ibatullin** is Professor and Corresponding Member of the Kazakh Academy of Agricultural Sciences, and also Director of the International Teaching Center for Dam Safety in Taraz, Kazakhstan. He has an M.Sc. in Hydraulic Engineering from the Zhambyl Hydromeliorative and Construction Institute and a Ph.D. in Hydraulic Engineering from the Moscow Hydromeliorative Institute. After more than 25 years of working at the Taraz State University in different positions, Saghit Ibatullin led the Kazakh Research Institute of Water Economy in Taraz from 2004 to 2008. In 2008, he was appointed Chairman of the Executive Committee of the International Fund for Saving the Aral Sea for three years by the president of Kazakhstan. He served from 2013 until 2016 as Vice-Chair of the Implementation Committee under the UNECE Convention on the Protection and Use of Transboundary Watercourses and International Lakes. He is the author of more than 100 scientific articles, books and methodical recommendations on issues related to water management and the Aral Sea.

**Anvar Kamolidinov** is a specialist on IWRM, irrigation, institutional improvement and water law, mainly with experience in Central Asia. He worked for many years as a lecturer and researcher at the Tajik Agrarian University and later became Head of Department at the former Ministry of Land Reclamation and Water Resources of Tajikistan, participating in activities of regional institutions of transboundary water management, such as IFAS and ICWC, and in negotiations on the development of inter-state draft agreements on water- and hydroenergy use. He was included as senior water resources expert in the national expert group for the development of the Rogun HPP design under World Bank support, and participated in the development of proposals for water-sector reform in Tajikistan. He is currently working as a consultant and Deputy Project Manager of the EU Zarafshan River Basin Project, which supports water-sector reform. Anvar obtained the degree of technical candidate (equivalent to Ph.D.) at Moscow in 1988, has written more than 50 articles and technical reports and is responsible for five inventions in the area of irrigation technologies.

**Ulan Kasymov** is Post-Doctoral Researcher at the Resource Economics Group, Humboldt University, Berlin, Germany. His research interests include environmental governance, institutional economics and rural development. Ulan is currently coordinating the 'CLIENT II-Definition Project – SmartCaDEF: How Smart Can Climate and Soil-Sensitive Resource Use in Central Asia Be? Innovative Technologies for Water and Biomass Reuse', funded by the German Federal Ministry of Education and Research (BMBF). He is also involved in the CARB-ASIA project Development of Methods to Assess Carbon Stocks and Improve Climate Reporting of Agricultural Ecosystems in Central Asia, financed by the German Federal Ministry for the Environment, Nature Conservation and Nuclear Safety. He also teaches the Advanced Empirical Methodology for Social-Ecological Systems Analysis master's course at the Humboldt University, Berlin.

**Jusipbek Kazbekov** is Water Management Specialist at CAREC, Uzbekistan. He has more than 15 years of academic, research and development experience in water management in Central Asia. He holds an M.Sc. from the Colorado State University and a Ph.D. from the Tashkent Irrigation and Melioration Institute (TIIM). Jusipbek has teaching experience with TIIM and research experience with the International Water Management Institute (IWMI). He has published more than 30 peer-reviewed publications on social and technical aspects of water management in Central Asia.

**Bjørn Kløve** is Professor in Water Engineering and Director of the Water, Energy and Environmental Engineering Research Unit as well as of the Kvantum Research Institute at the University of Oulu, Finland. He is an expert on hydrology, water resources and environmental engineering, with broad research and project experience in Finland, Norway and developing regions such as the Middle East, Central Asia and Indonesia. He has an M.Sc. in Civil Engineering from Helsinki University of Technology, Finland, and a Ph.D. in Water Resources Engineering from Lund University, Sweden. He has coordinated several large-scale European projects, supervised more than 20 Ph.D. candidates, has led a national doctoral school on Land and Water Resources Research and has published more than 150 peer-reviewed journal papers. He is Editor-in-Chief of the *Hydrology Research* journal of the International Water Association.

**Marton Krasznai** is Associate Professor and Scientific Director of the Center for Central Asia Research, Corvinus University, Budapest. He has 12 years of experience supporting regional cooperation in Central Asia as inter-regional adviser of the United Nations Economic Commission for Europe. He worked ten years in the Organization of Security and Cooperation in Europe (OSCE), first as Head of Delegation of Hungary and then as Director of the Conflict Prevention Center. He is a graduate of the Moscow Institute of International Relations, with 18 years of diplomatic experience in the Hungarian foreign service.

**Anastasia Kvasha** is a researcher in the field of earth observation applications, focusing particularly on community resilience and water security. She was a civil servant in the Moscow City government, and was also engaged in activities of international (UNOOSA) and regional (Asian Disaster Reduction Center) organisations dealing with the promotion of space-based information and geospatial technologies for sustainable economic and social development. Anastasia is Project Manager and Research Assistant at the Environmental Systems Laboratory at Central European University (CEU), Hungary, organising workshops on the use of ICTs for sustainable water management and disaster risk reduction. She holds a specialist degree in Geoecology from the Lomonosov Moscow State University, an M.Sc. in Environmental Sciences and Policy from the CEU and is currently pursuing a Ph.D. at CEU exploring challenges in the diffusion and application of earth observations in building community resilience.

**Murodbek Laldjebaev** is Assistant Professor in Earth and Environmental Sciences at the University of Central Asia's (UCA) School of Arts and Sciences, and a Research Fellow at the UCA's Mountain Societies Research Institute. Prior to joining UCA, he was a post-doctoral fellow at Cornell University, where he had obtained his Ph.D. in Natural Resources. He was a research associate on undergraduate core-curriculum development for UCA and previously also headed curriculum development and programme evaluation for the Institute for Professional Development of Teachers in Khorog, Tajikistan. Murodbek consulted the World Bank on pre-service education and worked as Senior Specialist with the Ministry of Economic Development and Trade of Tajikistan. He holds an M.Sc. in Public Policy from the Lee Kuan Yew School of Public Policy, National University of Singapore. His research interests include energy security, energy poverty, energy sovereignty, water resources management, food sovereignty and the energy–water–food nexus.

**Shreedhar Maskey** is Associate Professor of Hydrology and Water Resources at the IHE Delft Institute for Water Education in the Netherlands. He has over 25 years of experience in graduate and postgraduate-level teaching, research supervision, civil engineering design and development of flood and drought forecasting models integrating remote sensing-based data products. His research focuses on the development of catchment-river models for applications in flood and drought monitoring and forecasting, the assessment of climate change impacts and prediction uncertainty. Over the years he has worked on a wide range of projects, gaining experience on many river basins around the world, notably in Africa, Asia and Europe. He is also involved in IAHS Panta Rhei initiatives and serves as Associate Editor of the *Journal of Hydrology* and the *Journal of Flood Risk Management*.

**Hamid Mehmood** is Senior Researcher in Hydro-Informatics and Information Technology at the United Nations University Institute for Water Environment and Health (UNU-INWEH), based in Hamilton, Ontario, Canada. He holds M.Sc. and Ph.D. degrees in Remote Sensing and GIS from the Asian Institute of Technology (AIT) in Bangkok. Mehmood worked in research and technical programmes on hydro-informatics and information technology for over 20 years. He has worked on technology transfer projects in more than 15 developing countries and was involved in the implementation of the Hyogo Framework for Action of the UN Economic Commission for Asia and the Pacific (UNESCAP). He also served as an assistant professor at leading universities in Pakistan. His areas of interest and expertise include the application of hydro-meteorological sensors as a component of the internet of things (IoT); deploying blockchain in developing countries as a primary sensor data sharing and storing mechanism; solving hydro-meteorological data gaps using new information technologies; cloud computing for earth observations; and the use of open-source online learning platforms as part of a hybrid capacity building approach.

**Roman Mendelevitch** has been Research Associate at the Öko-Institut e.V. since January 2019. He was previously appointed as post-doctoral fellow in the Resource Economics Group of the Humboldt University, Berlin, and was also the managing editor of the *Economics of Energy & Environmental Policy* (EEEP) journal of the International Association of Energy Economists (IAEE). Until September 2016, he worked as a research assistant at the German Institute for Economic Research (DIW Berlin) and in the Working Group for Infrastructure Policy (WIP) at the Technical University of Berlin (TU Berlin). He holds a doctorate from the TU Berlin and the DIW Berlin Graduate Center in Economics (*summa cum laude*). He previously studied

Industrial Engineering (Dipl.-Ing.) at the TU Berlin and at the University of Maryland, USA. He has worked on several research projects on the economics of climate change that have combined technological and economic areas. His research focuses on medium- and long-term strategies of exporters and consumers of fossil fuels. He developed numerical models for the analysis of international coal markets, CCTS infrastructure and for high-carbon electricity markets in the energy mix. He focuses on alternative climate policy and the interface between economics and political economy on climate change.

**Alisher Mirzabaev** is Senior Researcher at the Center for Development Research (ZEF), University of Bonn, Germany. Before joining ZEF, he worked as a socio-economist at the International Center for Agricultural Research in Dry Areas (ICARDA). He has been leading ZEF's work on economics of land degradation since 2013. He is also serving as a coordinating lead author of the chapter on Desertification of the IPCC Special Report on Climate and Land. He has a Ph.D. in Agricultural Economics from the University of Bonn, Germany.

**Veruska Muccione** completed her Ph.D. in Astrophysics at the University of Geneva, Switzerland, in 2006 and obtained a Master of Advanced Studies (MAS) in Sustainable Development at SOAS, University of London, UK, in 2014. She did her post-doctorates at the University of Geneva and the University of Zurich and was Visiting Scientist at the University of Canterbury, Christchurch, New Zealand. Since 2011, she has been senior researcher at the Department of Geography of the University of Zurich, Switzerland. Her current research focuses on climate change impacts and risks in mountain areas, climate change adaptation and loss and damage. She has been involved in several projects on climate change and adaptation and has also a worked as programme coordinator in an interdisciplinary research programme on global change and biodiversity, as a project manager and consultant for different NGOs and as a natural risk analyst for the reinsurance sector. She is the lead author of the Europe chapter of the IPCC Working Group II Sixth Assessment Reportas well as for the cross-chapter paper 'Mountains'.

**Makhliyo Murzaeva** has worked as a consultant for IWMI since June 2018. She is currently involved in research activities of the EU-GIZ Project 'Sustainable Management of Water Resources in Rural Areas in Uzbekistan – Component 1: National Policy Framework for Water Governance and Integrated Water Resources Management and Supply Part'. Her work focuses on developing and strengthening water basin management plans; the demonstration of rational water use; and cost-effective water and land management practices. Makhliyo earned her B.Sc. in Ecology and Environmental Science from the Institute of Irrigation and Agricultural Mechanization Engineers, Tashkent, in 2016 and obtained additional experience as an Erasmus exchange student at the Humboldt University, Berlin, Germany. She obtained her M.Sc. in environmental protection in the water and agricultural sector at the Institute of Irrigation and Agricultural Mechanization Engineers in 2018.

**Asel Murzakulova** is Senior Research Fellow at the Mountain Societies Research Institute of the University of Central Asia (UCA) in Bishkek, Kyrgyzstan. She was a visiting scholar at the Davis Center at Harvard University in 2018 and at the Institute of Slavic, East European, and Eurasian Studies at the University of California, Berkeley in 2013. Between 2006 and 2015, she worked at the International Institute of Strategic Studies and as Associate Professor and Director of Academic Development at the Bishkek Humanities University (BHU) and as consultant and expert in projects of international organisations. Her research covers conflicts, migration, natural resource management, religion and nationalism. She is currently engaged in security and conflict-related research, with a focus on resource management challenges across the borders of Kyrgyzstan–Tajikistan and Kyrgyzstan–Uzbekistan.

**Serik Orazgaliyev** is Assistant Professor at the Graduate School of Public Policy in Nazarbayev University, Nur-Sultan, Kazakhstan. Prior positions include as a visiting Assistant Professor at Lee Kuan Yew School of Public Policy (at the National University of Singapore) and visiting Research Fellow at the Cambridge Central Asia Forum (Jesus College, University of Cambridge). Before joining academia, Serik gained professional experience in the public policy sector through his employment as a civil servant in the Parliament of Kazakhstan. He received his doctoral degree in Development Studies from the University of Cambridge (2016) and an M.Sc. in Public Law and Global Governance from King's College London (2010). His teaching portfolio includes graduate-level courses such as Governance and Public Policy, Institutions and Development Policies. His main research areas include governments and multinational enterprises (MNEs), institutions and development policies, international political economy and Central Asian Studies.

**Manzoor Qadir** is Assistant Director at the United Nations University Institute for Water, Environment and Health (UNU-INWEH), Hamilton, Ontario, Canada. He is an environmental scientist with a focus on policy, institutional and biophysical aspects of unconventional water resources, water recycling and safe reuse, and water quality and environmental health under a changing climate.

**Jay Sagin** is Assistant Professor in the Civil and Environmental Engineering Department of Nazarbayev University, Nur-Sultan, Kazakhstan. He holds a B.Sc. in Computational Mathematics and Cybernetics (CMC) from Moscow State University, an M.Sc. in Natural Resources and Environment Management (NREM) from Ball State University, Indiana, USA, and a Ph.D. in Geosciences from Western Michigan University, Kalamazoo, USA. He was a Post-Doctoral Fellow at the Global Institute for Water Security at the University of Saskatchewan, Canada. His academic background covers geomatics, GIS, hydrology, hydrogeology, disaster management and transboundary river basins. He is currently working on a number of projects: the Afghanistan–Pakistan transboundary river basin project 'Satellite Enhanced Snow-Melt Flood and Drought Predictions for the Kabul River Basin with Surface and Groundwater Modeling', the Burabay National Park project in Kazakhstan, the Korean Middle Latitude Region Network project, and the USA–Kazakhstan Hydro Meteorological Radar Technologies cooperation project.

**Tomas Saks** is Researcher at the University of Fribourg, Switzerland, currently employed in the SDC-funded Cryospheric Climate Services for Improved Adaptation (CICADA) project. The project is dedicated to enhancing the systematic exchange of water- and climate-related data to improve modelling and scenario-based forecasting of water flows and the reduction of related disaster risks in Central Asia. His research interests cover climate change impacts on the alpine cryosphere (Tien-Shan and Pamir), particularly glacier monitoring, glacier-related hazards and the paleoglaciology of Tien Shan and Pamir.

**Jenniver Sehring** is Senior Lecturer in Water Diplomacy at IHE Delft, with more than 15 years of experience in multi-level water governance in Central Asia, working both for academic as well as policy-making organizations. Prior to her current position, she was Environmental Affairs Adviser at the OSCE. She also worked as a political adviser on water affairs to the EU Special Representative for Central Asia, and as consultant for the GIZ Transboundary Water Management in Central Asia Programme, as well as for the German Foreign Office. In the academic field, she worked as a researcher at the Universities of Giessen and Hagen, and was Assistant Professor at the University of Wuerzburg, Germany. Jenniver Sehring is a member of the editorial board of the *Central Asian Journal of Water Research* (CAJWR) and a visiting scientist at the University of Helsinki.

**Maria Shahgedanova** is Professor of Climate Science at the University of Reading, UK. Her research focuses on changes in climate, glacier, hydrological and dust storm activity in Central Asia. Her work includes climate and glaciological monitoring at high elevations, remote sensing to assess glacier change and desert dust emissions and their impacts on glacier melt, as well as climate and hydrological modelling. She is a member of the Scientific Leadership Council of the Mountain Research Initiative, the largest international network focusing on mountain research, with responsibility for the Mountain Observatories project. She is also the editor of *Physical Geography of Northern Eurasia* (2003).

**Dietrich Schmidt-Vogt** is a geographer and Honorary Professor at the Faculty of Environment and Natural Resources, Freiburg University, Germany. From 2015 to 2017 he was also Director of the Mountain Societies Research Institute, University of Central Asia, in Bishkek, Kyrgyzstan. Prior positions include Full Professor at the Kunming Institute of Botany, Chinese Academy of Sciences, in Kunming, China, from 2009 to 2015; Associate Professor at the Asian Institute of Technology, Thailand, from 2002 to 2009; and Senior Lecturer at the South Asia Institute of Heidelberg University, Germany, from 1998 to 2002. Dietrich studied geography and English in Germany and Canada, and obtained his doctoral and post-doctoral degrees at Heidelberg University, Germany. Spanning more than 30 years of research experience in Asia, his interests include forest-farming interactions, multifunctional landscapes, integrated land use systems, land use change and natural resources management.

**Frank Schrader** is a senior consultant who worked in water resources and environmental projects from 1991 to 1994 in Germany, and since 1994 mainly in Central Asia and in the Caspian Sea and Caucasus regions. As a physical geographer specialising in hydrology and coastal sedimentology, he holds a Ph.D. (1973) and post-doc (1985). He worked from 1970 until 1991 as a researcher and lecturer, finally as Professor of Hydrology, and became a consultant for water resources management and environmental impact assessment in 1991. Working from 2008 for the German Society for International Cooperation (GIZ), in 2008 he established the Transboundary Water Management in Central Asia Programme, and presently supports the EC IFAS in Ashgabad. He also currently works for the Swiss Development Corporation (SDC) in Tajikistan, supporting the Tajik Ministry of Energy and Water Resources in the facilitation of water-sector reform, especially in the northern Tajik part of the Syr Darya River Basin.

**Vladimir Smakhtin** is Director of the United Nations University Institute for Water Environment and Health (UNU-INWEH) based in Hamilton, Ontario, Canada. Vladimir holds a Ph.D. in Hydrology and Water Resources from the Russian Academy of Sciences. He has over 35 years of experience in hydrology and water resources and has initiated, managed or contributed to numerous research initiatives in over 20 countries worldwide, including the state programme for mitigating the consequences of the Chernobyl Accident in Russia and Ukraine, a first global analysis of ecosystem water requirements and various projects focusing on managing water resources variability through enhanced surface and groundwater storage. He worked at Rhodes University and Council for Scientific and Industrial Research (CSIR) in South Africa, and as Research Program Director at the International Water Management Institute (IWMI) in Sri Lanka. His experience spreads across agricultural and environmental water management, low-flow and drought analyses, assessment of basin development and climate change impacts on water availability, the provision of hydrological information for data-poor regions and water-related disaster risk management. Vladimir has authored over 200 publications and has worked as a consultant for many national governments and international organisations.

**Thijs Stoffelen** is a Ph.D. candidate at Wageningen University, the Netherlands. He completed his M.Sc. in International Land and Water Management with a specific focus on transboundary water cooperation and water diplomacy in Central Asia. His research aims to identify the strategy and tactics of international actors in their pursuit of sustainable and peaceful transboundary water management. Thijs currently focuses his Ph.D. research on the regional coordination of water issues related to climate change adaptation in coastal zones, and also works on this subject as an independent consultant.

**Lucia de Strasser** is Consultant for the Water Convention of the UN Economic Commission for Europe, where she applies the water–energy–food nexus approach to advance and broaden cooperation in transboundary basins. Within this project, she has worked on inter-sectoral and transboundary cooperation projects in Europe, Central Asia, the Caucasus and North Africa. An energy analyst by background, Lucia has worked as researcher at Fondazione Eni Enrico Mattei (FEEM), where she investigated issues related to the energy sector transformation of the Sub-Saharan African region, as well as at the Royal Institute of Technology (KTH), where she focused on integrated water-energy system analysis and modelling.

**Janez Sušnik** is Senior Lecturer at IHE Delft Institute for Water Education, the Netherlands. He is an expert in global water–food–energy systems analysis, with over eight years of experience in the field. His career has developed around understanding the behaviour of entire systems, from volcano-rainfall systems to global water–food–energy systems. Present areas of research include local, national and global nexus analysis, focussing on quantifying complex relationships and future pathways; understanding in greater depth water–energy relationships in cities and exploring future (mega-)city water demand, and how this demand can be sustainably met. Janez has published several peer-reviewed articles and presented at conferences worldwide. He holds a B.Sc. from Lancaster University and a Ph.D. from the University of East Anglia, UK.

**Stella Tsani** is a visiting faculty at the Athens University of Economics and Business. She holds a Ph.D. in Economics and Business from the University of Reading, UK. Her research interests focus on resource economics, energy, public policy, development, political economy and socio-economic analysis. She has led research in projects funded by the European Commission, the World Bank, Revenue Watch Institute USA etc. She has worked for the Centre for Euro Asian studies in the UK, Europrism in Cyprus, the Public Finance Monitoring Centre in Azerbaijan, the Bank of Greece, the Institute of Energy for South East Europe and the E3MLab at the National Technical University of Athens in Greece. She has published in leading peer-reviewed journals, including *Economics Letters*, *Resources Policy*, *Energy Economics*, *Economic Systems* and others.

**Kai Wegerich** has over 15 years of experience in research on irrigation governance, water institutions and policy, and allocation at basin level in all five Central Asian countries, as well as in Afghanistan. He received his Ph.D. from the School of Oriental and African Studies, London, UK, with a thesis on the institutional reform of water management in Uzbekistan. He was Assistant Professor at Wageningen University, the Netherlands, and Senior Researcher at the International Water Management Institute (IWMI).

**Stefanos Xenarios** is an Environmental Economist with a focus on water resources and climate change aspects. He acquired three years of post-doctoral experience in India and Ethiopia as a staff member of the International Water Management Institute (IWMI), and subsequently did research on climate change and agriculture in South Asia, based at the Norwegian Institute of Bioeconomy (NIBIO). He also worked as Head of the Water and Energy Security Unit at the Organization for Security and Cooperation in Europe (OSCE) office in Tajikistan. Before joining Nazarbayev University as Associate Professor in the Graduate School of Public Policy, he was Senior Researcher in the University of Central Asia (UCA), leading activities on the water–energy–food and environment nexus in the region. He is Co-Editor-in-Chief of the *Central Asian Journal of Water Research* (CAJWR) and a committee member of the International Water Association's Specialist Group on Statistics and Economics.

**Vadim Yapiyev** received his B.Sc. from the Eurasian National University majoring in Chemistry and Ecology in 1997, and his M.Sc. from the Central European University in Environmental Sciences and Policy in 1999. He holds a Ph.D. from the School of Engineering at Nazarbayev University, Nur-Sultan, Kazakhstan. Vadim's research activities are centred on interdisciplinary water resources research, with particular emphasis on hydrology, soil science, environmental geochemistry and micrometeorology. His Ph.D. project is devoted to lake hydrology in Burabay National Nature Park, Kazakhstan, with a focus on open-water evaporation and water balance.

**Dinara R. Ziganshina** is Deputy Director at the Scientific Information Centre of the Interstate Commission for Water Coordination in Central Asia in Tashkent, Uzbekistan. She has more than 15 years of experience in national and transboundary water resources management in Central Asia and holds a Ph.D. in International Water Law from the IHP-HELP Centre for Water Law, Policy and Science under the auspices of UNESCO, University of Dundee, UK. She served as Alternate Governor of the World Water Council and is currently a member of the Implementation Committee under the UNECE Convention on the Protection and Use of Transboundary Watercourses and International Lakes and Associate Professor at Tashkent Institute of Irrigation and Agricultural Mechanization Engineers.

# FOREWORD

The book that you hold is neither yet another call to save the Aral Sea, nor focuses on the Aral Sea itself. It does focus though on the two major rivers—Amu Darya and Syr Darya—that together form what is referred to in the book as the Aral Sea Basin—the area from where the Aral Sea essentially collects its water, and which is a home for over 60 million people.

The book attempts to create a comprehensive narrative of the water management issues in the Basin from biophysical, environmental and socio-economic perspectives. While a lot has been written and said about water management (and mismanagement) in this region before, there is hardly an analogue that brings all these issues under one umbrella. This is essentially the first multidisciplinary volume on the subject.

The Aral Sea Basin is truly unique in many ways, encompassing a range of natural features from glaciers to deserts, boasting a rich history of water developments, possessing some of the most complex water infrastructure in the world, being a place of continuing political transformations and complex transboundary challenges. The socio-economic development in the Basin has historically been dependent on natural resources such as water and land. Accelerated development of irrigated agriculture has led to over-exploitation of water resources and caused widespread environmental degradation and long-term sustainability challenges that remain today. The management of Basin water resources became even more complex after the break-up of the Soviet Union in 1990, when each independent riparian country adopted separate national water policies and strategies.

The separate chapters of this book intend to cover the variety of issues listed above, also examining future prospects for the sustainable management of natural resources in the Basin. A characteristic feature of the book is that it considers all water-inter-related development aspects in the context of the UN Sustainable Development Goals (SDGs), and analyses where riparian countries stand in the SDG context and what they need to do to reach various water-related SDG targets.

It is published in English, unlike many other sources on the subject, which have been primarily in Russian—at least in the recent past. The book is also a product of the large team of authors both from riparian countries and international experts—an approach that enriches the book with diverse information and perspectives in each thematic area. Each chapter generally has a large reference list to provide the interested reader with many further leads.

This book will hopefully become a reference source on the Aral Sea Basin for a wide range of stakeholders from different disciplines and professional backgrounds, from within the Central Asia region and internationally.

**Guyzgeldi Bayjanov**
Chairman of the Executive Committee of the International Fund
for Saving the Aral Sea

# 1

# Introduction

*Stefanos Xenarios, Dietrich Schmidt-Vogt, Manzoor Qadir,
Barbara Janusz-Pawletta, Iskandar Abdullaev and
Vladimir Smakhtin*

The Aral Sea Basin encompasses a unique array of diverse ecosystems, from alpine forests and glacial lakes in its eastern parts to steppes and deserts in the west. The Basin includes two large transboundary rivers—the Amu Darya and Syr Darya—and is home to some 60 million people. The hydrological boundaries of the Aral Sea Basin enclose nearly the entire territory of Kyrgyzstan, Tajikistan and Uzbekistan, and to a lesser extent Turkmenistan, Kazakhstan and the northern fringes of Afghanistan (Figure 1.1). The Amu Darya is formed by the Pyanj and Vakhsh Rivers originating in the Pamir Mountains of Afghanistan, Kyrgyzstan and Tajikistan, and then flows through Uzbekistan and Turkmenistan, discharging into the Aral Sea in Uzbekistan. The Syr Darya is formed in the Tien Shan Mountains of Kyrgyzstan by the confluence of the Naryn and Kara Darya Rivers, and then flows through Uzbekistan, Tajikistan and Uzbekistan, also discharging into the Aral Sea in Kazakhstan. All the above countries except for Afghanistan were formerly part of the Soviet Union and are commonly referred to as the Central Asia region.

All of the countries are located at the core of the Eurasian landmass, blocked off by mountain ranges (except for northern Kazakhstan) that prevent or hinder the inflow of humid air from the south, thus creating the conditions for an arid to semi-arid continental climate. A distinctively uneven water allocation exists within the Basin between the water-producing uplands of Afghanistan, Tajikistan and Kyrgyzstan, and the water-consuming lowlands of Uzbekistan, Turkmenistan and Kazakhstan. More than 80 per cent of water in the Basin originates in the mountains of Kyrgyzstan, Tajikistan and Afghanistan, but more than 80 per cent is consumed in the lowlands of Uzbekistan, Turkmenistan and Kazakhstan.

Until the conquest of Central Asia by Tsarist Russia and the subsequent establishment of the Soviet Union, traditional agricultural practices were serving local economies through farming in the southern parts and through nomadic pastoralism in the northern parts of the Basin. During the Soviet Union period, land use in the Basin was transformed fundamentally from traditional to commercial agriculture based on irrigated cotton monoculture. The excessive use of freshwater sources resulted in an unparalleled environmental disaster, for which the drying-out of the Aral Sea became emblematic. Before 1961, the average water surface area of the Aral Sea was close to 68,000 $km^2$ and the estimated volume of water in the sea was 1,064 $km^3$. The expansion of irrigation caused the Aral Sea to shrink to 17 per cent of its original surface area, and 9 per cent of the original water volume (Figure 1.2).

The collapse of the Soviet Union and the emergence of the independent countries of Central Asia led to a change from centralized to inter-state transboundary water management. Despite this change, the intensive water withdrawal for agriculture continued until the present, or may even have been further aggravated by cross-border competition. The Central Asian countries are among the most water-intensive economies in the world, with mean annual water withdrawals of 2,200 $m^3$ per capita and nearly 90 per cent of water used for irrigation in all the five countries.

The gradual desiccation of the Aral Sea has caused the North Aral Sea (or the Small Aral Sea) to fragment into several parts. As part of Kazakhstan's efforts to save the northern parts of the Aral Sea, the Kok-Aral dike and dam were constructed in 2005, leading to a gradual replenishment (Figure 1.3). The southern part of the Aral Sea suffered the most from diminishing water inflow, with only 60 $km^3$ of water left. Some initiatives have been taken by Uzbekistan through the establishment of a Multi-Partner Human Security Trust-Fund for the Aral Sea region (MPHSTF) in 2017 under the auspices of the United Nations to reverse the desiccation trend.

Currently, the technocentric approach to managing water through large-scale hydropower and irrigation schemes goes hand-in-hand with the perception of water resources as a natural capital for economic development. The securitization of water as a national endowment has often resulted in disputes and conflicts between upstream and downstream countries, prioritizing their divergent agricultural and energy needs.

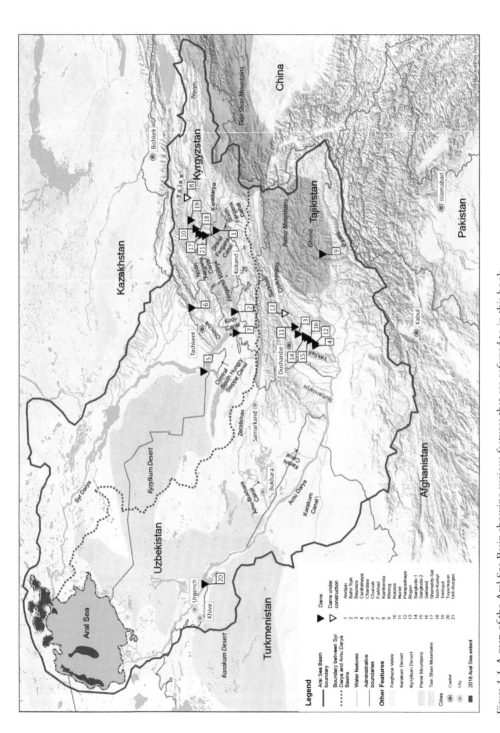

*Figure 1.1* A map of the Aral Sea Basin showing various features that are referred to in this book

Source: Compiled by Hamid Mehmood; UNU-INWEH

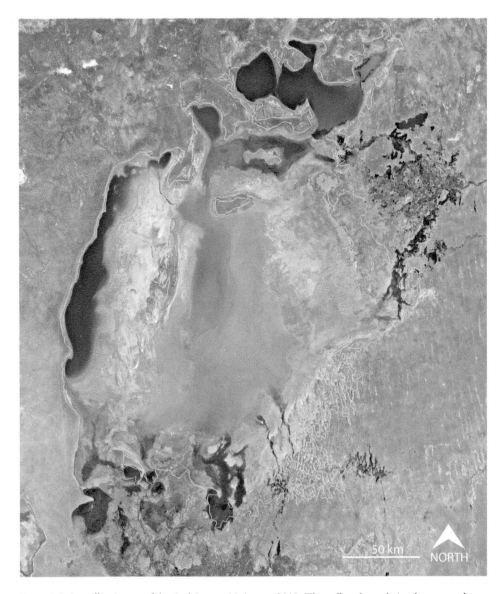

*Figure 1.2* A satellite image of the Aral Sea on 21 August 2018. The yellow boundaries demarcate the approximate shoreline in 1960

Source: NASA Earth Observatory, https://earthobservatory.nasa.gov/world-of-change/aral_sea.php.

New challenges are posed in the Basin by global environmental change, with major impacts on water resources, hydropower, food production and the natural environment in the region. Geopolitical challenges are also emerging through the consortium of the Eurasian Economic Union, led by Russia, the Belt Road Initiative, driven by China, and economic initiatives of Western allies in the region, which may stimulate economic growth but may also exert pressure on natural resources.

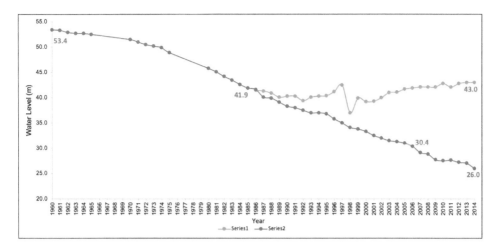

*Figure 1.3* The dynamics of water level in the Aral Sea. Combined time series (blue) of the water level in the Aral Sea (1960–1985), Big Aral (1986–2006) and Western Sea (2007–2014); and the water level of the Small Aral (1986–2014; green).

Source: UNU-INWEH, compiled based on data from CAWa Regional Research Network, www.cawa-project.net/

This book offers analytical insights into geophysical, hydrological, environmental and socio-economic challenges pertaining to water management in the Aral Sea Basin. Regional and international experts with experience from different disciplines present, for the first time, a comprehensive review of water resources opportunities and constraints in the Basin in one volume. Each chapter is a standalone component of the book, and all chapters are complementary. Each chapter lists, upfront, its key messages—the most substantial points to take home.

Chapter 2 presents the historical perspective of the water management evolution in the Basin from pre-historic times until today through milestones that changed the route of water management over time. The shift from subsistence and small-scale farming to industrialized and water-intensive agriculture during Soviet times is described, and the current transition to national water management and more integrated approaches is explained.

Chapter 3 provides an overview of the status and trends in surface water resources, covering specifics such as observational networks in the Basin, human-induced impacts on river flows, uncertainty associated with the climate change impacts and the status and developments in water infrastructure and use.

The role of groundwater management is presented in Chapter 4 by underlining that although water demand in the Basin is increasing, the overall groundwater consumption remains relatively low when compared to surface water use. The seasonal peaks of groundwater abstraction for irrigation in summer are noted and the risk of soil salinization due to groundwater depletion is stressed. The chapter also addresses the increasing deterioration of groundwater quality due to disintegrating drainage systems and salt depositions in aquifers.

Chapter 5 examines the role of hydropower as an underutilized energy source and the economic stimuli to realize its potential in the region. The hydropower potential of Tajikistan and Kyrgyzstan could improve energy security in both countries, while the agricultural needs of downstream countries could be preserved through multi-purpose dam constructions. The dilemma of large versus small hydropower facilities is underscored by addressing the drawbacks and comparative advantages in each case.

Environmental aspects, mainly reflected in land and water degradation in the Basin, are reviewed in Chapter 6. Dramatic economic annual losses of about US$3 billion are attributed to land degradation with direct and indirect effects to human livelihoods. There are currently technologies available to reverse land use management, which also have synergetic effects in the improvement of water management. The return on investments in sustainable land use and water management could offer up to a fivefold profit within a period of 6 to 30 years, depending on the type and scale of investment.

Chapter 7 covers the socio-economic aspects pertinent to water resources management in the Basin, with a focus on the role of agriculture in national economies and rural livelihoods. Attention is given to the water shortages mainly in semi-arid and arid areas, which are ascribed mainly to poor management coupled with intra-regional and inter-sectoral competition for water. The lack of a coordinated use of water for agriculture is presented as an outcome of the transition from commercial to individual farming practices after independence in 1991.

The status and impacts of climate change on the alpine cryosphere in the Basin are reviewed in Chapter 8. The alpine cryosphere, including snow, glaciers and permafrost, plays a crucial role in the water balance of the Basin. There is evidence that most cryosphere components are close to a melting point, which could temporarily increase water availability in July and August for the next 25–30 years. The situation will probably reverse once the glaciers will have shrunk below a critical threshold unless higher precipitation trends compensate the loss. More monitoring stations, especially in high altitudes, are needed for better assessment of natural hazards associated with cryosphere changes in the Basin.

The transboundary complexities of water management in the Basin are analyzed in Chapter 9, with reference to the broader economic and geopolitical changes in the Eurasian region. The needs for long-term regional strategies for transboundary water management, rather than an opportunistic approach, and for reform or replacement of the current regional mechanisms are emphasized. The participation of downstream countries in the development of large-scale hydropower projects upstream would be an essential element for the mitigation of hydro-political frictions in the Basin.

Chapter 10 focuses on the local and national institutions and policies governing water resources management in the Basin by providing an overview of the current situation in each Central Asian country. The water reforms currently undertaken by each country are outlined and the approaches and initiatives adopted towards a more coordinated water management within and between the countries are presented. The chapter acknowledges that technical and financial support of the authorities and organizations implementing water reforms is imperative, and that communities and water users at the local level should be engaged actively.

The international context of sustainable water management in the Basin is presented in Chapter 11 by reviewing the initiatives adopted by multilateral organizations, individual countries and other development partners. The impact of international support, although considerable, seems to be lower than expected due to limited understanding of local conditions and lack of coordination, while the socio-political context in Central Asia complicates the matter. Capacity building through longer-term programmes with a focus on local development should also consider the nuances of water politics in the region.

Chapter 12 is devoted to the future of water resources in the Basin. It points to potential pressures on water consumption due to economic and population growth in Central Asia by making it clear that piecemeal and short-term planning can no longer resolve existing and future needs and challenges. Evidence-based policies supported by sound research and ground-work should be adopted instead in the national and regional mechanisms.

Finally, Chapter 13 deals with the status of Sustainable Development Goal (SDG) 6 on 'Clean Water and Sanitation' by pointing out that the attainment level is uneven between the Central Asian countries but also within the rural and urban populations in each country. It is mostly foreign development partners and, to a lesser extent, national governments that invest in drinking water supply and sanitation. The need for water-efficient practices and cost-effective technologies is apparent in all the countries. The data for the measurement of achievement in SDG 6 have recently improved in terms of availability and reliability; however, more and better information is needed on the drinking water and sanitation conditions mostly in rural areas.

Transboundary water management in the Aral Sea Basin is a political priority for all Central Asian countries and has strongly affected many business sectors all over the region. Indicatively, foreign companies involved in agricultural and hydropower activities in Central Asia would like to learn more about the status of the Basin. Similarly, international organizations and donors are eager to further understand the situation in the Aral Basin and better plan and weigh the impact of their current and future activities. It is envisaged that the international research community has interest in this large basin, for which information on the various water-related aspects, e.g., hydrological, environmental, agricultural and socio-economic, has not been updated to a sig-nificant extent since the countries acquired independence at the end of the twentieth century.

The book intends to fill this vacuum by presenting a scientifically sound and comprehen-sive source of information in English, which reflects the current state of knowledge on water resources and water management in the Aral Sea Basin for a diverse group of readers, including the global community of water professionals that are interested in this region.

# 2

# History of water management in the Aral Sea Basin

*Iskandar Abdullaev, Kai Wegerich and Jusipbek Kazbekov*

## Key messages

- The Aral Sea Basin has a long and rich water history and water management has been the key element in the socio-economic development of the region.
- Water history of the basin may be divided into three periods: 1) pre-historic times until the 1930s, 2) the Soviet hydraulic mission of the 1930s to the 1980s and 3) the post-Soviet period, from the end of the 1980s until today.

- Water development during the first period was mainly a small-scale, subsistence-oriented and non-commercial activity for the local population. It was managed and governed by local, customary rules. The state had a limited role, although a specially appointed person from the highest authority supervised water affairs.
- During the second period, water management and governance was centralized, and pressure on water resources increased. This period is characterized by the transition from small-scale subsistence farming to large-scale commercial farming that increased both agricultural output and water use.
- Current water systems are in a state of transition, keeping attributes from the past while following new trends. New states in the Basin are building national water systems, which differ both on management and governance aspects. Water management in the Basin needs to tackle both supply and demand aspects, serving the economy of the states, but also taking care of the degraded water environment.

# Introduction

Central Asia (CA) is home to the oldest water-dependent civilizations in the world. The region covered by this chapter, which is identified by the Aral Sea Basin (ASB), is fed by two large rivers, the Amu Darya and the Syr Darya. Historically, water use in the region has always been a key element for socio-economic development. The central element for survival of any civilization in the ASB was its access to water. Great cities were erected near the banks of major rivers, such as Samarkand and Bukhara on the Zerafshan River, Urgench and Khiva on the Amu Darya and Kokand and Khujand on the Syr Darya. Historically, governments of the Aral Sea civilizations have played a very strong role in water governance, regulating water at the highest level. Until the middle of the nineteenth century, before Russian colonization, each governor had a deputy, responsible for water matters, including the use of water for irrigated agriculture (Bartold 1970; Mukhamedjanov, 1978; O'Hara 2000; Donish 1977; Kadirov 2007).

In the late 1980s, the Aral Sea region became known for the drying up of the world's fourth-largest inland lake as a result of excessive irrigation development for cotton production (Abdullaev and Atabaeva 2012; Rakhmatullaev *et al.* 2010, 2012). Although there are examples of misuse of water from other parts of the world (Rakhmatullaev *et al.* 2003; Abdullaev 2004; Abdullaev and Atabaeva 2012; Abdullaev *et al.* 2009a), the Aral Sea's degradation has become emblematic for failed water management.

The history of water development in the ASB is a paradigm of the hydraulic mission. The hydraulic mission refers to the ideology of conquering water resources, constructing and enhancing nature for the needs of human society by engineering infrastructure (capturing, delivering and using) and other means (Mollinga 2014; Molle *et al.* 2009). The contemporary history of water resources development in CA can be divided into three periods.

1 Local water management and governance, which includes the period starting from the early development of civilizations, focusing on subsistence irrigation and small-scale water diversions, until the 1930s, the early period of collectivization in the Soviet Union. During this period, the scale of water development did not disturb the environmental balance of water systems in the Basin.

2 The great hydraulic mission of the 1930s to the mid-1980s, beginning with the early Soviet construction period of large-scale water diversions and ending just before the collapse

of the Soviet Union. This was due to different reasons, including economic stagnation, change of political environment (Soviet leader Mikhail Gorbachev's *perestroika* and *glasnost* policies), environmental concerns and the rise of nationalism (O'Hara 2000; Abdullaev 2012). Large-scale water development projects, which often diverted water resources away from the river through monumental canals or pump stations (such as the Karakum and Dustlic canal and pump stations in Bukhara and Kashkardarya) totally changed both the landscape and the social fabric of the communities.

3    The post-Soviet period, starting from 1990s, which commenced after the collapse of the Soviet period and continues until now. This period has mainly been marked by institutional changes at the national level (with ministries of water resources being merged with other ministries such as energy or agriculture), as well as at the local level (collapse of the former state and collective farming management and introduction of water user associations). In addition, the collapse of financial support from the former Soviet Union was partly replaced by a piecemeal approach based on donor involvement.

This chapter presents the socio-technical, institutional and policy aspects of water history in the ASB for the local, great and post-hydraulic mission periods. The chapter is based on the authors' research, a review of recent publications and the analysis of relevant archival documents.

## Local water management and governance: from pre-historic times until the 1930s

CA civilizations have developed a network of water systems and relevant water institutions over thousands of years (Gulyamov 1957; Bartold 1970; Mukhamedjanov 1978; Kadirov 2007). This period can be divided into three sub-periods: i) ancient irrigation before the colonial period; ii) the Russian colonial period and iii) the early Soviet period.

### *From the ancient period until Russian colonization of Central Asia*

Communities developed small-scale, canal-based irrigation systems aiming primarily at providing water for the subsistence production of food and fibre. Ancient agricultural activities have shown that irrigated agriculture was located mainly in valleys of foothills, floodplains and deltas of the Amu Darya, Syr Darya and Zerafshan Rivers as, for instance, in the Margiana, Sogd, Ferghana, Khorezm and Northern Baktria regions. The history of water in ancient times in the ASB can be divided into the following sub-periods.

*Sixth to fifth millennium* BC. The ancient form of estuary agriculture originated from the lowlands—naturally irrigated by seasonal floods and fertilized by water deposits. Fluvial plains in the lower reaches of rivers were used for agriculture in the foothill zone. A number of wells were drilled for small-scale irrigation development. The irrigation of mountain valleys started in this period.

*Fourth to second millennium* BC. Transition from an opportunistic use of different water sources to an increasing regulation of the seasonal water flow of rivers (e.g., diking and other technical interventions). Allocation of surpluses of water in lowlands and clearing of natural channels allowed more rational use of water and the expansion of irrigated areas. The ancient irrigating systems were small lateral channels that carried water to areas distant from river sources. Transition from estuary to irrigated agriculture developed in river deltas such as the Tedzhen, Surkhan-Darya, Amu Darya and Zerafshan Rivers situated in southern CA.

*First millennium BC to sixth century AD.* This period is characterized by the transition to main canal irrigation—the invention of head water intake structures and agricultural development of extensive deltoid plains of the large CA rivers. The irrigation infrastructure built in the different parts of the Basin differed considerably. For example, ancient irrigation systems in Khorezm were built in such a way that water in the river would flow into the canals without water lifting (Altman 1947). Meanwhile, the local population gained experience on river flood patterns and initiated the construction of large canals and more complex irrigation systems. The constructed canals were more lengthy systems with many intakes to the irrigated areas, as, for example, the ancient Qirk-Qiz canal in Karakalpakstan (nowadays northeast Uzbekistan), which extended up to 90 km. In this period, the irrigated area in the lower reaches of the Amu Darya, Syr Darya and Zerafshan Rivers reached 3.5–3.8 million ha (Lewis 1966).

*Sixth to nineteenth century AD* (Figures 2.1a, b and c). Water systems were owned by local landlords until the spread of Islam in the seventh century and consecutive Arab invasions thereafter, which changed water management governance. The core of the water governance was based on the rules of Islam (*sharia* law) and the water supply was supervised by water regulators (*mirabs*). However, the application of *sharia* law was neither consistent nor strict. The long history of irrigation in the region led to the adaptation of *sharia* law to the local situation, with the *mirabs* developing practical and local interpretation of *sharia* that was simple and easy to apply. The established canals of the previous periods were extended to form more complex irrigation networks, as, for instance, the Chermen-Yab canal, with a length of 150 km and a width of 8–12 m.

The public water works (*Khashar*) have been a crucial element of maintenance of the irrigation systems through mobilization of the local population in digging and cleaning of the irrigation systems (Figure 2.1b). Both before and after the invasion of the Russian Empire, this type of mass mobilization of the local population had been the major method of keeping the irrigation systems functioning.

## *The Russian colonial period (1865–1917)*

In the middle of the nineteenth century, the Russian Empire began the process of colonization of CA, and by the mid-1880s this process had been accomplished. The Russian Empire created the Turkestan region (*Kray*) as the ninth protectorate of Russia. The new regime did not change existing land and water relations for a fairly long time, until the early twentieth century. In 1877, the Turkestan general governor's office issued the Temporary Rules on Irrigation of the Turkestan Territory. A new department of agriculture in CA was created in the same period, responsible for land-water issues. This administrative arrangement was retained up until 1917, when all management functions were transferred to the National Commissariat of Agriculture.

By and large, the operation of the irrigation systems and water use were conducted according to the customs in existence before the Russian annexation. Indicatively, the irrigation practices in the Bukhara emirate, now located in southern Uzbekistan, and the Khiva khanate,

a) Construction of water distribution point at the river

b) Large-scale canal desilting and digging works

c) Water extraction, with the help of a *chigir* (mechanic water lifting system)

*Figure 2.1* a, b, c. Water development works during the local hydraulic mission period
Source: Benjaminovich and Tersitskiy (1975)

now located in southeast Uzbekistan, did not change during the colonial period. Each settlement (*kishlak*) or a conglomeration of settlements in Buhkara had independent canals with their own water intake from rivers (Figure 2.2). These canals were very long, with many tributary channels, and the distribution points were made of stone, straw and turf, which entailed water losses up to 55–60 per cent of the withdrawn water volume.

Although under Russian occupation, the appointed ruler of the province (*Beks*) and the settlement/village/district (*Khakims*) still followed water regulations set by the local ruler (*Emir*). The *Mirabs* (water masters) remained in charge of irrigation management in each village. The *Emir* delegated duties to the first deputy (*Kushbegi* or grand *Vezir*) for the governance of the water in the whole state of Bukhara. The influential and rich people (*arbobs*) of the emirate also maintained distant supervision of the implementation of irrigation practices. Rarely, water conflicts were taken up to the courts, where judges (*Kazi'z*) looked into the land–water issues through the *sharia* system.

In the Khiva khanate, the Amu Darya River was the single source of irrigation. The main canals were wide, long and in some cases navigable. Therefore, the maintenance of these canals through processes such as desiltation required major efforts and significant human resources, sourced from the participation of the entire local population. Each farm had to provide workers and equip them with utensils as well as food supplies for one or two months. Often, the workers were requested to dredge tens of kilometers at the heads of canals (Wood 1875).

*Figure 2.2* The water distribution point at the river intake, 1865–1917.
Source: Benjaminovich and Tersitskiy (1975)

In another area called Sogd, now situated in the Ferghana Valley and located between the countries of Uzbekistan and Tajikistan, irrigation comprised the main economic stimulus for the local population. The water from tributaries of the Syr Darya River was intensively used for irrigation purposes (Tolstov 1962).

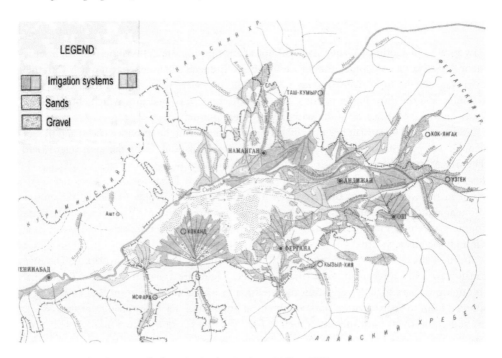

*Figure 2.3* Local utilization of tributaries in the Ferghana Valley, 1928.
Source: Benjaminovich and Tersitskiy (1975)

The high population density in the valley restricted extended irrigation use and the struggle for more access was significant, even before the Soviet regime was set in place in the early twentieth century.

## *The start of the 'new era': early Bolshevik rule (1917–1930)*

After the Bolsheviks took power in 1917, they announced major policy changes in land use by handing over plots to peasants and production plants to workers. In practice, the Soviets continued the hydraulic mission that had been interrupted by the beginning of World War I (WWI). Before the beginning of WWI, the Russian government had initiated the extension of irrigated lands in CA and the construction of large-scale canals in arid areas of the ASB—as in the 'Hungry Steppe', southwest of Tashkent in Uzbekistan (Figure 2.4).

Initially, the Bolsheviks were mainly concerned with the survival of their regime in the very difficult circumstances of Turkestan in that period. An order 'About the prohibition of bargains on sale, purchase, pledge, etc. of all immovable properties and grounds in cities in view of their forthcoming nationalization' was issued on 22 December 1917; all main channels and irrigation constructions were handed over to the National Commissariat on Agriculture, according to the decision of the Turkestan republic (*Turkrepublic*) *Sovnarkom* from 13 March 1918. A National *Commissariat* (ministry) on agriculture was assigned the task of restoring irrigated agriculture and the establishment of a cotton industry through the rehabilitation of irrigation canals and the construction of new ones. The decree 'On assignment of 50 million *roubles* for irrigation works in Turkestan and on organization of these works'

*Figure 2.4* Mass mobilization of the population for irrigation construction.

Source: Benjaminovich and Tersitskiy (1975)

was signed on 17 May 1918, stipulating the irrigation of 550,000 and 44,000 hectares in the *Golodni* and *Dal'verzinski* steppes on the left bank of the Syr Darya in Uzbekistan, and 12,000 hectares in the *Uchkurganski* steppe in the Ferghana Valley. A plan was also made to build a water reservoir on the *Zerafshan* River for the irrigation of about 120,000 hectares for cotton production. Other immediate interventions were to finalize the construction of irrigating systems in the valley of the Chu River and to improve the irrigation of 102,000 hectares (Matley 1970). These were actually unfinished projects from the Tsarist period that were to be completed in Soviet times.

In the next decade, from 1917 until the 1930s, the Bolsheviks tried to retain the main principles of water management of Tsarist Russia, or introduced only slight changes. The new government was focused on the professional training of the CA population in water management aspects. The National University in Tashkent was opened in 1918 with a technical faculty and a hydraulic engineering department. This department was later reorganized into an engineering-melioration faculty.

The Bolshevik state was also about to expand the cotton cultivation initiated by Tsarist Russia. In the mid-1860s, the Russian Empire had attempted to reduce its extensive cotton imports from the USA by promoting cotton farming in nearby regions, mostly in CA. The involvement of Russia in WWI enhanced the need for cotton fabric throughout the Empire.

Likewise, the Bolshevik state inaugurated the 3rd Congress of the Textile Industry Workers in 1920, with Lenin stating, 'Everyone knows that the textile industry experiences the greatest ruin because of the lack of cotton which is delivered from abroad. The only source is Turkestan' (Obertreis 2017). In order to intensify cotton planting, the Soviet regime also focused on restoring irrigation systems in Turkestan (Coates and Coates 1951).

A decree of the Council of National Commissioners of the Russian Soviet Federation of the Socialistic Republics (RSFSR) 'On restoration of cotton culture in Turkestan and Azerbaijan Soviet Socialist Republics' was issued in 1920 for enhancing cotton production. The decree obliged government bodies to be entrusted with land management (*Narkomzem*), the restoration and organization of new experimental fields and the introduction of crop rotation in cotton-growing areas. According to the decree, the government should provide cotton growers with seeds and initiate the restoration of irrigation facilities until the spring of 1921.

Further, in 1923 the Soviet government issued the 'Clause on Management of Water Economy of Turkrepublic'. The Clause included a plan for irrigation works for the organization of the labor force, the prevention of flooding and washouts, the production of scientific research, experimental and statistic-economic works, the use of old scientific research materials, professional development, training courses for workers of the irrigation sector, the development of the Water Law, water delimitation with the neighboring countries and more. The Soviets had reviewed all previous materials on water development and had prepared a programme for repair and construction, scientific research and specialist training. As a result of these measures, huge irrigation works were executed in 1923–1924; the Savay channel in the Ferghana Valley was restored, the Andizhan and Shakhrikhansai shores near to the village of Asaka were strengthened, the Sardobinski spillway in the Hungry Steppe was deepened, the Dargomski and Akkaradarya irrigation canals on the Zerafshan River were reconstructed, and the Gindukush dam on the Murgab irrigation network and the Kashutbeski and Sultanbentski dams on the Bairamaliski irrigation network were repaired, among quite a few other irrigation projects. The irrigated area was increased by 485,000 hectares from 1922 to 1923, and reached 1.7 million hectares in Turkestan (current Central Asia). By the end of 1924, the total cultivated area in Central Asia had reached 2.823 million hectares, which included 401,500 hectares under cotton cultivation. Most of the works during this period were accomplished by mass mobilization of the local workforce (Becker 2004).

While working on different aspects of state-building, the Soviets also developed plans on nation-state-building in Central Asia. In 1924, the government issued a decree 'On National Delimitation of Republics of Central Asia'. On the territory of Russian Central Asia, the Uzbek and Turkmen Soviet Socialist Republics, the Tadjik Autonomous Soviet Socialist Republic within the Uzbek SSR, the Kirghiz Autonomous Region, included in RSFSR, and the Karakalpak Autonomous Region in the structure of the Kazakh ASSR were formed. Water management was organized by the centralized authority for water management (*Sredazvodkhoz*), which was the umbrella organization of all research and construction organizations in CA. This was the beginning of the greatest hydraulic mission human society was to witness in the twentieth century, in terms of both socio-economic development and environmental disasters.

## The great hydraulic mission: from the 1930s to the 1980s

In the late 1930s, the collectivization of land and the formation of a new type of agricultural production (collective farms) was the crucial element of an emerging new water philosophy in the Aral Sea Basin. Formerly, land and water were interlinked and access to water was granted together with land ownership (Strikeleva *et al.* 2018). When land became collective property and the ownership was transferred *de facto* to the state, water became an instrument in the hands of the state to promote state interests. This period of the hydraulic mission brought new lands under irrigation. Most of the newly irrigated areas were purposively located in arid environments of desert and steppes in an attempt to cultivate new 'virgin lands'. Land area under irrigation in the CA region increased from 1.8 million hectares in 1939 to 7.8 million hectares in 1989 (Strikeleva *et al.* 2018).

In this period, the Soviets introduced both extensive collective farming (*Kolkhoz*) and the development of virgin lands. The virgin land concept was part of a socio-economic system that also included the construction of housing, hospitals and schools and the resettlement of the population from densely populated areas to newly irrigated areas. This period is also known for the great development of engineered irrigation systems (Figures 2.5a and b).

While some see the hydraulic mission as the quest for cotton production, others put emphasis on the irrigation bureaucracy (Thurman, 1999; 2001). The irrigation bureaucracy during this period was responsible for the planning and implementation of new irrigation systems, which also led to an increase in the budgets of the irrigation organizations (Abdullaev and Atabaeva 2012). Some authors highlight that water diversion schemes were developed to deal with the problem of population pressure and the avoidance of resettlement of the CA population into other parts of the Soviet Union (Gustafson 1980; Dukhovny and de Schuetter 2011). Improvements in the livelihoods of local people were distinctive, for instance in the health services and nutrition, which led to a birth rate increase in the region. In the period from 1970 to 1979, the population in the Uzbek, Turkmen and Tajik SSRs grew by about 30 per cent, compared to 6 per cent in the Russian Federative SSR. More recently, particularly looking at the expansion of the irrigated area along smaller tributaries within the Ferghana Valley, it has been reasoned that shifts to more costly irrigation infrastructure (from gravitation to lift irrigation) were justified to protect water allocation rights of the riparian states (Wegerich *et al.* 2016; Soliev *et al.* 2018).

*Figure 2.5a and b* Construction of irrigation systems during the great hydraulic mission, Ferghana Valley.
Source: Benjaminovich and Tersitskiy (1975)

During the Soviet period, water allocation within the larger basins, including their tributaries, was designed in complex basin planning procedures, considering drinking water, industrial and agricultural water needs. While in the early years of the Soviet period transboundary water infrastructure was implemented from the top down, in later years complex negotiations, compensation, property rights and obligations as well as benefit-sharing approaches were utilized (on dams/reservoirs, canals, pump stations and flood protection) (Wegerich *et al.* 2012; Pak and Wegerich 2014; Pak *et al.* 2014; Soliev *et al.* 2017).

In this period, the water resources of the two large river basins (Amu Darya and Syr Darya) were transferred through systems of canals and pump stations to other regions of the Aral Sea. The bulk of water was diverted to the former steppes in Karshi and Bukhara for the ambitious virgin land developments. The large water transfer from the Amu Darya began in the late 1950s through Karakum canal, where almost 50 per cent of river flow was directed into the Karakum Desert in southern Turkmenistan, also signalling the gradual desiccation of the Aral Sea.

In the mid-1950s, farming on collective and state farms was on a large scale, planned with the aim of establishing monocultures of specialized crops (such as cotton, wheat and rice). The infrastructure of the on-farm irrigation systems was designed to cater for such extended farming plots. On-farm water infrastructure within the collective farms differed significantly based on the geography of the agricultural areas as well as the time period in which the irrigation system was created. At the end of the great hydraulic mission, and as an outcome of heavy water withdrawals for the irrigation of 'white gold', as cotton was known, the Aral Sea had almost desiccated. The era of the great hydraulic mission ended with one of the greatest environmental impacts in human history.

The drying up of the Aral Sea led to the halting of many economic activities such as fishing, sea transportation and livestock grazing for more than 3 million people inhabiting the areas close to the Aral Sea. Around 5 million hectares of the former Aral Sea became a new desert—the Aral Kum. The Aral Sea disaster was well publicized during the *perestroika* years (Micklin 1988). One of the proposed solutions to the Aral Sea problem was the partial diversion of Siberian rivers into the Basin. Initial research and project preparation started at the end of the 1970s. Quite a few fact-finding and research expeditions took place and the Cabinet of Ministries was prepared for the implementation of this large project. However, in the mid-1980s the idea was heavily criticized by the Soviet *intelligentsia*, and the *perestroika* of Gorbachev stopped the implementation in the early 1990s. This marked the end of the hydraulic mission in the Aral Sea Basin.

## The post-Soviet period: the 1990s until present

In the mid-1990s the Soviet Union experienced a financial crisis, and at the same time political pressures were triggered by environmental degradation. During this period, the Soviet Union released many rules specifying the reduction of the irrigated area and the increase of water allocations to the Aral Sea. However, in 1991, the Soviet Union collapsed and the CA republics became five independent countries. The water allocation agreements after independence tried to retain the Soviet regulations (in 1984 for the Syr Darya and in 1987 for the Amu Darya), whereby the environment was given the lowest priority.

After independence, the formerly centralized and regulated freshwater systems became interstate or transboundary waters between the CA countries, with immediate effects on the existent infrastructures. In particular, new benefit-sharing agreements were formulated on critical transboundary infrastructure, as, for instance, the Tuyamuyun reservoir and Bukhara and Kashkardarya pump stations, which were situated in Turkmenistan but owned by Uzbekistan. However, other less-significant infrastructural assets were not settled through bilateral agreements.

Shortly after the collapse of the Soviet Union, Central Asian representatives agreed to set up inter-state water coordination institutions. The Interstate Commission on Water Coordination (ICWC) was the first inter-state organization set up by CA states to coordinate transboundary water management. The formerly strong and well-funded water institutions started to decline and ministries of water resources were merged with other ministries that were the main users of water (such as the Ministry of Agriculture or the Ministry of Energy) (Wegerich 2005; Yalcin and Mollinga 2007; Abdullaev and Atabaeva 2012). Due to these developments, the influence of water professionals on water-related decisions diminished (Abdullaev and Atabaeva 2012).

Socio-political and economical transformations of the post-Soviet period have resulted in an institutional and technical vacuum due to limited monetary support and degradation of human capacities in water management organizations of CA states (Wegerich *et al.* 2015). Along with de-collectivization in the agriculture sector (e.g., individual farming systems), water management systems were also transformed in CA to river basin/watershed and on-farm management level (Yalchin and Mollinga 2007; Abdullaev *et al.* 2009b; Wegerich *et al.* 2012). At the beginning of the twenty-first century, the river basin/watershed approach replaced territorial-administrative management in Kazakhstan, Kyrgyzstan and Uzbekistan. The same occurred with a slower pace in Tajikistan and stagnated in Turkmenistan (Sehring 2009; Duchovny and de Schutter 2011; Abdullaev 2012; Wegerich 2015).

The water sector in ASB countries is a very crucial part of their policy-making. The governments are coming back to the idea of establishing independent water-related ministries or agencies (Uzbekistan, Turkmenistan) by providing financial and technical support. Financing of the water sector in all countries of CA has been considerably increased since the mid-2000s. The revival of support to the water sector is related to both the improved economic outlook of CA countries and the increased importance of the water-related agenda in the region. However, water-sector reforms in CA are still ongoing. Currently, in Tajikistan, water-sector reforms are in progress with the aim of separating water governance and management aspects. The Ministry of Energy and Water is shaping water policies and governance while separate agencies are implementing water management. In 2018, Uzbekistan started to reform its water sector and established a new Ministry of Water Management. The focus of the reforms is to improve water productivity and efficiency of water control in the irrigation sector. In Turkmenistan, a new agency for water management was set up in 2018 with a focus on increasing water availability for irrigated agriculture. The CA countries are now moving towards a more a pragmatic approach on transboundary water issues than in previous years, when there was higher competition over water allocation. The change of political leader in Uzbekistan in late 2016 has led to improved water–energy cooperation between the riparian states of the Aral Sea.

In all countries of CA, land reforms have resulted in the formation of hundreds of thousands of individual farms. These reforms have also increased the need for water control between individual landowners and farming units. Two main principles of land reforms can generally be observed in the ASB: 1) equitable distribution of land to all former collective farm workers and their families and 2) selective land distribution, based on 'potential success' criteria like experience, human and financial capital and machinery that could improve agricultural productivity. The consequences of the land reforms have resulted in the formation of two different agricultural systems in the region: one aiming at poverty reduction and food self-sufficiency, with many smaller farming units, and the other aiming at commercial farming, with larger-scale farms. With the exception of Uzbekistan, the land reforms have followed the principle of allocating land titles based on family members working on the state and collective farms (Kazakhstan, Kyrgyzstan, Tajikistan), which has caused fragmentation of the irrigated area into many small plots, mainly for subsistence farming (in Kyrgyzstan and Tajikistan). Uzbekistan allocated land leases (with specific crops to be grown) according to the aforementioned criteria. In addition, in Uzbekistan lease agreements are currently reviewed on a regular basis, leading to a further decrease in the number of farms and an enlargement in plot sizes.

In parallel, water reforms have been taking place in an attempt to decentralize management and governance, as well as transferring the costs of maintenance and rehabilitation of irrigation systems to community level. To this aim, 'water users' associations' (WUAs) have been gradually established in all the four CA countries except Turkmenistan (known as water consumers' associations in Uzbekistan, with practically the same duties). The establishment of WUAs seemed to be initially designed for large-scale farms and infrastructures. Such non-governmental institutions were advocated to replace former collective farming systems for the operation, maintenance and management of irrigation and drainage infrastructures (Kazbekov and Qureshi 2011; Yakubov 2012). However, the creation of a multitude of small farming units has required the establishment of new water systems at farm level. Furthermore, the boundaries of WUAs (administrative or hydraulic) have created some confusion in the management to be applied in each case. All the above have resulted in the redesigning and reestablishment of many WUAs so as to serve smaller-scale individual farming systems (Yakubov and Hassan 2007).

## Conclusions

The history of water management in the ASB is closely linked with the socio-political history of the Central Asia region. Water has been a key social and economic policy catalyst in the pre-Russian, Russian and Soviet history of the Basin. The region has a rich experience of both great water developments and failures of 'man's fight against nature'. The great developments are closely linked to the histories of the states in the Aral Sea. Great Central Asian empires have developed irrigation systems to feed their cities and populations.

Before colonization by Russia in 1865, water management in the region supported subsistence farming for food production, with multi-cropping supplied by numerous small irrigation canals. Later on, the share of commercial cropping slowly increased and the water system was administered by a multi-layer governance system. The states of the Basin were able to maintain water systems mainly through the mobilization of both financing support and workforce from the communities. The irrigated area in the Aral Sea Basin before Russian conquest was rather small, at around 3.5–3.8 million hectares.

In the initial years of the Russian conquest, no changes were made to the state of water affairs. However, the region was gradually facing an increase of cotton cultivation in the cropping structure. Commercial and private interests coupled with the imperial interests of the Russian state drove the demand for cotton at the end of the era of Russian domination. The funding of new irrigation systems was initiated in the early twentieth century, and the governance of the water system, originally in private hands, was gradually taken over by the state under the Soviet regime.

The Soviet system continued the water management methods of imperial Russia until the mid-1920s, tailoring their policies to the existing land-water systems in the Basin. However, this was only a tactical retreat, and in the 1930s the region witnessed large-scale land and water use changes. Collective farming became a major production system and large-scale water developments were major elements of the Soviet policy in the Basin. Around 8.5 million hectares were irrigated by the mid-1980s, with almost 75 per cent devoted to cotton production.

From the mid-1980s the region witnessed serious transformations—the retreat of the hydraulic mission, which coincided with the collapse of the Soviet system, and environmental disasters, in the form of the drying of the Aral Sea. The former republics became independent states and started to develop national self-interests. Formerly common, centrally controlled water resources became national, and the states of the ASB immediately set up systems for regional water-sharing for preserving the status quo of the Soviet period. Although the current policy framework is heavily criticized for its inefficiency and curtailed functioning, the Aral Sea region has not been an area of major water conflicts for the last 30 years. The CA countries were able to set up institutions to oversee water-environmental issues between the countries. This has prevented many potential inter-state conflicts. However, the water sectors of the Basin countries are in continuous transformation.

Water management and governance throughout the history of CA also reflects the governmental systems that succeeded one another in the Basin. Before the Russian Empire took over CA, rulers were focussed on increased tax collection and the stability of their territories. The Russian Empire and early Soviet periods were concerned with the increased production of cotton and other crops, keeping local water systems intact. During the period of the 1930s until the 1980s, the region experienced a hydraulic mission on an unprecedented scale. Development plans of the USSR in Central Asia have been both grandiose and water-intensive. Cotton irrigation increased manifold, irrigated areas were extended and production increased at record rates. Since the collapse of the Soviet Union, former CA countries have turned into transboundary riparian states of the Aral Sea Basin. The water sector is still undergoing reforms and changes. Currently, water is a key dimension of the inter-state political relations of Central Asian countries.

The water-sector reforms in CA are an outcome of the nation-building process after the Soviet Union's dismantlement. Historically, water has been a crucial element of statehood in the Aral Sea Basin. In each period, water administrations were a central component of the governmental policies. Effective and efficient water governance and management was a pre-requisite for the success of states in this region. Therefore, while post-Soviet transformations at the national, Basin and on-farm levels were aimed at improving water governance and management in the region, a series of both externally and internally led reforms have not brought the expected results. State-owned water management organizations have changed 'nametags', but the underlying ideology of water governance and management has remained nearly identical. The challenge of decentralizing fiscal, technical and political responsibilities remains largely unresolved. At present, the biggest challenges for the water sector in the region come from weak decision-making capacities of state water management organizations, and from un-coordinated and competing interests of different stakeholders and public actors across all levels of water management. These problems are the most important root causes and bottlenecks of the failure of integrated water management approaches at river basin level. Current water systems are in transformation, having the attributes of both past and present systems. However, an effective water administration needs to tackle both supply and demand needs, taking care of both the developing economy and the degraded environment.

# References

Abdullaev I (2004) The analysis of water management in Bukhara oasis of Uzbekistan: historical and territorial trends, *Water International* 29(1): 107–111

Abdullaev I (2012) *Socio-Technical Aspects of Water Resources Management in Central Asia*. Lambert Academic Publishing, Saarbrücken, Germany

Abdullaev I, Atabaeva S (2012) Water sector in Central Asia: slow transformation and potential for cooperation, *International Journal of Sustainable Society* 4(1/2): 103–112

Abdullaev I, Kazbekov J, Manthritilake H, Jumaboev K (2009a) Water user groups in Central Asia: emerging form of collective action in irrigation water management, *Water Resources Management.* doi:10.1007/s11269-009-9484-4

Abdullaev I, Kazbekov J, Manthritilake H, Jumaboev K (2009b) Participatory water management at the main canal: a case from South Ferghana canal in Uzbekistan, *Agricultural Water Management* 96(2): 317–329

Altman V (1947) Ancient Khorezmian civilization in the light of the latest archaeological discoveries (1937–1945), *Journal of the American Oriental Society* 67(2): 81–85

Bartold VV (1970) *History of Irrigation in Central Asia* [in Russian]. Nauka Publishing, Moscow, Russia, pp 23–65

Becker S (2004) *Russia's Protectorates in Central Asia: Bukhara and Khiva, 1865–1924*. Central Asian Studies, Taylor & Francis, London, UK

Benjaminovich ZM, Tersitskiy DK (1975) *Irrigation of Uzbekistan II* [in Russian]. Fan Publishing House, Tashkent, Uzbekistan

Coates WP, Coates ZK (1951) *Soviets in Central Asia*. Lawrence & Wishart, London, UK

Donish A (1977) *History of Bukhara and Samarkand Oasis* [in Uzbek]. Fan Publishing, Tashkent, Uzbekistan, pp 5–68

Dukhovny VA, de Schuetter JLG (2011) *Water in Central Asia: Past, Present, Future*. CRC 899 Press/Balkema, Leiden, the Netherlands

Gulyamov YG (1957) *Irrigation History of Khorezm from Ancient Times to the Present* [in Russian]. Publishing of Academy of Sciences of the Soviet Uzbek Republic, Tashkent, Uzbekistan, pp 5–74

Gustafson T (1980) Technology assessment, Soviet style, *Science*, New Series 208(4450): 1343–1348

Kadirov AA (2007) *From Historical Water Dividing Bridges in Zerafshan River up to IWRM*. Global Water Partnership Central Asia and the Caucasus, Tashkent, Uzbekistan

Kazbekov J, Qureshi AS (2011) *Agricultural Extension in Central Asia: Existing Strategies and Future Needs*. IWMI Working Paper 145. International Water Management Institute, Colombo, Sri Lanka. doi:10.5337/2011.211

Lewis RA (1966) Early irrigation in West Turkestan, *Annals of the Association of American Geographers* 56(3): 467–491

Matley IM (1970) The Golodnaya Steppe: a Russian irrigation venture in Central Asia, *Geographical Review* 30(3): 328–346

Micklin PP (1988) Dessication of the Aral Sea: a water management disaster in the Soviet Union, *Science* 241: 1170–1176

Molle F, Mollinga PP, Wester P (2009). Hydraulic bureaucracies and the hydraulic mission: flows of water, flows of power, *Water Alternatives* 2(3): 329–349

Mollinga P (2014) Canal irrigation and the hydrosocial cycle. the morphogenesis of contested water control in the Tungabhadra Left Bank Canal, South India, *Geoforum* 57: 192–204

Mukhamedjanov AR (1978) *History of Irrigation in the Bukhara Oasis (From Ancient Times to the Beginning of the XX Century)* [in Russian]. Fan Publishing, Tashkent, Uzbekistan, pp 23–45

O'Hara SL (2000) Lessons from the past: water management in Central Asia, *Water Policy* 2: 365–384

Obertreis J (2017) *Imperial Desert Dreams: Cotton Growing and Irrigation in Central Asia, 1860–1991*. Vandenhoeck & Ruprecht Gmbh & Co, Berlin, Germany

Pak M, Wegerich K (2014). Competition and benefit sharing in the Ferghana Valley: Soviet negotiations on transboundary small reservoir construction, *Central Asian Affairs* 1(2) 225–246

Pak MK, Wegerich K, Kazbekov J (2014) Re-examining conflict and cooperation in Central Asia: a case study from the Isfara River, Ferghana Valley, *International Journal of Water Resources Development* 30(2): 230–245

Rakhmatullaev S, Bazarov DR, Kazbekov J (2003) *Historical Irrigation Development in Uzbekistan from Ancient to Present: Past Lessons and Future Perspectives for Sustainable Development*. In Proceedings of the Third International Conference of International Water History Association, Alexandria, Egypt, pp 79–80

Rakhmatullaev S, Huneau F, Celle-Jeanton H, Le Cousumer P, Motelica-Heino M, Bakiev M (2012) Water reservoirs, irrigation and sedimentation in Central Asia: a first-cut assessment for Uzbekistan, *Environmental Earth Sciences* 68: 985–998

Rakhmatullaev S, Huneau F, Le Cousumer P, Motelica-Heino M (2010) Sustainable irrigated agricultural production of countries in economic transition: challenges and opportunities (a case study of Uzbekistan, Central Asia). In: Wager FC (ed.), *Agricultural Production*. Nova Science Publishers, New York, pp 139–161

Sehring J (2009) Path dependencies and institutional bricolage in post-Soviet water governance, *Water Alternatives* 2(1): 61–81

Soliev I, Theesfeld I, Wegerich K, Platonov A (2017) Dealing with "baggage" in riparian relationship on water allocation: a longitudinal comparative study from the Ferghana Valley, *Ecological Economics* 142: 148–162

Soliev I, Wegerich K, Akramova I, Mukhamedova N (2018) Balancing the discussion of benefit sharing in transboundary water governance: stressing the long-term costs in an empirical example from Central Asia, *International Journal of Water Governance* 6(2): 19–42

Strikeleva E, Abdullaev I, Reznikova T (2018) Influence of land and water rights on land degradation in Central Asia, *Water* 10: 1242

Thurman M (1999) *Modes of Organization in Central Asian Irrigation: The Ferghana Valley, 1876–Present.* Dissertation, University of Indiana, Bloomington, Indiana, USA

Thurman M (2001) *Irrigation and Poverty in Central Asia: A Field Assessment.* World Bank Group, Washington, DC, USA

Tolstov SP (1962) *Along the Ancient Deltas of the Oks and Yaksart Rivers* [in Russian]. Oriental Literature Publishing, Moscow, Russia, pp 9–19

Wegerich K (2005) What happens in a merger? Experiences of the State Department for Water Resources in Khorezm, Uzbekistan, *Physics and Chemistry of the Earth*, 30(6–7): 455–462

Wegerich K (2015) Shifting to hydrological/hydrographic boundaries: a comparative assessment of national policy implementation in the Zerafshan and Fergana Valleys, *International Journal of Water Resources Development* 31: 88–105

Wegerich K, Kazbekov J, Kabilov F, Mukhamedova N (2012) Meso-level cooperation on transboundary tributaries and infrastructure in the Ferghana Valley, *International Journal of Water Resources Development* 28(3): 525–543

Wegerich K, Kazbekov J, Lautze J, Platonov A, Yakubov M (2012) From monocentric ideal to polycentric pragmatism in the Syr Darya: searching for second-best approaches, *International Journal of Sustainable Society* 4(1&2): 113–130

Wegerich K, Soliev I, Akramova I (2016) Dynamics of water reallocation and cost implications in the transboundary setting of Ferghana Province, *Central Asian Survey*, doi:10.1080/02634937.2016.1138739.

Wegerich K, Van Rooijen D, Soliev I, Mukhamedova N (2015) Water security in the Syr Darya basin, *Water* 7(9), 4657–4684

Wood H (1875) Notes on the lower Amu-Darya, Syr-Darya and Lake Aral, in 1874, *Journal of the Royal Geographical Society of London* 45: 367–413

Yakubov MA (2012) Programme theory approach in measuring impacts of irrigation management transfer interventions: the case of Central Asia, *International Journal of Water Resources Development* 28(3): 507–523

Yakubov MA, Ul Hassan M (2007) Mainstreaming rural poor in water resources management: preliminary lessons of a bottom-up WUA development approach in Central Asia, *Irrigation and Drainage* 56: 261–276

Yalcin R, Mollinga PP (2007) *Institutional Transformation in Uzbekistan's Agricultural and Water Resources Administration: The Creation of a New Bureaucracy.* Working Paper Series No. 22. Center for Development Research (ZEF) University of Bonn, Germany

# 3

# Surface water resources

*Kakhramon Djumaboev, Oyture Anarbekov, Bunyod Holmatov,*
*Ahmad Hamidov, Zafar Gafurov, Makhliyo Murzaeva, Janez Sušnik,*
*Shreedhar Maskey, Hamid Mehmood and Vladimir Smakhtin*

## Key messages

- The Amu Darya and the Syr Darya are the two primary rivers draining into the Aral Sea, with a total basin area of over 1,737,000 km$^2$, which includes parts of six states. The estimated (pre-development) mean annual flow the two rivers used to discharge into Aral Sea was 115 km$^3$, but the estimated current total flow is around 10 per cent of that, due to massive water resource developments in the upstream parts of the Basin, which took place primarily in the second half of the twentieth century.

- The existing observational meteorological and hydrological networks in the Basin are not sufficient to support informed water management. The observational networks have declined since the 1990s and have not been improved since then. Regional water data sharing is also suboptimal at present.
- The flows of both main rivers in the Basin are primarily meltwater-dependent. Meltwater from snow and glaciers together contributes close to 70 per cent and 80 per cent of the mean annual river flow of the Amu Darya and Syr Darya Basin, respectively. Snowmelt runoff significantly outweighs that of glaciers.
- The Basin contains some of the world's most complex and largest water management infrastructure, including the world's longest canal, the Kara Kum in Turkmenistan, and what is soon to be the world's highest reservoir, the Rogun dam in Tajikistan (currently under construction). Water infrastructure serves primarily two economic sectors—irrigated agriculture and hydropower. The network of irrigation canals in the Basin is dense and complex, covering thousands of kilometres. There are over 80 reservoirs in the Basin with an individual capacity of over 10 million m$^3$. Upstream riparian countries—Kyrgyzstan and Tajikistan—are implementing ambitious development plans of new hydropower facilities. A characteristic feature of water infrastructure in the region is its ageing, causing it to require costly maintenance.
- Tajikistan, Kyrgyzstan, Uzbekistan and Turkmenistan are among the ten most vulnerable states to climate change in Europe and Central Asia. There is high uncertainty about possible future impacts on water availability, although warming trends are already clear and flow reduction is very likely. Future reduction of glaciers and seasonal snow cover due to climate change will most likely affect mainly the seasonality of river flow and only marginally impact mean annual flow itself in both large basins.

## Introduction

The Aral Sea Basin (ASB; Figure 1.1) includes the basins of the Syr Darya and Amu Darya Rivers, which cover parts of six states in the Central Asia region. The characteristic features of the upstream—south and southwest—parts of the ASB are the Tian Shan and Pamir Mountains, with peaks exceeding 7,000 metres above sea level (masl) and with high mean annual precipitation ranging from 800 to 1,600 mm (Kirmani and Moigne 1997). The mountains are home to forest reserves and national parks. Low precipitation (under 100 mm/year) and high evaporation rates characterize the downstream lowland parts of the ASB—the deserts of Karakum and Kyzylkum—which cover most of the ASB area (Figure 1.1).

The Amu Darya, the largest of the two ASB rivers, has a total length of some 2,400 km. The estimates of its total basin area vary from some 612,000 km$^2$ to over 1,385,000 km$^2$, depending on a range of factors including difficulties in delineating basin areas in flat and arid regions (e.g., those to the north and northwest of the Aral Sea), approach to basin area delineation, i.e., whether certain tributaries or rivers that naturally discharged to inland sinks are included (e.g., Murghab and Tedzhen Rivers) or if tributaries that no longer reach the Amu Darya (i.e., Zerafshan) are included (www.fao.org/nr/water/aquastat/basins/aral-sea/index.stm; UNECE 2011). In this book, the basin area of the Amu Darya is considered to be without the Murghab and Tedzhen Rivers, which discharge naturally to inland sinks not reaching the Amu Darya, but with the Zerafshan River, which was disconnected from the Amu Darya due to anthropogenic factors. The total area is therefore similar to that of the FAO's AQUASTAT estimate at 1,023,610 km$^2$ (www.fao.org/nr/water/aquastat/basins/aral-sea/index.stm).

The Amu Darya originates from the confluence of the Pyanj and Vakhsh Rivers and generates a long-term mean annual flow of just under 80 km$^3$ (UNEP *et al.* 2011; Frenken 2013). Other significant tributaries of the Amu Darya are the Surkhan Darya and the Kafirnigan. The upper Amu Darya passes through parts of Afghanistan, Kyrgyzstan and Tajikistan, while its middle and downstream reaches pass through Turkmenistan and Uzbekistan before discharging into the Aral Sea.

The Syr Darya, the second-largest river in the ASB, has a total length of over 2,210 km. Its basin area estimates also vary (for similar reasons as for the Amu Darya) between 402,760 km$^2$ and close to 800,000 km$^2$. Again, the FAO's AQUASTAT estimate of 531,650 km$^2$ is accepted in this book. The Syr Darya originates from the confluence of the Naryn and Kara Darya Rivers, and generates a flow of around 37 km$^3$ (Frenken 2013)—about half that of the Amu Darya. The upstream parts of the Naryn and the Kara Darya are located in the mountain ranges of Kyrgyzstan. The Syr Darya then flows through Uzbekistan and Tajikistan and back to Uzbekistan before continuing into Kazakhstan and eventually discharging into the Aral Sea. In both the Amu Darya and Syr Darya Basins, only the upstream part of each basin contributes significantly to river flow generation.

## Hydro-meteorological data sources

The observational network in the ASB was developed primarily during the Soviet period. The development of observation networks was particularly intensive from the 1970s until the 1980s (Chub 2000). At its peak in 1985, Central Asia as a whole was endowed with almost 560 meteorological stations. In the post-Soviet period, many stations were abandoned, and their number dropped to under 400 by 1996 (Chub 2000). The results of observations at each station during Soviet period were published annually by the State Hydro-meteorological Service, and these data were distributed through libraries of republics and specialized research institutes.

Information on the exact number of currently operational hydrological and meteorological stations in the ASB, as well as in the entire CA region, varies. These data are dispersed among the CA countries, and no common regional database seems to exist. An attempt is made in this chapter to compile the maps of current meteorological and hydrological observation station networks in the ASB based on consultations with local experts and publicly available sources (Figure 3.1).

The hydro-meteorological observational networks are currently operated by the hydro-meteorological services of each riparian country. For example, in Uzbekistan, *Uzhydromet* is the only body authorized to collect, process and disseminate hydro-meteorological data from the country's observational network. The website of the Scientific Information Center of Interstate Commission of Water Coordination (SIC-ICWC) is essentially the data platform for the Aral Sea Basin and contains a large amount of hydro-meteorological data (CAWa 2018). There are attempts to improve regional hydro-meteorological networks (e.g., Olsson and Bauer 2010; CAWa 2018). However, not all of the data collected are available in the public domain. Overall, the status of hydro-meteorological data collection and sharing between the riparian countries is suboptimal, which, naturally, hampers the development of sustainable regional and national water resources management strategies. Hydrological and water resources models are a conventional global practice for simulating required water data at either ungauged locations or for 'ungauged scenarios' (i.e., various options of water resources development/ water allocation and/or impacts of climate change).

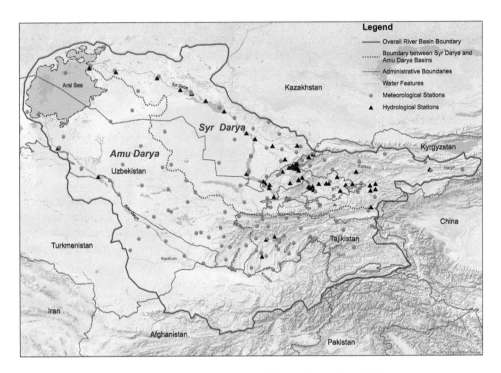

*Figure 3.1* Meteorological and hydrological observational networks in the ASB

Source: Based on Zaitov, SIC-ICWC, personal communication, and NOAA (2018)

Various models were used for various purposes in the ASB and the larger CA region over decades (Wang *et al.* 2016; McDermid and Winter 2017; Omani *et al.* 2017). Cai *et al.* (2003) used an integrated hydro-agronomic-economic model to simulate water management alternatives in the Syr Darya Basin. The STREAM model (the Spatial Tool for River Basins and Environmental Analysis of Management, which combines GIS and the water balance model) was applied for the Amu Darya and Syr Darya River Basins to simulate long-term discharge over the Holocene and future periods (Aerts *et al.* 2006). The WEAP (Water Evaluation and Planning) model was used to simulate water availability and use in the Syr Darya Basin under climate change scenarios (Savoskul *et al.* 2003; Savoskul *et al.* 2004). Immerzeel *et al.* (2012) used the 'Aral-Mountain' raster-based detailed distributed cryospheric-hydrological model to simulate hydrology and water availability in both the Syr Darya and Amu Darya Basins. A high-resolution version of the PCR-GLOBWB hydrological model was used to simulate mass balance of glaciers and their contribution to inflows of major rivers in Central Asia (Punkari *et al.* 2014), suggesting 20–35-per cent flow reduction for the Syr Darya and Amu Darya by 2050. The applications and limitations of hydrological modeling in the glacierized catchments of Central Asia are summarized in Yaning *et al.* (2017).

The above are just a few examples of modelling initiatives that were carried out in the region and in the ASB. Other similar initiatives will inevitably continue. The questions that need to be asked in this regard are whether and how these initiatives have contributed to better water policy formulation and better water management practices in the ASB. With a number of local and international actors in the Basin, modelling effort needs to be better coordinated, as in many other large river basins of the world (Johnston and Smakhtin 2014).

## Climate and hydrology

The overall climate in the ASB is continental (Figures 3.2 and 3.3). However, local climatic conditions created by high mountain ranges vary considerably. Average temperatures in the Basin range from 28 to 32°C during summer and from 0 to 4 °C during winter. Maximum temperatures may reach 50°C, while minimum temperatures may be as low as −30°C. The annual precipitation ranges between 80 and 200 mm in valleys and lowlands, increasing to 300–400 mm in the foothills and reaching 600–800 mm on the southern slopes of the mountains. Annual precipitation in the mountains, e.g., central Tajikistan, may exceed 1,800 mm. The highest precipitation occurs in spring.

Seasonal distribution of annual runoff is uneven, with around 80 per cent of it occurring from April to September (Figure 3.4). Most of the river flow is generated in the high mountains, while the lowland arid and desert plains do not generate much flow. Most of the surface water flowing in the basin is generated in the mountains of Tajikistan and Kyrgyzstan, while Kazakhstan, Turkmenistan and Uzbekistan supply just over 10 per cent of the total basin flow. The downstream areas are also characterized by high natural transmission losses due to high annual evaporation, which exceeds annual precipitation by more than an order of magnitude.

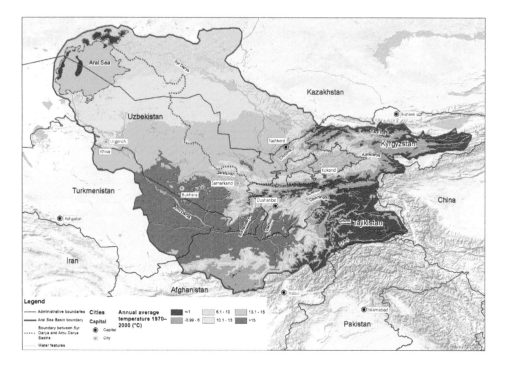

*Figure 3.2* Mean annual temperature in the Aral Sea Basin

Source: Compiled by Hamid Mehmood (UNU-INWEH) based on WorldClim data (http://worldclim.org/version2, with 2.5' spatial resolution)

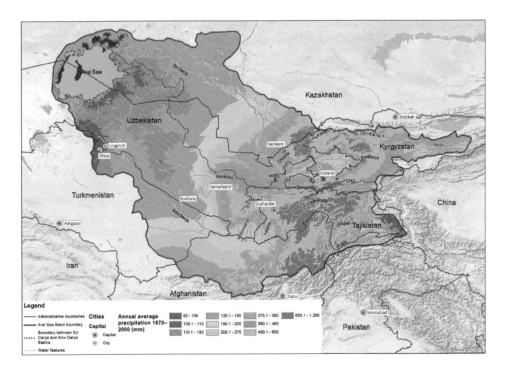

*Figure 3.3* Mean annual precipitation in the Aral Sea Basin

Source: Compiled by Hamid Mehmood (UNU-INWEH) based on WorldClim data (http://worldclim.org/version2, with 2.5' spatial resolution)

Meltwater components (snow and glaciers) play the dominant role in the hydrological regime of both the Amu Darya and Syr Darya Basins. Estimates suggest that meltwater contributes close to 70 per cent and 80 per cent of the mean annual river flows of the Amu-Darya and Syr Darya Basins, respectively. The ratios of snowmelt to glacier-generated runoff are 65:35 and 89:11 in the two Basins respectively, which indicates that snowmelt significantly outweighs that of glaciers (Savoskul and Smakhtin 2013b). Detailed estimates of glaciated areas and glacier runoff were made in the 1990s for all major mountainous sub-catchments of both the Amu Darya and Syr Darya Basins (reviewed by Savoskul and Smakhtin 2013b).

## Anthropogenic impacts on surface water resources

Together the Amu Darya and Syr Darya Rivers generate an estimated (pre-development) mean annual flow of around 115 km³, but the actual flow reaching the Aral Sea at present is much lower. The hydrographs of many rivers in the ASB reflect the impacts of various upstream developments over decades (Figures 3.5 and 3.6).

Figure 3.5 reflects the impacts of water resource developments from the early 1960s onwards. Expansion of irrigated land in the basin from the 1960s stimulated the development of large storage facilities for river flow regulation, and by the late 1980s the total reservoir capacity in the upstream parts of the Syr Darya River Basin alone exceeded 30,000 Mm³.

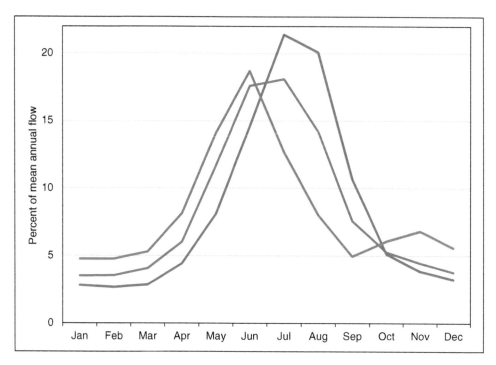

*Figure 3.4* Example of seasonal flow distribution based on observed flow data over 1936–1985 for Naryn at Mouth Kekirim, upstream basin area 34,600 km² (blue); Vaksh at Surk-Ob, upstream basin area 20,000 km² (red); and Syr Darya at Kal, upstream basin area 90,000 km² (green)

Source: Compiled by authors based on UNESCO data (https://wayback.archive-it.org/10611/20160803175443/http://webworld.unesco.org/water/ihp/db/shiklomanov/index.shtml)

Water withdrawal from the river for irrigation amounted to 85 per cent of the total long-term annual river flow against 40 per cent before the 1960s, which significantly reduced the river inflow to the Aral Sea (Rubinova 1987).

Figure 3.6 reflects the impact of more recent developments, when the states of the ASB became independent. The operation of the Toktogul reservoir on the Naryn River changed to maximize hydropower production for Kyrgyzstan rather than meeting the downstream irriga-tion needs in Uzbekistan. This resulted in a shift of the main releases from the reservoir from summer to winter months, and in the 1990s the ratio of summer to winter flow dropped from 2:3 to below 1. While this is not a reflection of the absolute flow decrease, the change in flow seasonality is evident.

Overall, the combined inflow of the Syr Darya and Amu Darya Rivers into the Aral Sea reduced from its 'pre-development' levels prior to the onset of large irrigation infrastruc-ture in the 1960s, to some 22 km³ in the 1970s. It dropped further to only 6.0 km³ in the 1980s, increased slightly in the 1990s up to 15 km³, but then declined further in the 2000s (Micklin 2010). The Aral Sea's water level is a clear cumulative indicator of these changes. The sea is actually a lake with no outflow, while its water level is determined by the interplay of inflow and net evaporation. Historical fluctuations of the Aral Sea water level (by up to 36 meters, likely due to glaciation in the Quaternary period) were periodically

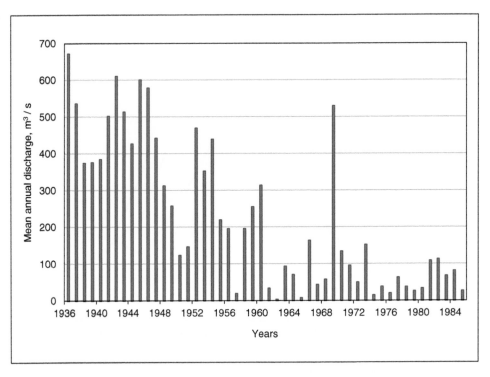

*Figure 3.5* A time series of observed mean annual discharge in the Syr Darya at Bekabod (upstream basin area of 142,000 km²)

Source: Compiled by authors based on UNESCO data (https://wayback.archive-it.org/10611/20160803175443/ http://webworld.unesco.org/water/ihp/db/shiklomanov/index.shtml; data after 1986 for this gauged location are not available from this source)

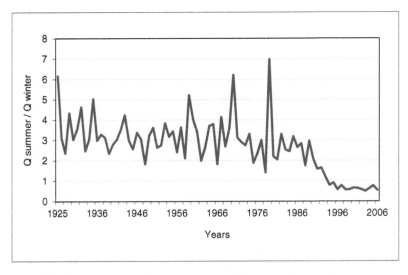

*Figure 3.6* Ratio of summer to winter discharge (Q) in the Naryn River at Uchkurgan station (upstream basin area 58,400 km²)

Source: Karimov et al. (2010)

observed (Severskiy *et al.* 2005). However, since the mid-1950s, upstream water resource development decreased the Sea's water level by more than 20 meters (Figures 1.2 and 1.3 in Chapter 1), creating one of the planet's most well-known environmental disasters, with a range of social and economic consequences.

## Climate change impacts on surface water resources

According to the World Bank (2013b), Tajikistan, Kyrgyzstan, Uzbekistan and Turkmenistan are among the ten most vulnerable countries to climate change among the countries of Europe and Central Asia. This is primarily due to their low adaptive capacity.

There are numerous sources of information on possible impacts of climate change on water resources of the Amu Darya and Syr Darya (e.g., Aerts *et al.* 2006; Lioubimtseva 2015; McDermid and Winter 2017; Milanova *et al.* 2018; Olsson and Bauer 2010; Savoskul *et al.* 2006; Chevallier *et al.* 2014; Xenarios *et al.* 2019). One common observation of the above literature is that climate change is already the reality globally, and in the Basin specifically. The already-observed warming rates since the early 1950s in Uzbekistan, for example, exceed the average rates around the world by more than double; warming rates in the mountains are slightly lower (World Bank 2013c). Although the exact degree of warming that will occur in the near future is uncertain, the overall warming trend appears to be clear and projections suggest that by 2050 a temperature increase of between 1.7 and 4.7°C is expected in Uzbekistan (Milanova *et al.* 2018). In Tajikistan, an increase in mean annual temperature by 0.2–0.4°C is expected in most areas by 2030 (in comparison with the period 1961–1990; World Bank 2013b). Projected future precipitation in the ASB is expected to show large variations in terms of means, intensity and geographical distribution, but overall climate models do not seem to reach consensus with regard to future precipitation projections for the region. Qualitatively, it is most likely that an overall increase in precipitation in lowland areas and some decrease in the highlands will occur. Annual river flow is anticipated to decrease by 2–5 per cent in the Syr Darya River Basin and by 10–15 per cent in the Amu Darya River Basin by 2050 (World Bank 2013c).

Climate change already has and will have an impact on Basin glaciers and snow, and, consequently, on river flow (Hagg *et al.* 2006; Siegfried *et al.* 2012). Estimates suggest that glacier-covered area and maximum seasonal snow area extent in the Amu Darya Basin have decreased by 15 per cent and 7 per cent respectively in the first decade of the twenty-first century compared to the 1960–1990 period. In the Syr Darya Basin, these reductions were 22 and 15 per cent (Savoskul and Smakhtin 2013a). Meltwater contributions to mean annual flow already decreased in the compared periods by 5 per cent for the Amu Darya and by 20 per cent for the Syr Darya. In the short term melting glaciers may increase runoff, but in the long term the runoff is likely to decline (Lioubimtseva 2015). Yet, future reduction of glaciers and seasonal snow cover due to climate change will most likely affect mainly the seasonality of river flow and only marginally impact mean annual flow itself in both large basins (Smakhtin and Savoskul 2013b). At the same time, the impacts may be more pronounced in some upstream sub-catchments of both basins, and in upstream countries overall. Many small glaciers of Tajikistan, for example, may disappear in 30 years, and the glacial area may be reduced by 15–20 per cent compared to present if the current rate of glacial degradation continues; this may adversely affect the river regimes of such rivers as the Kafirnigan, Karatag and Obihingou (World Bank 2013b).

Overall, the uncertainty associated with future surface hydrology and water availability throughout the ASB is very high. In such conditions, a 'no-regrets' or 'low- regrets' approach to future water planning and management is justified. Some initiatives (e.g., Olsson and Bauer 2010) suggest that three future scenarios for water availability need to be adopted by the riparian states of the ASB: a 10 per cent, 20 per cent and (the worst case) 30 per cent reduction in Basin water resources (represented primarily by surface water) by 2085. Understanding the implications of such reductions to the economies of the riparian states now will help minimize adverse socio-economic impacts and ensure stability of the region in the future.

## Water use and infrastructure

Irrigated agriculture plays the key role in the economy of the riparian countries, with the agricultural sector contributing from 10 to 45 per cent of GDP and employing 20 to 50 per cent of the rural population, depending on the state (Qushimov *et al.* 2007). Most water infrastructure in the ASB was built to expand and improve irrigated agriculture, and the major part of this infrastructure, still operational today, was developed during the Soviet period in the early 1960s, primarily for cotton production.

The irrigated area has expanded by 150 per cent in the Amu Darya Basin and by 130 per cent in the Syr Darya Basin between 1970 and 1989 (World Bank 2003). In more recent times, more than 90 per cent of ASB crops were produced on almost 10 million hectares of irrigated land (Horsman 2008). The irrigated area in the Amu Darya Basin is 6 million hectares (1.3 million hectares in northern Afghanistan, 0.1 million hectares in Kyrgyzstan, 0.5 million hectares in Tajikistan, 1.8 million hectares in Turkmenistan and 2.3 million hectares in Uzbekistan (Sokolov 2009; Horsman 2008)). The water infrastructure was built to deliver water to large cooperative farms owned by the state. Today the network serves small, medium and large farms, where size and ownership status may vary between countries.

Irrigation and drainage facilities in the ASB are among the most complex in the world. Before 2000, the overall length of the main and inter-farm irrigation networks (larger-size water conveyors that deliver water to individual farms) in the basin was 47,750 km and the on-farm irrigation network (smaller-size water conveyors that deliver water to irrigation sites) length was 268,500 km. Table 3.1 lists the salient features of some of the main canals in the ASB, such as the world's longest, the Kara Kum canal in Turkmenistan, constructed in the 1950s. The canal conveys water to Turkmenistan's capital Ashgabat and oases in the south, and abstracts 10–12 $km^3$ from the Amu Darya every year (Orlovsky and Orlovsky 2002). Approximate locations of most of these canals are shown in Figure 1.1 in Chapter 1.

A significant part of water infrastructure in the region is currently reaching the end of its design life (CAWaterInfo 2011a; Bucknall *et al.* 2003) and needs costly maintenance. This has already led to excessive water losses, low irrigation efficiencies, waterlogging and widespread soil salinization, and declining crop yields. International financial institutions have made proposals for viable water infrastructure operations in the Basin. One example is the Asian Development Bank's proposal to consider proportional cost-sharing for the operation and maintenance of hydraulic facilities between the countries of the ASB.

Besides agriculture, water infrastructure in the Basin also extensively serves the energy sector. More than 80 reservoirs with individual capacity of over 10 million $m^3$ and 45 hydropower plants with individual capacity between 50 and 2,700 MW are located in the Aral Sea Basin (Frenken 2013). Most of the large dams (>1 $km^3$) are located in Uzbekistan, although the two existing flagship dams are in Kyrgyzstan and Tajikistan. These are, respectively, Toktogul—on

*Table 3.1* Major canals in the ASB

| Name | Capacity | Length | Function | Reference |
|---|---|---|---|---|
| Great Fergana canal | 150 m³/sec* | 249 km | Irrigation in the Ferghana Valley | Wegerich *et al.* 2012 |
| North Fergana canal | 113 m³/sec* | 133 km | Irrigation in the Ferghana Valley | Wegerich *et al.* 2012 |
| South Fergana canal | 92 m³/sec | 142 km | Irrigation of 96,215 ha in the Ferghana Valley | Abdullaev *et al.* 2009 |
| Karshi Main canal | 350 m³/sec | 86 km | Irrigation of 212,500 ha in the Karshi Steppe | Anarbekov *et al.* 2018 |
| Amu-Bukhara canal | 150 m³/sec | 186 km* | Irrigation of the land in Bukhara and Navoi Oases in Uzbekistan from the Amu Darya | Abdullaev 2004 |
| South Hungry Steppe canal | 310 m³/sec* | 127 km* | Irrigation of 300,000 ha of the Golodnaya Steppe from the Syr Darya | Anarbekov *et al.* 2018 |
| Kara Kum canal | 630 m³/sec | 1,300 km | Irrigation (over 1.2 million hectares) and fishing. The canal unites the Amu Darya, Murghab and Tedzhen Rivers | Frenken 2013 |

Note: * Source: Anarbekov *et al.* 2018

the Naryn—with a capacity of 19.5 km³, and Nurek—on the Vakhsh—with a capacity of 10.5 km³ (Frenken 2013; World Bank 2013a). Tajikistan and Kyrgyzstan each control about 60 per cent of the total storage capacities in the Amu Darya and Syr Darya Basins (O'Hara 2000). Nurek provides over 75 per cent of Tajikistan's electricity (World Bank 2013a).

After independence from the Soviet Union, the upstream riparian countries set hydropower development as a national priority (Government of Tajikistan 2011). Tajikistan relaunched the construction of the Rogun dam in 2016, which will be the world's highest dam, with a height of 335 m. This will provide another 3,600 MW of generating capacity, doubling electricity production in Tajikistan. Initial construction of the Rogun dam began during the Soviet era (1976) but was later halted. After independence in 1991, Uzbekistan expressed repeated concerns about the dam's construction, but more recently it showed interest in supporting the dam subject to adherence to international norms and standards (MFA of Uzbekistan 2018). The Rogun dam is expected to operate in incremental stages until its full capacity is installed. The first stage was launched in 2018, and full capacity is expected to be realized in 2028 (Sputnik 2018).

Kyrgyzstan—another upstream CA state—has also set forth major plans for the improvement of its hydropower potential. Kyrgyzstan has initiated the construction of the Kambarata 1 and Kambarata 2 dams in the eastern part of the country, close to the source of the Naryn River. Once constructed, the first unit of the Kambarata 2 will allow Kyrgyzstan to produce an additional 500 to 700 million kWh/year, increasing the current production of 14 billion kWh/year and enhancing energy export potential (Dzyubenko 2010). The exporting option is planned through the Central Asia–South Asia electricity network (CASA-1000), which will connect Kyrgyzstan, Tajikistan, Afghanistan and Pakistan through extensive transmission lines. CASA-1000 will enable the supply to Afghanistan and Pakistan of up to 5 billion kWh once completed (SNC-Lavalin 2011). The network is are already connecting Tajikistan and Kyrgyzstan, while extension to Pakistan and provisional connection of Uzbekistan is foreseen within the next five years (www.casa-1000.org/index.php).

# References

Abdullaev I (2004) The analysis of water management in Bukhara Oasis of Uzbekistan: historical and territorial trends, *Water International* 29(1): 20–26

Abdullaev I, Kazbekov J, Manthritilake H, Jumaboev K (2009) Participatory water management at the main canal: a case from South Ferghana canal in Uzbekistan, *Ag. Water Management*, 96: 317–329

Aerts J, Renssen H, Ward P, De Moel H, Odada E, Bouwer L, Goosse, H (2006) Sensitivity of global river discharges under Holocene and future climate conditions, *Geophysical Research Letters* 33, L19401

Anarbekov O, Gaipnazarov N, Akramov I, Djumaboev K, Gafurov Z, Solieva U, Khodjaev S, Eltazarov S, Tashmatova M (2018) *Overview of Existing River Basins in Uzbekistan and the Selection of Pilot Basins* [Project Report of the Sustainable Management of Water Resources in Rural Areas in Uzbekistan. Component 1: National policy framework for water governance and integrated water resources management and supply part]. International Water Management Institute (IWMI), Colombo, Sri Lanka, DOI: 10.5337/2018.203

Bucknall L, Klytchnikova I, Lampietti J, Lundell M, Scatasta M, Thurman M (2003) *Irrigation in Central Asia: Social, Economic and Environmental Considerations.* World Bank, Washington DC, USA

Cai X, McKinney DC, Lasdon LS (2003) Integrated hydrologic-agronomic-economic model for river basin management, *Journal of Water Resources Planning and Management* 129: 4–17

CAWa (2018) Regional research network 'Central Asia Water'. Retrieved 10 March 2019, www.cawa-project.net/ru/

CAWaterInfo (2011a) The Aral Sea Basin: irrigated lands. Retrieved 10 March 2019, www.cawater-info.net/aral/land_e.htm#irrigation

CAwaterInfo (2011b) The Aral Sea Basin: dams and hydropower. Retrieved 10 March 2019, www.cawater-info.net/aral/groundwater_e.htm#dams

CAWaterinfo (2018) Central Asian water information. Retrieved 10 March 2019, www.cawater-info.net/

Chevallier P, Pouyaud B, Mojaïsky M, Bolgov M, Olsson O, Bauer M, Froebrich J (2014) River flow regime and snow cover of the Pamir Alay (Central Asia) in a changing climate, *Hydrological Sciences Journal* 59(8): 1491–1506

Chub VE (2000) *Climate Change and its Impact on Natural Resources Potential of the Republic of Uzbekistan*, Central Asian Hydro-Meteorological Research Institute, Tashkent, Uzbekistan

Dzyubenko O (2010) Corrected – Updated 1 – Kyrgyzstan launches new hydroelectric power plant. Retrieved 10 March 2019, www.reuters.com/article/kyrgyzstan-hydro/corrected-update-1-kyrgyzstan-launches-new-hydroelectric-power-plant-idUSLDE67T0LK20100830

Frenken K (2013) *Irrigation in Central Asia in Figures.* AQUASTAT Survey 2012, FAO, Rome, Italy

Government of Tajikistan (2011) *Resolution of the Government of Tajikistan.* Programme for the Efficient Use of Hydropower Resources and Energy 2012–2016

Hagg W, Braun L, Weber M, Becht M (2006) Runoff modelling in glacierized Central Asian catchments for present-day and future climate, *Nordic Hydrology* 37: 93–105

Horsman S (2008) Afghanistan and transboundary water management on the Amu Darya: a political history. Retrieved 3 March 2019, www.researchgate.net/publication/252580222_Afghanistan_and_transboundary_water_management_on_the_Amu_Darya_a_political_history

Immerzeel WW, Lutz A, Droogers P (2012) *Climate Change Impacts on the Upstream Water Resources of the Amu and Syr Darya River Basins*, Wageningen, the Netherlands, 1–103

Johnston R, Smakhtin V (2014) Hydrological modeling of large river basins: how much is enough? *Water Resources Management.* DOI: 10.1007/s11269-014-0637-8

Karimov A, Smakhtin V, Mavlonov A, Gracheva I (2010) Water 'banking' in Fergana valley aquifers—a solution to water allocation in the Syrdarya river basin? *Agricultural Water Management* 97: 1461–1468

Kirmani S, Moigne GL (1997) *Fostering Riparian Cooperation in International River Basins: The World Bank at its Best in Development Diplomacy.* World Bank Technical Paper 349, World Bank, Washington, DC, USA

Lioubimtseva E (2015) A multi-scale assessment of human vulnerability to climate change in the Aral Sea basin, *Environmental Earth Sciences* 73: 719–729

McDermid SS, Winter J (2017) Anthropogenic forcings on the climate of the Aral Sea: a regional modelling perspective, *Anthropocene* 20: 48–60

Micklin P (2010) The past, present, and future Aral Sea, *Lakes & Reservoirs: Science, Policy and Management for Sustainable Use* 15(3): 193–213

Milanova E, Nikanorova A, Kirilenko A, Dronin N (2018) Water deficit estimation under climate change and irrigation conditions in the Fergana Valley, Central Asia. In: Mal S, Singh R, Huggel C (eds), *Climate Change, Extreme Events and Disaster Risk Reduction*, Sustainable Development Goals Series, Springer, Cham, Switzerland

MFA (Ministry of Foreign Affairs of the Republic of Uzbekistan) (2018) Statement by the President of the Republic of Uzbekistan Shavkat Mirziyoyev and the President of the Republic of Tajikistan Emomali Rahmon on the strengthening of friendship and good neighborly relations. Retrieved 3 March 2019, https://mfa.uz/uz/press/news/2018/03/14187/ [in Uzbek]

NOAA (National Oceanic and Atmospheric Administration) (2018) *Global Historical Climatology Network – Daily (CHCN-Daily)*, Version 2 (Version Superseded). Retrieved 3 March 2019, www.ncei.noaa.gov/metadata/geoportal/rest/metadata/item/gov.noaa.ncdc:C00838/html

O'Hara SL (2000) Central Asia's water resources: contemporary and future management issues, *International Journal of Water Resources Development* 16(3): 423–441

Olsson O, Bauer M (eds) (2010) *Interstate Water Resource Risk Management: Towards a Sustainable Future for the Aral Basin*. (JAYHUN) IWA Publishing, London, UK

Omani N, Srinivasan R, Karthikeyen R, Smith PK (2017) Hydrological modeling of highly glacierized basins (Andes, Alps and Central Asia), *Water* 9(2): 111

Orlovsky N, Orlovsky L (2002) Water resources of Turkmenistan: use and conservation. Retrieved 3 March 2019, http://citeseerx.ist.psu.edu/viewdoc/download?doi=10.1.1.167.2605&rep=rep1&type=pdf

Punkari M, Droogers P, Immerzeel W, Korhonen N, Lutz A, Venalainen A (2014) *Climate Change and Sustainable Water Management in Central Asia*. ADB Central and West Asia Working Paper Series, No 5

Qushimov B, Ganiev IM, Rustamova I, Haitov B, Islam KR (2007) Land degradation by agricultural activities in Central Asia. In: Lal R, Suleimenov M, Stewart BA, Hansen DO, Doraiswamy P (eds), *Climate Change and Terrestrial Carbon Sequestration in Central Asia*. Taylor and Francis, New York, USA, pp 137–146

Rakhmatullaev S, Abdullaev I (2014) Central Asian irrigation sector in a climate change context: some reflections, *Journal of Water and Climate Change* 5(3): 341–356

Rubinova FE (1987) *Land Development and River Flow Quality in the Aral Sea Basin*. Gidrometeoizdat, Moscow, Russia [in Russian].

Savoskul OS, Shevnina EV, Perziger F, Barburin V, Danshin A (2004) How much water will be available for irrigation in the future? The Syr Darya Basin (Central Asia). In: Aerts JC, Droogers P (eds), *Climate Change in Contrasting River Basins: Adaptation Strategies for Water, Food and Environment*. Institution for Environmental Studies (IVM), Free University, Amsterdam, the Netherlands; CABI, Wallingford, UK, pp 93–113

Savoskul OS, Shevnina EV, Perziger FI, Vasilina LY, Baburin VL, Danshin AI, Matyakubov B, Murakaev RR (2003) *Water, Climate, Food, and Environment in the Syr Darya Basin*. Contribution to the project ADAPT, Adaptation Strategies to Changing Environments. Retrieved 10 March 2019, www.weap21.org/downloads/AdaptSyrDarya.pdf

Savoskul OS, Smakhtin V (2013a) *Glacier Systems and Seasonal Snow Cover in Six Major Asian River Basins: Water Storage Properties under Changing Climate*. International Water Management Institute (IWMI), Colombo, Sri Lanka

Savoskul OS, Smakhtin V (2013b) *Glacier Systems and Seasonal Snow Cover in Six Major Asian River Basins: Hydrological Role under Changing Climate*. International Water Management Institute (IWMI), Colombo, Sri Lanka

Severskiy I, Chervanyov I, Ponomarenko Y, Novikova NM, Miagkov SV, Rautalahti E, Daler D (2005) *Global International Water Assessment: Regional Assessment 24 – Aral Sea*. United Nations Environment Programme. Retrieved 3 March 2019, https://iwlearn.net/resolveuid/2cd20b92e5b24f2f6a1f34b9d131074c

Siegfried T, Bernauer T, Guiennet R, Sellars S, Robertson AW, Mankin J, Bauer-Gottwein P, Yakovlev A (2012) Will climate change exacerbate water stress in Central Asia? *Climatic Change* 112: 881–899

SNC-Lavalin (2011) Central Asia–South Asia electricity transmission and trade (CASA-1000) project feasibility study update. Montreal. Retrieved 3 March 2019, www.casa-1000.org/1)Techno-EconomicFeasbilityStudy_MainRep_English.pdf

Sokolov V (2009) *Future of Irrigation in Central Asia. IWMI-FAO Workshop on Trends and Transitions in Asian Irrigation. What are the Prospects for the Future?* 19–21 January 2009, Bangkok, Thailand

Sputnik (2018) Oʻzbekiston suv xoʻjaligi vaziri Rogʻun GESini ishga tushirish marosimida ishtirok etadi, Retrieved 3 March 2019, https://sputniknews-uz.com/world/20181112/9933082/zbekiston-suv-khzhaligi-vaziri-Roun-GESi-ishga-tushirilish-marosimida-ishtirok-etadi.html [in Uzbek].

UNECE (United Nations Economic Commission for Europe) (2007) Drainage basin of the Aral Sea and other transboundary surface waters in Central Asia. In: *First Assessment of Transboundary Rivers, Lakes and Groundwaters.* United Nations, Geneva, Switzerland, pp 68–91

UNECE (United Nations Economic Commission for Europe) (2011) Drainage basins of the Aral Sea and other transboundary waters in Central Asia. In: *Second Assessment of Transboundary Rivers, Lakes and Groundwaters.* United Nations, Geneva, Switzerland, pp 107–130

UNEP, UNDP, UNECE, OSCE, REC, NATO (2011) *Environment and Security in the Amu Darya Basin.* Retrieved 3 March 2019, www.envsec.org/publications/AmuDarya-EN-Web.pdf

Viviroli D, Weingartner R (2004) The hydrological significance of mountains: from regional to global scale, *Hydrological and Earth System Sciences Discussions* 8(6): 1017–1030

Wang X, Luo Y, Sun L, He C, Zhang Y, Liu S (2016) Attribution of runoff decline in the Amu Darya River in Central Asia during 1951–2007, *Journal of Hydrometeorology* 17: 1543–1560

Wegerich K, Kazbekov J, Mukhamedova N, Musayev S (2012) Is it possible to shift to hydrological boundaries? The Ferghana Valley meshed system, *International Journal of Water Resources Development* 28(3): 545–564

World Bank (2003) *Irrigation in Central Asia: Social, Economic and Environmental Considerations.* World Bank, Washington, DC, USA

World Bank (2013a) *Tajikistan's Winter Energy Crisis: Electricity Supply and Demand Alternatives.* World Bank Study 79616, Washington, DC, USA

World Bank (2013b) *Tajikistan: Overview of Climate Change Activities.* Washington, DC, USA. Retrieved 3 April 2019, https://openknowledge.worldbank.org/handle/10986/17552

World Bank (2013c) Uzbekistan: overview of climate change activities. Retrieved 3 March 2019, http://documents.worldbank.org/curated/en/777011468308642720/text/855660WP0Uzbek0Box382161 B00PUBLIC0.txt

Yaning C, Weihong L, Gonghuan F, Zhi L (2017) Review article: hydrological modelling in glacierized catchments of Central Asia – status and challenges, *Hydrology and Earth System Sciences Journal* 21: 669–684

Xenarios S, Gafurov A, Schmidt-Vogt D *et al.* (2019) Climate change and adaptation of mountain societies in Central Asia: uncertainties, knowledge gaps, and data constraints, *Regional Environmental Change* 19: 1339 DOI: https://doi.org/10.1007/s10113-018-1384-9

# 4

# Groundwater resources

*Abror Gafurov, Vadim Yapiyev, Mohamed Ahmed, Jay Sagin,*
*Ali Torabi Haghighi, Aigul Akylbekova and Bjørn Kløve*

**Key messages**

- Groundwater in the Aral Sea Basin (ASB) is used as a rural water supply, for irrigation and as a supply for some major cities. Recently, groundwater use has been increasing due to growing demand in agriculture. Exploitation of aquifers is favored because of lower infrastructural investment in comparison to surface water use.

- The proportion of groundwater use in the ASB varies between countries. Tajikistan and Afghanistan are the largest groundwater users, with 19.7 per cent and 15 per cent, respectively, of their total country-wide annual water withdrawal; groundwater use in Turkmenistan is only 1.1 per cent of the total national withdrawal. These numbers, however, show that the proportion of groundwater use of total use remains small within the Basin.
- In many regions, the seasonal variation of groundwater levels is linked to water consumption during irrigation season. Higher groundwater levels due to irrigation carry the risk of increased soil salinity. Reduced groundwater levels are a serious threat to ecosystems and are conducive to desertification.
- The impacts of land use change and climate change on groundwater are not well understood in the Basin at present. Groundwater recharge in the Basin seems to be reduced due to reduced soil moisture.
- Due to the lack of *in-situ* data on groundwater, the total water storage variation is estimated by remote sensing (GRACE). This indicates that the ASB is experiencing a total water storage depletion of $-3.35 \pm 0.45$ mm/year. The depletion is largely due to increase in evaporation rates and groundwater extraction given the fact that the average annual rainfall increased by 11 per cent during the investigated period (2002–2017: 333 mm) compared to that of the previous two decades (1979–2000: 299 mm).
- Groundwater quality has deteriorated in many parts of the ASB where intensive irrigation for agricultural production takes place. This also occurs because drainage systems are no longer maintained, entailing salt depositions in the soil.

## Introduction

Groundwater in the Aral Sea Basin (ASB) has a special role as a large water storage system, which interacts with the surface water systems. Major rivers originating in the mountains are mainly fed by groundwater during winter when precipitation falls and is stored in the snowpack with only little runoff. In the summer months, melting snow and river discharge replenish groundwater in the region. Despite the importance of groundwater, the observations and quantification methods in the ASB are not developed as much as for surface water. For integrated and sustainable water management the surface and groundwater resources should be considered jointly.

Groundwater resources in the ASB are important for agricultural production and for drinking purposes. Moreover, groundwater acts as a buffer for environmental flow during drought years, when the availability of surface water resources becomes limited (Zhiltsov *et al.* 2018). The population increase in the ASB and the possibility of reduced freshwater volume due to climate change in the forthcoming years is expected to augment the pressure on groundwater resources. This may result in the degradation of groundwater quality and its gradual depletion. Historically, the groundwater status in the ASB after the 1960s has deteriorated. Understanding the processes related to groundwater resources is essential also to ensure food security in the region.

This chapter initially overviews groundwater demand and supply in the ASB, followed by an analysis of inter-annual groundwater variation and trends through observed and gravimetry-based time series. The quality of groundwater in the region is then presented while the economic importance of groundwater for the ASB and the challenges to be met are demonstrated.

## Groundwater supply and demand

The land cover types of the ASB include the high mountainous areas of Pamir and Tien Shan, where the two major rivers of the Amu Darya and the Syr Darya are sourced and freshwater is formed from glaciers and snowmelt sources. The very flat lowlands in the downstream areas of the Amu Darya and Syr Darya Rivers support intensive irrigation in agricultural areas (Figure 4.1).

The amount of rainfall in lowlands is less than 200 mm/year, while the actual evapotranspiration is seasonal and may well exceed 700 mm from April till October (Forkutsa 2009). In arid regions without irrigation, groundwater recharge would occur mainly through rivers and ephemeral streams and often during episodic events (Cuthbert *et al.* 2016). Recently, Micklin (2016) has estimated the renewable usable groundwater resources in the ASB to be as high as 44 km³, of which 16 km³ is freshwater that may be connected to surface water resources. In the same study, it was shown that important ecosystems have been downgraded while desertification increased due to surface water and groundwater depletion.

Irrigation of the lowlands in the ASB leads to groundwater recharge in the summer months. Winter runoff can also become an important source of shallow groundwater recharge as well as river bank infiltration due to modified reservoir operation in upstream countries (Tajikistan, Kyrgyzstan). These changes have influenced the intra-annual hydrological regimes within the entire Basin.

Groundwater is often the main source for domestic water supply in rural areas and a major resource in some cities (Rakhmatullaev *et al.* 2010; UNESCO-IHP 2016). Some studies have

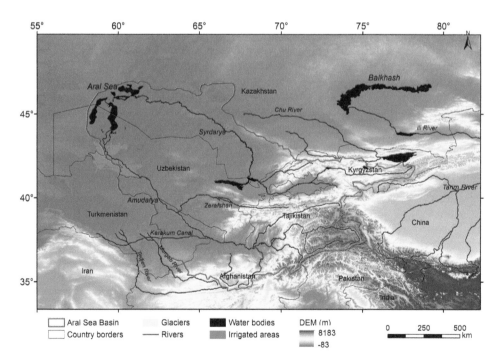

*Figure 4.1* Irrigated areas in the ASB
Source: After Conrad *et al.* 2016

indicated that groundwater resources in some of the arid areas of Central Asia (CA) could be sufficient to satisfy the demands for crop irrigation (Ostrovsky 2007). While groundwater has been considered to represent a strategic resource since Soviet times, surface water (such as rivers, canals, lakes and reservoirs) was the primary source of water. Groundwater exploitation in the ASB has, however, been increasing in recent years due to its good quality and quantity as an alternative to surface water sources. More extensive use of groundwater for irrigation purposes began in the drought years of 1998–2001 (FAO 2013). Since then, groundwater has been considered a reliable source of irrigation conjunctively used with surface waters.

The trends of irrigated areas with groundwater withdrawal vary from country to country and over time. For example, in Afghanistan the irrigated areas with groundwater withdrawal increased from 492.646 ha in 1993 to 577.156 ha in 2002. In Tajikistan, on the other hand, the proportion decreased from 68,000 ha in 1994 to 32,500 ha in 2009 (FAO 2013). As shown in Table 4.1, Tajikistan has the highest share of groundwater use (19.7 per cent) followed by Afghanistan (15 per cent) and Uzbekistan (8.9 per cent). Kyrgyzstan (3.8 per cent) and Turkmenistan (1.1 per cent) present the lowest groundwater withdrawal, while in Kazakhstan the share is about 4.9 per cent. However, only a relatively small part of Kazakhstan belongs to the ASB, so the country is less relevant to this study.

However, in volumetric usage, groundwater extraction is highest in Uzbekistan among all ASB countries, with about 5km$^3$ (5,000 million m$^3$; 8.9 per cent of total water consumption), followed by Afghanistan, with an extraction rate of about 3 km$^3$ (15 per cent of total water consumption). The groundwater extraction rate in Tajikistan is about 2.3 km$^3$ (19.7 per cent of total water consumption), while in Kazakhstan it is distinctively lower, at about 1 km$^3$ (4.9 per cent of total water consumption) (FAO 2013). Turkmenistan and Kyrgyzstan are among the lowest groundwater users, with about 0.3 km$^3$ each (3.8 per cent in Kyrgyzstan; 1.1 per cent in Turkmenistan).

Groundwater resources are extremely important for Afghanistan since nearly 18 per cent of agricultural area in the country is irrigated by groundwater withdrawal, which corresponds to the intensity of groundwater extraction as illustrated in Figure 4.2. Uzbekistan, with about 6.4 per cent of its agricultural area irrigated by groundwater withdrawals, is the second major consumer in the ASB, followed by Tajikistan (6.4 per cent). Kyrgyzstan, Kazakhstan and Turkmenistan use groundwater for less than 1 per cent of the irrigated area.

Extensive groundwater use can decrease surface water availability. In Uzbekistan, the threshold for actual groundwater withdrawal with no impact on surface water is estimated at 7.5 km$^3$/year. The groundwater extraction potential in the entire ASB region is estimated at 13.1 km$^3$/year (CAWaterInfo 2018).

*Table 4.1* Share of surface (SF) and groundwater (GW) used for irrigation in Central Asia

| Country | Annual water withdrawal | | | | | Area (thousand ha) irrigated using | | | |
|---|---|---|---|---|---|---|---|---|---|
| | Total (km$^3$) | SF (km$^3$) | % | GW (km$^3$) | % | SF | % | GW | % |
| **Afghanistan** | 20.4 | 17.3 | 85.0 | 3.1 | 15.0 | 2,631 | 82.0 | 577 | 18.0 |
| **Kyrgyzstan** | 8.0 | 7.4 | 92.4 | 0.3 | 3.8 | 1,011 | 99.0 | 10.2 | 1.0 |
| **Tajikistan** | 11.5 | 8.9 | 77.7 | 2.3 | 19.7 | 696 | 93.9 | 32.5 | 4.4 |
| **Uzbekistan** | 56.0 | 44.2 | 78.9 | 5.0 | 8.9 | 3,929 | 93.6 | 269 | 6.4 |
| **Turkmenistan** | 27.9 | 27.2 | 97.4 | 0.3 | 1.1 | 1,981 | 99.5 | 9.6 | 0.5 |
| **Kazakhstan** | 21.1 | 18.9 | 89.7 | 1.0 | 4.9 | 1,197 | 99.8 | 2.0 | 0.2 |

Source: FAO 2013

## Regional and transboundary aquifers

The regional groundwater reserves in the ASB are estimated to be about 31.17 km³, of which 14.7 km³ are located in the Amu Darya Basin and 16.4 km³ in the Syr Darya Basin. About 30 per cent of these reserves are of transboundary nature (CAWaterInfo 2018). Groundwater recharge occurs by rainfall and river flow originating from the Pamir and Tien Shan Mountains in the eastern parts of the Basin. Recently, researchers have focused on identifying large aquifers on the territory of CA (Lee *et al.* 2018; Karimov *et al.* 2015). In addition, international partners have carried out groundwater mappings in some regions.

The largest aquifer in the region, with an area of about 400,000 km², is the Syr Darya aquifer (Figure 4.2), located between Kazakhstan and Uzbekistan (Lee *et al.* 2018). The second-largest transboundary aquifer in the ASB is the Amu Darya artesian basin. It is located between Kazakhstan, Turkmenistan and Uzbekistan and covers an area of about 260,000 km². The third-largest transboundary aquifer system is the Birata-Urgench artesian basin, located along the Amu Darya River. The Birata-Urgench artesian system covers about 80,000 km² and is located between Turkmenistan, Uzbekistan and Kazakhstan.

All of these aquifer systems are connected to the Aral Sea. The Syr Darya artesian system is connected to the southeastern part of the Aral Sea, the Amu Darya artesian system to the

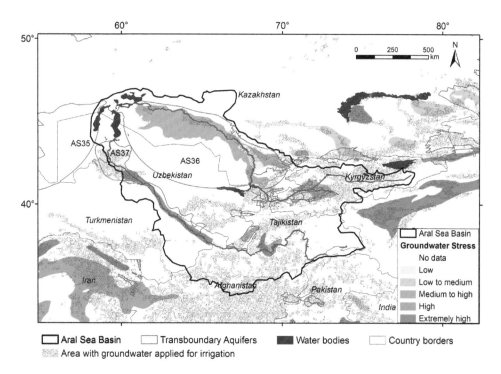

*Figure 4.2* Map of the Aral Sea Basin with transboundary aquifers (TBAs), areas with groundwater applied for irrigation and groundwater stress index. The largest TBAs with international codes of AS35, AS36 and AS37 are located in the Amu Darya, Syr Darya and Birata-Urgench artesian basins

Source: UNESCO-IHP-WINS (n.d.)

northern part and the Birata-Urgench artesian system to the southern part (the Amu Darya Delta). Aquifers with a connection to the Aral Sea also contribute to the water balance of the lake. Jarsjö and Destouni (2004) have estimated the total groundwater contribution to the water balance of the Aral Sea from 1960 to 1996. They concluded that the role of groundwater has increased drastically during this period from 12 per cent of all the rivers' runoff in 1960 to approximately 100 per cent in 1996 due to the well-known streamflow reductions from the Amu Darya and Syr Darya Rivers. Oberhänsli *et al.* (2009) performed an extensive stable water isotope (hydrogen and oxygen) survey (2004–2006) for the characterization of the Aral Sea water. One of their goals was to assess the interactions between the Aral and groundwater bodies in several locations. They concluded that groundwater flow contribution to the Aral Sea is significant in the spring and the autumn. The stable water isotopic composition of both artesian water, shallow groundwater and the lake indicates the increased importance of fossil (old) subsurface water flow input, most probably due to lowering of the water level of the Aral Sea over many years.

During Soviet times, the aquifers were studied with regards to their potential for water supply in the region. For example, a study was conducted to analyze aquifer potential for water supply of the Pretashkentskiy aquifer, located between Kazakhstan and Uzbekistan (near Tashkent) and the groundwater extraction capacity was estimated. The amount was 1,464 $m^3$/ day for Kazakhstan and 2,044 $m^3$/day for Uzbekistan. The total groundwater volume of this transboundary aquifer has been estimated to be as high as 97.6 $km^3$ (UNESCO 2016). Figure 4.2 also shows the areas with groundwater used for irrigation. Groundwater extraction for irrigation mainly occurs along the rivers of CA. Moreover, intensified groundwater use for irrigation can be observed in the Amu Darya Delta. The groundwater stress index presented in Figure 4.2 shows areas where groundwater extraction is much higher than groundwater recharge. The groundwater stress along the Amu Darya River is extremely high and along the Syr Darya River is moderate to high.

## Seasonal groundwater variation in irrigated lowlands

In the Amu Darya River Delta, about 1,200 mm surface water is used seasonally for cotton production alone (Conrad 2006). Out of this amount, 430 mm is used for pre-season soil leaching (Djanibekov 2008), which is essential in many parts of CA to cope with salinity, and eventually feeds into the groundwater (Ibrakhimov *et al.* 2007). Water losses to groundwater occur also from unlined irrigation canals (Forkutsa 2009).

The overall effects of irrigation may contribute to seasonal groundwater level fluctuation over 100 mm in some places (Schettler *et al.* 2013). In the Khorezm region, for instance (Amu Darya Delta), the groundwater level rises in spring with the start of cropping activities (pre-season leaching) and reaches a peak in March–April (Figure 4.3). The groundwater levels are the lowest during July–August when irrigation is at its maximum (Ibrakhimov 2004).

## Groundwater variability by GRACE satellite data

As data availability from groundwater observations in Central Asia is limited, remote sensing-based observation is used to reveal the spatial variation of groundwater in the region. Gravity Recovery and Climate Experiment (GRACE) data are used as an integrated dataset of total water storage (TWS) variations, representing the total, vertically integrated, variations of

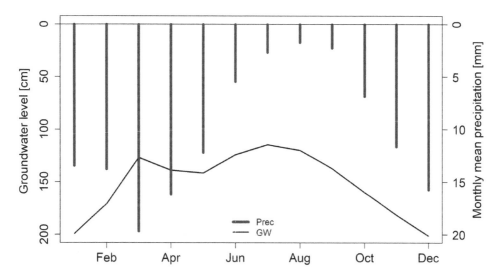

*Figure 4.3* Seasonal variations of the groundwater table in the Amu Darya Delta. Monthly values correspond to long-term means derived from decadal observations during the period from 2000 to 2005 in the Khorezm region. The precipitation data is the long-term means derived from daily precipitation observations at Urgench meteorological station in the period from 1936 to 2006

Source: Authors

water stored as ice, snow, soil moisture, groundwater and surface water bodies. GRACE is a joint satellite mission between the National Aeronautics and Space Administration (NASA, the USA) and the German Aerospace Center (DLR, Germany) launched in March 2002 to map the global, temporal and static gravity fields of the Earth (Tapley *et al.* 2004). GRACE data have been successfully applied to monitor groundwater resources in several parts of the world (e.g., Ahmed *et al.* 2011, 2014, 2016; Fallatah *et al.* 2017, 2018; Ahmed and Abdelmohsen 2018; Niyazi *et al.* 2018) including the CA region (Ebead *et al.* 2017; Deng and Chen 2016).

As the other components of TWS (e.g., surface water, snow and ice) are negligible in the lowland areas of CA, we attribute the spatiotemporal variations in TWS data obtained by GRACE satellites mainly to spatiotemporal variations in soil moisture and groundwater storages. Monthly GRACE mass concentration (mascon) solutions, generated by the University of Texas Center for Space Research (UTCSR), spanning April 2002 to June 2017 have been used in this study. GRACE mascon solutions do not require spectral (e.g., destriping) and spatial (e.g., smoothing) filtering and provide higher signal-to-noise ratio, higher spatial resolution and reduced error compared to other GRACE approaches (e.g., Save *et al.* 2016). UTCSR mascon solutions have been generated over hexagonal grids (1° at the Equator, total 41,000) and resampled at 0.5° × 0.5° TWS grids. TWS time series were generated by spatially averaging TWS results over the spatial domain of these areas. Trends in TWS data were extracted by simultaneously fitting annual (sine and cosine), semiannual (sine and cosine) and a trend parameter to each TWS time series. The TWS trend solutions over CA Asia are shown in Figure 4.4.

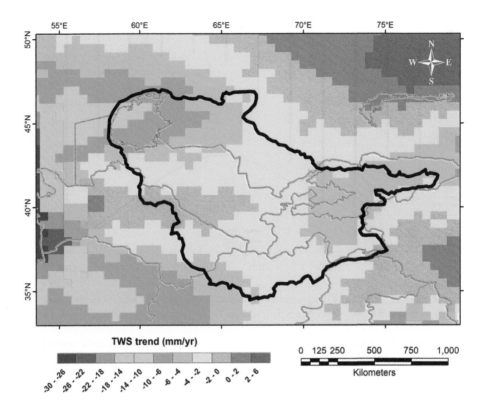

*Figure 4.4* Secular trends (mm/yr) in GRACE-derived TWS over the Central Asia region and surroundings. Also shown is the spatial distribution of the ASB (black polygon)

Source: Authors

The blue areas in Figure 4.4 are experiencing TWS depletion trends, whereas red areas are experiencing TWS wetting trends. In particular, it seems that the areas to the northeast and southwest of the ASB are subject to wetting TWS trends, whereas those to the west of the ASB are witnessing TWS depletion trends. Within the ASB, TWS trends indicate that depletion increases from the center to the margins of the ASB. Given the fact that GRACE data have no vertical resolution, independent sources for soil moisture, reservoirs and snow/ice storage should be utilized to quantify the GRACE-based groundwater storage. However, in this investigation the TWS was used as a proxy for the 'accessible water' storage given the scale of the study area. Accessible water storage variations are defined in this study as changes in the combination of groundwater, soil moisture, surface water and snow/ice.

Figure 4.5 shows the temporal variations in GRACE-derived TWS averaged over the ASB. As presented, ASB is experiencing a TWS depletion of $-3.35 \pm 0.45$ mm/year, which is equivalent to $-5.82 \pm 0.87$ Gt/year. It is worth mentioning that TWS depletion rates during the period from April 2002 to March 2016 were estimated at $-4.46 \pm 0.52$ mm/year ($-7.75 \pm 0.90$ Gt/year) and that these rates are quite similar to those ($-7.31 \pm 1.68$ Gt/year) reported by Wang *et al.* (2013) during the same period.

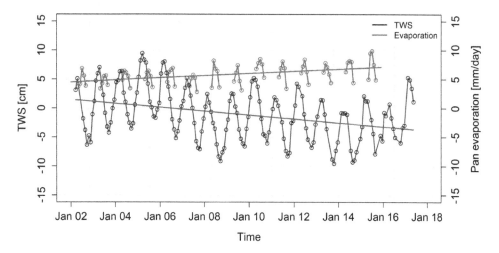

*Figure 4.5* Temporal variations in monthly (April 2002–June 2017) GRACE-derived TWS (averaged over the ASB) and observed mean daily pan evaporation from May to September (May 2002–September 2015) at Shalkar observation point

Source: Authors

Surface water storage depletion accounts for more than 80 per cent of the observed TWS depletion in the ASB (Wang *et al.* 2013). The decline in surface water storage could be attributed to a decline in rainfall rates and/or an increase in evapotranspiration rates. However, the ASB is witnessing an increase in rainfall rates during the investigated period. Analysis of the Global Precipitation Climatology Project (GPCP) rainfall data (Adler *et al.* 2018) indicated that the average annual rainfall over the ASB increased by 11 per cent during the investigated period (2002–2017: 333 mm) compared to that of the previous two decades (1979–2000: 299 mm). The northern part of the ASB is witnessing increased evaporation rates. Figure 4.5 also shows an evaporation time series for the Shalkar pan evaporation observation station (northwest of the ASB, 47°49'59.00"N 59°37'14.00"E, 171 msl). In this area, the pan evaporation rate approximately equals open-water evaporation. The increasing trend in pan evaporation indicates a growth in atmospheric demand for water that can lead to soil moisture desiccation in the ASB. The decreasing trend in soil moisture is also supported by recent satellite observations in the ASB (Jin *et al.* 2017). The sharp increase in evaporation rates over the ASB provides a potential explanation for the TWS decline. TWS decline could be also attributed to groundwater extraction. The central parts of the ASB are lowland areas and areas of increased groundwater extraction for agricultural activities.

## Groundwater quality

Groundwater quality in the ASB is commonly affected by agriculture, municipal wastewater and industry (Bekturganov *et al.* 2016). The quality of groundwater varies from region to region. In the areas close to mountains and in the western part of the ASB, the quality of groundwater is better when compared to the deltas of the Amu Darya and Syr Darya.

In the deltas, there is much return flow from irrigated areas, which degrades groundwater sources (Abuduwaili *et al.* 2019).

In areas with extensive irrigation, waterlogging causes pollution from chemicals used in agriculture (World Bank 2003). The pollution of shallow groundwater is of particular concern in irrigated areas of the ASB since nitrates, ammonium and persistent organics permeate to groundwater used for drinking purposes (Panichkin *et al.* 2017). Shallow and fossil groundwater has good potential to satisfy the needs of the local population if adequate water treatment technologies are implemented.

The total dissolved content (TDS) depends on the aquifer features and may vary considerably in the ASB. The salt content of groundwater varies from 1 to 3 g/L (CAWaterInfo 2018). The shallow groundwater is more saline in the lowlands where intensive evaporation leads to salt accumulation in the top soil. For example, in the southeastern part of the Syr Darya River Basin near the Karatau Ridge, groundwater (the aquifer of the Upper-Turon-Senonian deposits) is fresh (0.3–0.4 g/L) with a dominance of bicarbonate and calcium ions (Panichkin *et al.* 2011). The groundwater salinity in ASB increases in aquifers from south to north and from east to west with domination of chloride-sulfate ions. It is highest near the Aral Sea bed with a peak value of 50–100 g/L (Panichkin *et al.* 2011). Recently, Kazhydromet reported that in the lowlands of the Syr Darya River (Kyzylorda town) open water has a salinity ranging from 1.0 to 1.3 g/L with sulphate content of approximately 500 mg/L. At the same time, it is mentioned that the salinity of groundwater potable water sources in Kazakhstan meets drinking water standards with a sulphate level comparable to surface water bodies (Ministry of Energy 2017).

Previously, Friedrich (2009) reported that ASB was contaminated in the Syr Darya watershed from uranium mining activities, and groundwater was also affected. Panichkin *et al.* (2011) reported that near the Aral Sea the local population increasingly relies on groundwater as a main source of potable water due to the degradation of surface water quality. Proper surface and subsurface quality monitoring and mapping is needed in the lowland areas of the endorheic ASB in particular, as these are the final sinks for contaminants.

## Groundwater mapping and data availability

Groundwater maps are outdated and data sharing is still scarce between the ASB countries. The publicly available geodatabases are still deprived of hydrogeological maps due to restriction policies in each country. Most maps were developed during Soviet times and some countries have not yet updated the data since this period.

The International Groundwater Resource Assessment Center (IGRAC) has tried to conduct studies related to groundwater in the transboundary Kazakhstan–Uzbekistan Pretaskent aquifer, but has only been able to access data for a part of Kazakhstan. The previous government of Uzbekistan had imposed restrictions on access to groundwater data, maps and hydrogeological studies. Only recently have Uzbek scientists been able to join Kazakh colleagues during an experts' meeting in Tashkent held in May 2018, in a common effort to model groundwater resources of the Pretashkent aquifer (UN-IGRAC 2019). Currently, hydrogeological information and maps on the Syr Darya aquifer are publicly available only in Kazakhstan. Hydrogeological information and maps for the Amu Darya aquifer are still not available from the side of Uzbekistan. However, the new Uzbek government prioritizes regional and international cooperation in joint research and data transparency and more access to groundwater data is anticipated.

## Socio-economic implications of groundwater use

Groundwater use can directly influence the river flow regime by decreasing the baseflow of rivers and the drying up of springs and wells in headwaters. Change in surface water use can also impair the performance of current hydraulics. In addition, mining groundwater can affect groundwater-dependent ecosystems (Kløve *et al.* 2014) such as groundwater lakes. Elevated groundwater levels in regions with malfunctioning drainage systems may lead to the deterioration of soil quality (salinity), reduced agricultural production, and unemployment in the agriculture sector. About 70 per cent of irrigated land in downstream Amu Darya areas has a groundwater level of about 1.5 meters (NATO 2016), which may lead to severe land degradation in the ASB.

Groundwater resources are becoming more important as an alternative source of water for irrigation and drinking, due to frequent droughts in the last years (1998–2001, 2007–2008, 2018). In many parts of CA, depletion of groundwater was significant during these periods with a decreased amount of water to be pumped from below the ground due to insufficient groundwater recharge. Thus, groundwater contributes significantly to food security and economic stability in the ASB.

The use of fertilizers and pesticides for agricultural production in CA can contribute to reduced groundwater quality in aquifers. Saline and arsenic groundwater is common in the region. Information on pesticide or nitrate concentration is lacking. In some places, fresh groundwater is filled into bottles and sold as mineral drinking water. For example, in the Pretashkenstkiy TBA, 0.265 km$^3$ of fresh groundwater was used for bottling as drinking water (UNESCO-IHP 2016).

## Future challenges for groundwater use

Artificially recharged groundwater storage-water banking could be considered as one of the options to meet water issues within the ASB in the near future (Karimov *et al.* 2014). However, thorough research on the groundwater storage potential in the region must be conducted. Aquifer systems are superior to surface reservoirs in terms of evaporation losses and can also function as water storage during drought periods. However, increased water storage may lead to increased use of water and expansion of agricultural lands, which, in turn, may once again cause water scarcity. Elevated groundwater level due to inefficient irrigation approaches (e.g., surface irrigation) and a malfunctioning drainage system (Dukhovny *et al.* 2007) may further lead to deterioration of groundwater quality and salinization of agricultural land. Thus, more efficient irrigation methods should be applied in the ASB to mitigate the salinization of soil due to evaporation. Proper legislation on securing groundwater resources is still missing or outdated. A new legislative framework should be developed among countries sharing the ASB to jointly manage groundwater resources, especially transboundary aquifers.

## Acknowledgments

Khorezm Rural Advisory Support Service (KRASS) and the Ministry of Agricultural and Water Resources of Uzbekistan are acknowledged for providing observed groundwater level data.

# References

Abuduwaili J, Issanova G, Saparov G (2019) *Hydrology and Limnology of Central Asia.* Singapore, Springer Singapore. DOI: https://doi.org/10.1007/978-981-13-0929-8

Adler R, Sapiano M, Huffman G, Wang J-J, Gu G, Bolvin D *et al.* (2018) The Global Precipitation Climatology Project (GPCP) monthly analysis (new version 2.3) and a review of 2017 global precipitation, *Atmosphere* 9(4): 138

Ahmed M, Abdelmohsen K (2018) Quantifying modern recharge and depletion rates of the Nubian Aquifer in Egypt, *Surveys in Geophysics* 39(4): 729–751

Ahmed M, Sultan M, Wahr J, Yan E, Milewski A, Sauck W *et al.* (2011) Integration of GRACE (Gravity Recovery and Climate Experiment) data with traditional data sets for a better understanding of the time-dependent water partitioning in African watersheds, *Geology* 39(5): 479–482

Ahmed M, Sultan M, Wahr J, Yan E (2014) The use of GRACE data to monitor natural and anthropogenic induced variations in water availability across Africa, *Earth-Science Reviews* 136: 289–300

Ahmed M, Sultan M, Yan E, Wahr J (2016) Assessing and improving land surface model outputs over Africa using GRACE, field, and remote sensing data, *Surveys in Geophysics* 37(3): 529–556

Bekturganov Z, Tussupova K, Berndtsson R, Sharapatova N, Aryngazin K, Zhanasova M (2016) Water-related health problems in Central Asia – a review, *Water* 8: 219

CAWaterInfo (2018) The Aral Sea Basin. Retrieved 5 January 2018, www.cawater-info.net/index_e.htm

Conrad C (2006) Fernerkundungsbasierte Modellierung und hydrologische Messungen zur Analyse und Bewertung der landwirtschaftlichen Wassernutzung in der Region Khorezm (Usbekistan). Ph.D. Thesis, University of Würzburg, Würzburg, Germany

Conrad C, Schönbrodt-Stitt S, Löw F, Sorokin, Paeth H (2016) Cropping intensity in the Aral Sea Basin and its dependency from the runoff formation 2000–2012, *Remote Sensing* 8(8): 630

Cuthbert MO, Acworth RI, Andersen MS, Larsen JR, McCallum AM, Rau GC, Tellam JH (2016) Understanding and quantifying focused, indirect groundwater recharge from ephemeral streams using water table fluctuations, *Water Resources Research* 52: 827–840

Deng H, Cheng Y (2016) Influences of recent climate change and human activities on water storage variations in Central Asia, *Journal of Hydrology* 544: 46–57

Djanibekov N (2008) *A Micro-Economic Analysis of Farm Restructuring in the Khorezm Region, Uzbekistan*, Ph.D. thesis, University of Bonn, Germany

Dukhovny V, Umarov P, Yakubov H, Madramootoo Ch (2007) Drainage in the Aral Sea Basin, *Irrigation and Drainage* 56: 91–100

Ebead B, Ahmed M, Niu Z, Huang N (2017) Quantifying the anthropogenic impact on groundwater resources of North China using Gravity Recovery and Climate Experiment data and land surface models, *Journal of Applied Remote Sensing* 11: 026029

Fallatah OA, Ahmed M, Save H, Akanda AS (2017) Quantifying temporal variations in water resources of a vulnerable Middle Eastern transboundary aquifer system, *Hydrological Processes.* DOI: https://doi.org/10.1002/hyp.11285

Fallatah OA, Ahmed M, Cardace D, Boving T, Akanda AS (2018) Assessment of modern recharge to arid region aquifers using an integrated geophysical, geochemical, and remote sensing approach, *Journal of Hydrology.* DOI: https://doi.org/10.1016/J.JHYDROL.2018.09.061

FAO (2013) *Irrigation in Central Asia in Figures.* AQUASTAT Survey–2012. FAO Water Reports 39. Rome, Italy

Forkutsa I, Sommer R, Shirokova I, Lamers J, Kienzler K, Tischbein B, Martius C, Vlek B (2009) Modeling irrigated cotton with shallow groundwater in the Aral Sea Basin of Uzbekistan: I. Water dynamics, *Irrigation Science* 27: 331

Friedrich J (2009) Uranium contamination of the Aral Sea, *Journal of Marine Systems* 76(3): 322–335

Jarsjö J, Destouni G (2004) Groundwater discharge into the Aral Sea after 1960, *Journal of Marine Systems* 47(1–4): 109–120

Jin Q, Wei J, Yang Z-L, Lin P (2017) Irrigation-induced environmental changes around the Aral Sea: an integrated view from multiple satellite observations, *Remote Sensing* 9: 900

Ibrakhimov M (2004) *Spatial and Temporal Dynamics of Groundwater Table and Salinity in Khorezm (Aral Sea Basin), Uzbekistan.* Ecology and Development Series 24, Cuvillier Verlag, Göttingen, Germany

Ibrakhimov N, Everett RS, Esanbekov U, Kamilov BS, Mirzaev L, Lamers JPA (2007) Water use efficiency of irrigated cotton in Uzbekistan under drip and furrow irrigation, *Agricultural Water Management* 90(1–2): 112–120

Karimov A, Smakhtin V, Mavlonov A, Borisov V, Gracheva I, Miryusupov F, Akhmedov A, Anzelm K, Yakubov S, Karimov A (2015) Managed aquifer recharge: potential component of water management in the Syr Darya River Basin, *Journal of Hydrologic Engineering* 20(3): B5014004 (1–12)

Kløve B, Ala-Aho P, Bertrand G, Gurdak JJ, Kupfersberger H, Kværner J *et al.* (2014) Climate change impacts on groundwater and dependent ecosystems, *Journal of Hydrology* 518: 250–266

Lee E, Jayakumar R, Shrestha S, Han Z (2018) Assessment of transboundary aquifer resources in Asia: Status and progress towards sustainable groundwater management, *Journal of Hydrology: Regional Studies* 20: 103–115

Micklin P (2016) The future Aral Sea: hope and despair, *Environmental Earth Sciences* 75(9): 844

Ministry of Energy (2017) *Information Bulletin on the Environment and Human Health in the Aral Sea Basin: 2017.* Retrieved 5 January 2018, https://kazhydromet.kz/upload/pdf/ru_1516180142.pdf

NATO (2016) *Proceedings of the NATO Advanced Research Workshop on The Socio-Economic Causes and Consequences of Desertification in Central Asia*, Bishkek, Kyrgyzstan

Niyazi BA, Ahmed M, Basahi JM, Masoud MZ, Rashed MA (2018) Spatiotemporal trends in freshwater availability in the Red Sea Hills, Saudi Arabia, *Arabian Journal of Geosciences* 11(22): 702

Oberhänsli H, Weise SM, Stanichny S (2009) Oxygen and hydrogen isotopic water characteristics of the Aral Sea, Central Asia, *Journal of Marine Systems* 76(3): 310–321

Ostrovsky VN (2007) Comparative analysis of groundwater formation in arid and super-arid deserts (with examples from Central Asia and Northeastern Arabian Peninsula), *Hydrogeology Journal* 15: 759

Panichkin V, Miroshnichenko O, Zakharova N, Satpayev A, Trushel L, Kalmykova N, Veselov V *et al.* (2011) ll. Eastern Priaralye groundwaters. Retrieved 14 February 2019, http://old.unesco.kz/water/ar_ch_2_r.htm

Panichkin V, Sagin J, Miroshnichenko O, Trushel L, Zakharova N, Yerikuly Z, Livinskiy Y (2017) Assessment and forecasting of the subsurface drain of the Aral Sea, Central Asia, *International Journal of Environmental Studies* 74(2): 202–213

Rakhmatullaev Sh, Huneau F, Kazbekov J, Le Coustumer Ph, Jumanov J, El-Oifi B, Motelica-Heino M, Hrkal Z (2010) Groundwater resources use and management in the Amu Datya River Basin (Central Asia), *Environmental Earth Sciences* 59(6): 1183–1193

Save H, Bettadpur S, Tapley B (2016) High resolution CSR GRACE RL05 mascons, *Journal of Geophysical Research: Solid Earth* 121: 7547–7569

Schettler G *et al.* (2013) Hydrochemical water evolution in the Aral Sea Basin. Part I: Unconfined groundwater of the Amu Darya Delta – Interactions with surface waters, *Hydrology* 495: 267–284

Tapley BD, Bettadpur S, Ries JC, Thompson PF, Watkins MM (2004). GRACE measurements of mass variability in the Earth system, *Science* 305(5683): 503–505

UNESCO-IHP (2016) *Governance of Groundwater Resources in Transboundary Aquifers (GGRETA), Main Achievements and Key Findings.* Retrieved 10 January 2019, https://unesdoc.unesco.org/ark:/48223/pf0000245266

UNESCO-IHP-WINS (n.d.) *The Water Information Network System.* Retrieved 25 January 2019, https://en.unesco.org/ihp-wins

UN-IGRAC (2019) *International Groundwater Resources Assessment Centre.* Retrieved 25 January 2019, www.un-igrac.org/

Wang P, Yu J, Zhang Y, Liu C (2013) Groundwater recharge and hydrogeochemical evolution in the Ejina Basin, northwest China, *Journal of Hydrology* 476: 72–86

World Bank (2003) *Irrigation in Central Asia. Social, Economic and Environmental Considerations.* World Bank, Washington, DC, USA

Zhiltsoz S, Zonn S, Kostinanoy I, Semenov A (2018) Water resources in Central Asia: international context. In: *The Handbook of Environmental Chemistry.* Springer, Berlin, Germany

# 5

# Hydropower

*Frank Schrader, Anvar Kamolidinov, Maksud Bekchanov,*
*Murodbek Laldjebaev and Stella Tsani*

## Key messages

- Hydropower can become a major source of energy and support socio-economic development in the entire Aral Sea Basin. Hydropower potential in the Basin is large but remains underutilised to date. The upstream riparian states—Kyrgyzstan and Tajikistan—are the lead producers of hydropower in the Syr Darya and Amu Darya Basins, respectively.

- In order to avoid regional conflicts over the use of water for food or electricity generation, riparian countries should develop viable forms of cooperation. There are solid examples of such cooperation in the Basin already. Legal and regulatory solutions like the 1998 Syr Darya Agreement should be introduced for supporting food production in summer for the upstream countries (Tajikistan, Kyrgyzstan) and electricity generation in the downstream countries (Kazakhstan, Uzbekistan, Turkmenistan).
- Large water storage and hydropower development projects in upstream river courses where hydropower potential is also the largest continue to emerge in the region (e.g., the Rogun Dam Project, which is currently under construction). If managed in a collaborative manner, they will bring multiple benefits to collaborating countries, including improved reliability of supply and availability of water for agriculture, domestic use and electricity generation.
- Small hydropower plants and other renewable sources represent another energy option that can be developed in the Basin to supplement carbon fuel-based and large hydropower facilities.

## Introduction

There is great hydropower potential in the Aral Sea Basin (ASB) due to large snowpack areas and glaciers in the mountains of the upstream countries. During the 1960–1980s, a complex and closely interconnected system of reservoirs and canals often associated with hydropower stations was developed with the objectives to regulate the Syr Darya and Amu Darya Rivers for irrigated agriculture on large alluvial plains, flood risk reduction and hydropower generation.

The seasonal fluctuations associated with hydropower were previously compensated by a Central Asian regional energy system. The Central Asia Power System (CAPS) was established in the 1970s and included all five former Central Asian Soviet republics: Kazakhstan, Kyrgyzstan, Uzbekistan, Turkmenistan and Tajikistan. During the Soviet period, internal borders were disregarded, and the CAPS could meet the needs of the whole region. In summer months, water-rich upstream countries Tajikistan and Kyrgyzstan were responsible for releasing water and generating electricity for the whole region. In return, they were receiving fossil fuels and surplus electricity in winter from the hydrocarbon-rich downstream countries, Uzbekistan, Kazakhstan and Turkmenistan. The high regional demand for irrigation was met throughout summer, while winter energy shortages in upstream countries were similarly compensated (Shenhav *et al.* 2019).

The reservoirs in the upstream and midstream locations of both the Syr Darya and Amu Darya Rivers became the backbone of the economic and social development mainly in steppe and arid lowland zones. Large irrigation schemes, large state (*sovkhoz*) and collective (*kolkhoz*) farms, cities and industrial infrastructure were established to improve the economic and social conditions in Central Asia (CA).

In the post-Soviet Union era, the newly established CA countries were challenged by maintaining the regional water and energy infrastructure that they had inherited while the regional water–energy trade was almost halted. The upstream countries of Kyrgyzstan and Tajikistan, with their lack of fossil fuels and strongly centralised governmental systems, developed a predominantly engineering and self-sufficiency approach to water resource management that focuses mainly on electricity production through water releases from hydropower stations during winter time (Valentini *et al.* 2004; Jigarev 2008).

Since independence, controversial views and disputes on principles, shares and annual limits of water resources mainly between upstream and downstream CA countries have partially impeded regional cooperation. There are, for instance, views in Kyrgyzstan that historical water distribution schemes should be abandoned, and upstream CA countries should have greater rights on the water resources originating from their territory. Kazakhstan and Uzbekistan have indicated in the past that such an approach is unacceptable, and that water should be managed as a common resource and distribution decisions should consider the interests of all riparian states (Rakhmatullaev *et al.* 2010).

Although regional cooperation structures and rules for common water resources use are in place, the practical implementation of shares and limits is insufficient, especially in dry periods at the end of the growing season, when national needs often supersede regional concerns.

During the short period from 1999 until 2003, Kazakhstan, Kyrgyzstan and Uzbekistan developed and established new cooperation and water sharing systems, defined by the 1998 Agreement on the Use of Water and Energy Resources of the Syr Darya Basin (later also joined by Tajikistan), which in 2004 unfortunately came to an end due to disagreements on water allocation and rights. Upstream–downstream conflicts in the last two decades were mainly caused by water release shortfalls in amount and time (Karimov *et al.* 2013).

This chapter describes the history, current state and prospects of hydropower development in the ASB. It also reviews the approaches that were or are attempted by riparian countries to find a common ground in hydropower development, including i) the development of legal and regulatory mechanisms, such as the 1998 Syr Darya Agreement and the UN Water Convention for balancing national water–food–energy interests, ii) construction and joint use of large water storage facilities in the upstream areas, using the Rogun hydropower plant (HPP) as an example and iii) increased development of renewable energy sources (RES), especially small HPPs based on run-off structures without reservoirs (ADB 2010).

## Reservoirs and hydropower: Basin overview

Most of the population, irrigation zones and industrial enterprises in CA are concentrated along the lower reaches of the Amu Darya and Syr Darya Rivers. High annual flow variation of both rivers increases the risks of inadequate supply of water for irrigated agriculture, industrial development and drinking water, primarily in downstream regions with higher dependence on freshwater.

The flow regulation and water distribution between upstream and downstream countries continued through Soviet times and is ongoing. Indicatively, the total storage capacity of all water reservoirs in the ASB has increased from 4 km³ to 76 km³ during the period between 1950 and 1990. The main leap—from 17 km³ to 56 km³—occurred in the late 1970s (Dukhovny and Schutter 2011), when large hydro-technical infrastructure was established in the mountainous zones of the upstream countries. The main goal was to distribute water flows more evenly within the ASB on a seasonal basis and to provide a stable water supply for irrigating croplands, especially in midstream and downstream areas. The reservoirs were unevenly distributed mainly due to physiographical reasons.

Since the 1990s, both Kyrgyzstan and Tajikistan have prioritised energy production over downstream irrigation needs. The reservoirs that were initially planned for irrigation downstream and secondary for hydropower purposes upstream have now inverted their operational mode by prioritising hydropower over agricultural needs. Both upstream countries were releasing more water from reservoirs for electricity generation during the winter period than had been agreed upon during the Soviet Union era. This significant change in the operational regime along the Syr Darya Basin especially has caused shortfalls of water for irrigation during the vegetation period, and often heavy floods and damages of settlements, infrastructure and irrigated land during winter in downstream countries. Nowadays, Tajikistan generates the highest portion of hydroelectricity in CA, at 58.2 per cent (21 out of 36 GWh; Table 5.1).

The Toktogul HPP situated in Kyrgyzstan in the Syr Darya Basin and the Nurek HPP situated in Tajikistan in the Amu Darya Basin are the two largest reservoirs operating as main regulators for both basins (Petrushkov 2018). The Toktogul HPP, with 1,200 MW, and the Nurek HPP, with 3,000 MW installed capacities, are the major national electricity providers in Kyrgyzstan and Tajikistan. Also, two large HPP projects are currently under construction in Tajikistan—the aforementioned Rogun dam in Tajikistan on the upper Vakhsh Basin, with 3,600 MW installed capacity, and the Kambarata 1 and 2 dams on the Naryn River (upper Syr Darya Basin) in Kyrgyzstan, with an installed capacity of 1,900 MW.

During the Soviet Union period, numerous other hydropower projects were planned (Table 5.2), particularly in the upper reaches of the Amu Darya River, which would have significantly increased storage capacity and hydropower generation in CA (Rizk and Utemuratov 2012).

*Table 5.1* Hydropower generation with reservoirs in the Aral Sea Basin

| Description | Kazakhstan | Kyrgyzstan | Tajikistan | Turkmenistan | Uzbekistan | Central Asia |
|---|---|---|---|---|---|---|
| **Full capacity (million m³)*** | 5,700 | 19,500 | 18,736 | 3,000 | 11,050 | 57,986 |
| **HPP installed capacity (MW)** | 100 | 1,200 | 5,217 | | 974 | 7,491 |
| **Total production (thousand MWh)** | 516 | 4,400 | 21,011 | | 10,170 | 36,097 |

*\*Note:* Includes reservoirs without HPPs and for Kazakhstan accounts only for reservoirs in the ASB

*Table 5.2* Planned HPPs with reservoirs during the Soviet period

| Planned HPPs with reservoirs in CA | Reservoir capacity (mill m³) | Installed capacity (MW) | Annual power generation (1,000 MWh) | River (Basin) |
|---|---|---|---|---|
| **Shurob (Tajikistan)** | 27 | 850 | 2,100 | Vakhsh |
| **Sangvor (Tajikistan)** | 1,500 | 800 | 2,000 | Obikhingou (Vakhsh) |
| **Jalbulak (Tajikistan)** | 1,400 | 600 | 2,000 | Surkhob (Vakhsh) |
| **Dashtijum (Tajikistan–Afghanistan)** | 17,600 | 4,000 | 15,600 | Pyanj (upper Amu Darya) |
| **Rushon (Tajikistan)** | 5,500 | 3,000 | 14,800 | Pyanj (upper Amu Darya) |
| **Jumar (Tajikistan)** | 1,300 | 2,000 | 8,200 | Pyanj (upper Amu Darya) |
| *Total* | *27,327* | *11,250* | *44,700* | |

# Hydropower in the Syr Darya Basin

Water resources of the Syr Darya are formed mainly in Kyrgyzstan (75.2 per cent), which is also the main hydropower producer on the Syr Darya, and to a much lesser extent in Uzbekistan (15.2 per cent), Kazakhstan (6.9 per cent) and Tajikistan (2.7 per cent). Several HPPs were built along the upstream courses of the Syr Darya and its tributaries, the largest being the Toktogul and Charvak HPPs (Table 5.3).

After the collapse of the Soviet Union, the regional energy network between CA countries was disrupted and Kyrgyzstan was cut off from energy during the winter season. Being without alternative sources of energy to provide for the heating needs in winter, Kyrgyzstan changed the operational mode of the Toktogul HPP from irrigation–energy to energy–irrigation status. This led to the release of large water volumes from the Toktogul reservoir in winter for electricity generation, while the summer releases where significantly decreased. As a result, a shortage of water for irrigation in downstream countries appeared during the vegetation period, especially on the irrigated lands of Kazakhstan and Uzbekistan in the *Golodnaya* (Hungry) Steppe.

Likewise, Tajikistan experienced great difficulties to provide its population with heat and electricity during winter in the 1990s and in the first decade of the twenty-first century. The operating mode of the reservoir in Kayrakkum (presently named Bahri Tojik HPP) was therefore switched to energy generation in winter and irrigation during the vegetation season in summer. The Bahri Tojik HPP in Tajikistan can play an important role for the country but also for the lower-lying irrigated lands of Uzbekistan and Kazakhstan. The plant has an operational storage capacity of 2.6 km³, which is almost five times less than the operational capacity of the Toktogul reservoir (14 km³). However, the volume of water passing through the reservoir (17–20 km³) is about twice as much as the annual discharge from the Toktogul reservoir. The mean seasonal and yearly water releases of the Bahri Tojik and Toktogul HPPs are presented in Figure 5.1. The Bahri Tojik HPP can seasonally regulate the Syr Darya flow, compensate for water shortages in summer caused by the Toktogul reservoir and retain water releases in winter and early spring for the avoidance of floods downstream. These contributions could be offered in agreement with Uzbekistan and Tajikistan on the hydropower and energy needs between the two countries.

*Table 5.3* Hydropower stations in the Syr Darya River Basin

| HPPs with reservoirs | Reservoir capacity (million m³) | | Installed capacity (MW) | Annual power generation (1,000 MWh) | River |
|---|---|---|---|---|---|
| | Full | Operational | | | |
| Chardara (Kazakhstan) | 5,700 | 4,700 | 100 | 516 | Syr Darya |
| Farhad (Uzbekistan–Tajikistan) | 350 | 20 | 126 | 870 | Syr Darya |
| Bahri Tojik (Tajikistan) | 4,160 | 2,600 | 126 | 691 | Syr Darya |
| Charvak (Uzbekistan) | 2,000 | 1,580 | 621 | 2,000 | Chirchik |
| Andijan (Uzbekistan) | 1,750 | 1,600 | 140 | 435 | Karadarya |
| Toktogul (Kyrgyzstan) | 19,500 | 14,000 | 1,200 | 4,400 | Naryn |
| Other stations | 650 | 87 | 1,955 | 7,028 | Various tributaries |
| Total | *34,110* | *24,587* | *4,268* | *15,940* | |

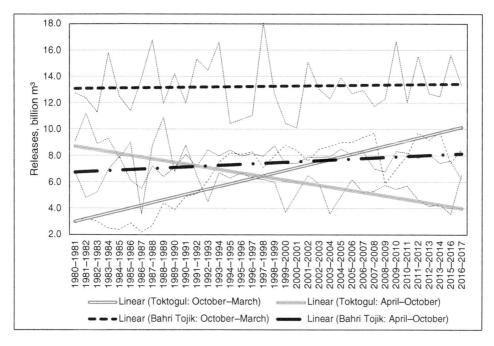

*Figure 5.1* Water releases from Toktogul and Bahri Tojik reservoirs from 1980 to 2016
Source: Anvar Kamolidinov

More broadly, the diverse hydropower and irrigation demands could be managed by compensating for the shortage of electricity and hydrocarbons in Kyrgyzstan and Tajikistan through energy supply from Uzbekistan and Kazakhstan, in an exchange for water in summer, as was the practice during the Soviet period (Wegerich 2010). However, this practice was gradually phased out in the 1990s. In the late 1990s, the discrepancy of water and energy relations between the riparian countries of the Syr Darya became profound. The major impediments were the lack of a legal mechanism to resolve water and energy issues, the lack of political will to resolve heterogeneous needs, and little experience among the newly independent states to negotiate these issues and manage the socio-economic repercussions (Weinthal 2002).

A consensus between CA states on Syr Darya water allocation and use came with the Syr Darya Agreement on Use of Water and Energy Resources (1998). Its main objective was to define the use of water and energy resources primarily between upstream Kyrgyzstan and downstream Kazakhstan and Uzbekistan. Although officially not part of the Agreement, the Tajik government participated in the development of its final draft. The agreement adopted joint decisions on water supplying for the irrigation needs mainly of Kazakhstan and Uzbekistan by also ensuring the operation of hydropower facilities and reservoirs in Kyrgyzstan. Decisions were taken in a coordinated manner and touched upon issues of i) water releases, generation and transmission of electricity, and compensation for energy losses on an equivalent basis (Article 2) and ii) transfer of additionally generated electric energy associated with the regime of water releases in summer and long-term flow regulation in the Toktogul reservoir to the downstream countries (Article 4).

A compensation to Kyrgyzstan was foreseen for the volume of water transferred in the summer season for agricultural needs, through energy transfer upstream, other products and services or monetary compensation. The goal was also to create annual and long-term water reserves on the upstream course of the Syr Darya, mainly for irrigation purposes. The Agreement was signed in March 1998, entered into force in 1999 and was implemented unanimously until 2003. In 2003, the agreed deliveries of fuel and energy resources and electricity to Kyrgyzstan were not provided. The Kyrgyz authorities decided then to change the operational regime of the Toktogul reservoir by restricting summer releases. In 2004, Uzbekistan withdrew from the Agreement and began constructing reservoirs for securing water in the Ferghana Valley upstream of the Tajik reservoir of Bahri Tojik. Kazakhstan also built the Koksarai reservoir for flood prevention and water storage purposes for summer irrigation in the Kyzylorda agricultural zone. Although the Syr Darya Agreement ended, major experience was gained on basin regulation in a transboundary setting. It was also noticed that a unified information system for the operational regime of the Syr Darya River is of major importance.

## Hydropower in the Amu Darya Basin

The Amu Darya River is the largest river in CA, with an average annual flow of water of 78.5 km³, over 80 per cent of which is formed in Tajikistan. With total potential reserves of hydropower resources of 527 TWh/year, Tajikistan ranks eighth in the world. The estimated hydropower potential of the country is around 3,697 MWh/year/km², which is three times higher than the current electricity consumption in Central Asia (Statistical Agency under President of the Republic of Tajikistan 2018).

The total installed capacity of the HPPs in Tajikistan (including that in the Amu Darya—Table 5.4) is 5,217 MW, which accounts for 98 per cent of the national electricity generation. In the 1990s and during the first decade of the new millennium, Tajikistan endured severe power shortages. The shortfall was about 4–5 TWh in winter, although an equivalent electricity surplus could be produced in summer but not stored for the energy-deficient periods in winter. The development of small HPPs has helped to provide remote rural areas with

*Table 5.4* Hydropower stations in the Amu Darya River Basin in 2018

| HPPs | Reservoir capacity (million m³) | | Installed capacity (MW) | Annual power generation (1,000 MWh) | River |
|---|---|---|---|---|---|
| | Full | Operational | | | |
| **Nurek (Tajikistan)** | 10,500 | 4,500 | 3,000 | 11,200 | Vakhsh |
| **Baypaza (Tajikistan)** | 84 | 80 | 600 | 2,900 | Vakhsh |
| **Sangtuda 1 (Tajikistan)** | 2,700 | 20 | 670 | 2,500 | Vakhsh |
| **Sangtuda 2 (Tajikistan)** | 932 | 5 | 220 | 665 | Vakhsh |
| **Sarband (Tajikistan)** | | | 240 | 960 | Vakhsh |
| **Perepadnaya (Tajikistan)** | | | 30 | 250 | Vakhsh |
| **Centralnaya (Tajikistan)** | | | 18 | 110 | Vakhsh |
| **Khorog (Tajikistan)** | 10 | | 250 | 1,300 | Vakhsh |
| **Tuyamuyun (Uzbekistan)** | 7,300 | 5,100 | 150 | 550 | Ghund (Pyanj) |
| *Total* | *21,526* | *9,705* | *5,178* | *20,435* | |

electricity. However, it has become clear that a decrease in the flow of small rivers during winter time can significantly reduce the efficiency of small plants. Some 283 small HPPs were built between 1990 and 2016 in Tajikistan (Statistical Agency under President of the Republic of Tajikistan 2018).

The most important contribution to covering the winter power shortages was the construction of a thermal power and heating plant (2012–2016) fuelled by coal. This coal-fired power plant has a capacity of 400 MW and can cover the energy deficit in winter, which is still apparent in rural areas through erratic electricity power cuts that may last for many hours during the day.

The major part of the hydropower potential of Tajikistan is concentrated along the Vakhsh River—a large transboundary tributary of the Amu Darya, formed by the confluence of the Surkhob and Obikhingou Rivers. The total length of the Vakhsh is 524 km, and the catchment area is 39,100 km², of which 31,200 km² lies in Tajikistan. The discharge peak occurs in July–August, while the minimum flows in the winter season are due to low ice and snow melting. The average annual discharge at the confluence with the Amu Darya is about 650 m³/s, and the average annual flow volume is 20.2 km³. Periods of high discharges were observed in 1950–1958 and 1992–2008, while low discharges appeared in the periods 1962–1966 and 1980–1990 (World Bank 2014).

The existent cascade of hydropower stations in the Vakhsh River can be seen as a model for improved regulation, and sustainable use of water for electricity generation and irrigation

*Figure 5.2* HPP Nurek in Tajikistan

Source: Frank Schrader

in the ASB. The first hydropower station in the Vakhsh Basin was commissioned in 1958. The Golovnaya (1962), Centralnaya (1964), Nurek (1972), Boyghozy (1986), Sangtuda 1 (2008) and Sangtuda 2 (2011) hydropower plants were constructed thereafter.

The operation of the largest HPP on the Vakhsh—Nurek—began in 1972 and its full capacity was reached in 1979, when the reservoir filled up. However, Nurek could not meet the winter shortages in Tajikistan (Petrushkov 2018). The gradual siltation in the Nurek HPP forced the Tajik government to look for additional energy sources by constructing the Rogun HPP, also on the Vakhsh River, about 100 km upstream of the Nurek HPP. A Feasibility and Environmental and Social Impact Assessment Study was prepared between 2008 and 2014 based upon independent international expertise, with funding from the World Bank (Word Bank 2014). The dam construction was initiated with national funds of the Tajikistan government and is still ongoing while the operation of the first turbine took place in December 2018. Once completed, Rogun will become the largest HPP in CA, with an embankment dam of 335 m height, installed capacity 3,600 MW (6 units of 600 MW) and an average annual output of 13.1 billion kWh (Table 5.5; World Bank 2014).

The filling of the Rogun reservoir will be carried out within 13–18 years (baseline year 2018), depending on water usage and availability by setting the dam in full operation. The main objectives of the Rogun HPP are to ensure energy independence in Tajikistan, to reduce the siltation process in downstream reservoirs, and especially the Nurek HPP, and to mitigate floods. Despite different opinions of CA experts on the construction of the Rogun HPP, the facility is of strategic importance for all countries in the region, as well as for those that may benefit from electricity trading with Tajikistan, such as, for instance, Afghanistan and Pakistan (Sulton 2014).

In particular, Tajikistan, Kyrgyzstan, Afghanistan and Pakistan have proposed the implementation of the Central Asia–South Asia 1000 kV project, known as CASA-1000. This concerns a transmission line project that will become a unique facility for supplying surplus electricity during the summer season from Kyrgyzstan and Tajikistan to Afghanistan and Pakistan. The CASA-1000 project is supported by the World Bank Group, the Islamic Development Bank, the United States Agency for International Development (USAID), the US Department of State, the Department for International Co-operation of the United Kingdom, the Australian Agency for International Development and several other donors and organisations (CASA-1000 2018). CASA-1000 will provide electricity supply via high-voltage power lines, initiating from Kyrgyzstan to Tajikistan with a length of 477 km, then from Tajikistan to Afghanistan and Pakistan with a 750-km power line. In Tajikistan and Kyrgyzstan, there is excess electricity

*Table 5.5* Technical features of the Rogun HPP in Tajikistan

| Parameters | Unit | Quantity |
| --- | --- | --- |
| **Total volume of reservoir (total storage)** | million m$^3$ | 10.3 |
| **Operational volume of reservoir (live storage)** | million m$^3$ | 13.3 |
| **Height** | million m$^3$ | 335.0 |
| **Height of dam crest** | masl | 1,296.25 |
| **Maximum operational water level** | masl | 1,296.0 |
| **Minimum operational water level** | masl | 1,185.0 |
| **Average annual flow** | km$^3$ | 20.1 |
| **Level of first-stages dam** | Masl | 1.11 |

*Note:* masl = meters above sea level

in the summer period to be exported to Afghanistan and Pakistan through lucrative energy trading. This US$1.2-billion project connects Tajikistan, Kyrgyzstan, Afghanistan and Pakistan through massive transmission lines, enabling large power flows from north to south (CASA-1000 2018). Tajikistan and Kyrgyzstan are expected annually to supply up to 5 billion kWh of summer electricity to Afghanistan and Pakistan via this transmission line (SNC-Lavalin 2011). In September 2018, the first contracts were signed for the construction of converter stations in Tajikistan and Pakistan, while the energy exchange should reach its full capacity by the summer of 2021 (World Bank 2018). The very lucrative energy trading could provide significant revenues to the upstream economies of Tajikistan and Kyrgyzstan and contribute to rehabilitating vulnerable energy systems in both countries (Shenhav *et al.* 2019).

The Rogun reservoir, together with Nurek on the Vakhsh River and the Tuyamuyun reservoir on the Amu Darya are also expected to provide an annual water yield of 5.0–5.9 km³ for irrigation in Afghanistan, Uzbekistan and Turkmenistan.

The environmental impacts of the cascading HPPs on the Vakhsh River could be comparatively lower than those of similar facilities in lowland areas. For example, one of the main indicators of the degree of influence of reservoirs on the environment can be considered the ratio of the area of flooding of the reservoir to its total volume. This indicator for mountain reservoirs is much smaller than for plain reservoirs when comparing different reservoirs in CA, as shown in Figure 5.3.

In view of the relatively smaller surface area of water in mountain reservoirs, relatively high elevation of location and therefore low water temperatures, evaporation from the water surface is much less. Also, there is less loss of water for filtration and waterlogging below the reservoir.

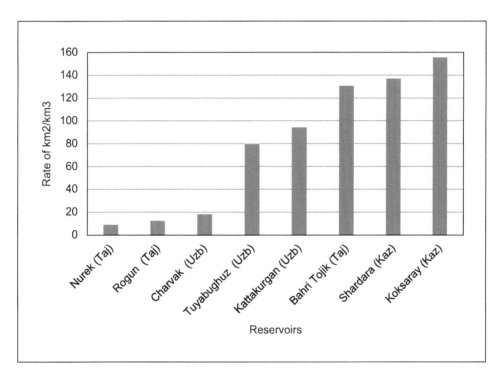

*Figure 5.3* Ratio of flooding area of selected reservoirs in Central Asia

Source: Anvar Kamolidinov

The fish population and diversity in the Vakhsh and Amu Darya Rivers are also considered in the Rogun HPP through the planned construction of fish spillways. Although the mountain rivers of Tajikistan are not popular destinations for commercial fishing and angling, they contain a number of species like the *Oncorhychus mykiss* (rainbow trout), which is originally from North America but has been distributed widely around the globe for angling activities. Other fish species are the *Salmo trutta* (brown trout) and the *Salmo trutta oxianus*, the latter of which is endemic in the Amu Darya Basin.

The fauna surrounding the Vakhsh Basin is rich in various mammals like bears, wolves, foxes, hares and rodents, but also in ornithofauna and reptiles. There is some consideration for the establishment of a national park in the adjacent area of the cascade reservoirs, which could conserve the local fauna, as proposed by the IUCN (2013).

## Small hydropower plants and renewable sources

According to the World Small Hydropower Report (2016), a variety of definitions of small HPPs are found in CA. In Kazakhstan, the upper limit of small HPPs is set at 35 MW, in Kyrgyzstan and Tajikistan at 30 MW and in Uzbekistan at 10MW of installed capacity; Turkmenistan provides no definition. Tajikistan further defines mini (0.1–1 MW) and micro (<100 kW) hydropower stations (UNIDO & ICSHP 2016). Karatayev and Clarke (2016) define small (1–10 MW) and medium-scale (10–50 MW) plants for Kazakhstan. Due to the terminology variation and the absence of definitions in many relevant reports, the numbers discussed herewith should be interpreted with caution, especially when aggregate comparisons among countries are made. Small HPPs may be advantageous over larger ones in that they 'make it possible to use the potential of small rivers and watercourses; place less load on the river ecosystem; make it possible to build small hydropower plants without significant flooding of land and without completely damming up the river; promote the development of local industry; make it possible to resolve the region's social problems; require less initial major spending, [and] operational expenses' (Baum 2008).

The CA region is endowed with substantial potential for small hydropower. In Kazakhstan, there is an estimated potential of 4,800 MW for plant capacity up to 35 MW, and 2,707 MW for less than 10 MW. In Kyrgyzstan, the potential is estimated at 900 MW for plants up to 30 MW, and 275 MW for up to 10 MW of installed capacity (Temiraliev 2015). Tajikistan boasts a potential capacity of 30,000 MW for HPPs up to 30 MW. The estimated potential for small HPPs in Turkmenistan is 1,300 MW (though clear definition is lacking). In Uzbekistan, the potential is estimated at 1,180 MW for HPPs up to 10 MW (Avezova *et al.* 2017), while other sources estimate it at 5,931 GWh/year (Kochnakyan *et al.* 2013) and 8,000 GWh/year (Jorde and Biegert 2009).

The existing capacity of small HPPs in CA comprises a tiny fraction of the potential hydropower to derive from the further development of such facilities. In Kazakhstan, the installed capacity for plants up to 35 MW is 119 MW and for plants less than 10 MW is estimated at 78 MW. In Kyrgyzstan, 41.5 MW is installed, whereas in Tajikistan only 25 MW is installed. Turkmenistan has the lowest installed capacity, at 5 MW, while Uzbekistan has 71 MW installed for plants up to 10 MW (Embassy of Uzbekistan to New Delhi, 2017). In Tajikistan, power generated by small HPPs in 2016 accounted to 27.9 million kWh (Statistical Agency under President of the Republic of Tajikistan 2018).

CA countries have made plans to exploit the potential for small hydropower. In Kazakhstan, plans were put forward to construct 41 small hydropower plants with a total capacity of 539

MW by 2020 (UNIDO & ICSHP 2016). Key factors for driving commercial interest are the 'low cost, reliability and environmental friendliness' (Karatayev and Clarke 2016). In Kyrgyzstan, building and rehabilitation of 132 plants totalling 275 MW was planned between 2010 and 2025 (UNIDO & ICSHP 2016). In 2008, 24 plants (totalling 200 MW) were planned to undergo reconstruction (Baum 2008). According to the Long-term Programme for the Construction of Small Hydropower Plants for 2007–2020, in Tajikistan, 189 plants with a total capacity of 103 MW were to be constructed between 2009 and 2020 (Doukas *et al.* 2012). By 2013, about 300 plants were added, increasing the country's small hydro capacity to 132 MW. A further addition of up to 3 GW of small hydropower is envisaged, particularly for remote areas (UNIDO & ICSHP 2016). In Turkmenistan, nine plants with 57 MW of potential capacity were proposed, but without a clear construction and development horizon. Uzbekistan has planned the construction of four new plants with total capacity of 23.5 MW and the rehabilitation of 11 plants (with no capacity specified) between 2016 and 2020 (UNIDO & ICSHP 2016).

The development of small HPPs presents a set of important challenges. Four major barriers are identified, as noted by UNIDO and ICSHP (2016): 'Market barriers: . . . lack of awareness and information on the potential and possible application of SHP. Institutional and regulatory barriers: The existing institutional and regulatory frameworks in the energy sector are not fully considering the peculiarities of SHP. Technical barriers: Technical and market conditions are not supportive of implementation and operation of SHP. Financing barriers: . . . lack of functioning and affordable financing mechanisms (credits) available for developers of SHP projects'.

The present situation regarding the use and development of small HPPs in CA encompasses a number of barriers. There is progress in some countries, but also major constraints that hinder developments in small HPPs and the installation of renewable energy resources (RES) more broadly. In the case of Kazakhstan, for instance, Karatayev and Clarke (2014, 102) provided further details such as 'a lack of awareness of the opportunities associated with renewable energy, a lack of technical expertise and capacity, insufficient governmental support to overcome high initial financial and capital requirements and investment disincentives due to subsidies of other energy sources (primarily fossil fuels), low price of electricity in the country, uncertainties with the long term power purchasing tariffs, difficulties in attracting foreign investment and a lack of access to credit for both consumers and investors, absence of a clear national program for renewable energy development, a lack of specific action plans and instruments, a lack of concrete competitive legislation and regulation relating to the newly developed renewable energy market'.

Moreover, the energy supply prices in nearly all CA countries are among the lowest in the world, which prevents investment in RES and particularly small HPPs (Balliyev *et al.* 2009; Eshchanov *et al.* 2013). An important challenge that is also frequently overlooked is the lack of reliable information on the energy potential and profitability of RES in the region, with the notable exception of small HPPs. Table 5.6 summarises the major impediments on RES development in CA.

## Conclusions

There are examples of successful cooperation and benefit-sharing from water and hydropower in the ASB. Sustainable future development of the ASB could be achieved on the basis of collaboration, higher efficiency of water use, more attention paid to multiple-use reservoirs and the introduction of environmental flows into water management practices. For this purpose, it is necessary to develop and approve clear rules for managing and discharging water from reservoirs along the main rivers (Khamraev 2005).

Table 5.6 Barriers to renewable energy in Central Asia.

| | Kazakhstan | Kyrgyzstan | Tajikistan | Turkmenistan | Uzbekistan |
|---|---|---|---|---|---|
| **Data and information** | Lack of updated information | Poor and outdated information | Lack of reliable data | Very limited data availability | Lack of feasibility studies |
| **Expertise and know-how** | Lack of project plan and expertise | Low technical capacity on construction and maintenance | Lack of expertise on project development and maintenance | No information | No information |
| **Regulatory framework** | Lack of regulations for technical specifications | Lack of framework for investors and regulations for enforcement | Uncertainty about private-sector participation | Absence of regulatory framework | No information |
| **Infrastructure** | Difficult to transfer RES electricity in the grid | Old electricity assets | Old electricity assets | No information | Old electricity assets |
| **Finances** | No information | Low tariffs, lack of support from state | Low tariffs, monopoly market, lack of mechanism to manage funds | Investment on RES is not promoted | Low tariffs, lack of finance and investment |
| **Awareness and support** | No information | Low awareness of citizens, state agencies and institutions | Lack of awareness of RES and energy security | No information | Lack of supportive mechanisms and awareness |

Source: Authors' compilation based on the World Small Hydropower Report 2017 (UNIDO & ICSHP 2016)

The hydropower potential in the ASB is currently underutilised while the low capital (investment) costs of hydropower and the provision of 'clean energy' without dependence on fossil fuels make hydropower a promising energy source for the Basin.

Hydropower development in CA is linked to the resolution of various technical and institutional challenges. For example, existing natural conditions in a river, especially in floodplains, need to be assessed prior to HPP construction. The assessment of potential real-location of communities in case of the development of reservoirs should be conducted prior to the design phase. The financial risk of large HPP constructions should be scrutinised. The domestic and international electricity demand should be also explored to ensure the financial sustainability of a hydropower project. There is also a need to simplify the procedures for issuing HPP permits and for the creation of public–private ventures by also updating the legislative documents for the improvement of hydropower.

The water–energy nexus in the ASB and the larger CA region is closely intertwined with the sustainability of economic growth of all riparian countries. The well-planned and agreed development of hydropower could offer mutual benefits to both upstream and downstream countries.

# References

Asian Development Bank (ADB) (2010) *Central Asia Atlas of Natural Resources.* Manila, Philippines. Retrieved 9 February 2019, www.adb.org/publications/central-asia-atlas-natural-resources

Avezova N, Khaitmukhamedov A, Vokhidov A (2017) Uzbekistan renewable energy short overview: Programs and prospects, *International Journal of Energy and Smart Grid* 2(2): 43–46

Balliyev K, Jorde K, Biegert A (2009) *Country Chapter: Republic of Turkmenistan.* In: Deutsche Gesellschaft für Technische Zusammenarbeit, GTZ (ed.), *Regional Reports on Renewable Energies. 30 Country Analyses on Potentials and Markets in West Africa (17), East Africa (5), Central Asia (8).* Deutsche Gesellschaft für technische Zusammenarbeit (GTZ) GmbH, Frankfurt/Eschborn, Germany, pp 155–170. Retrieved 19 March 2019, www.ecreee.org/sites/default/files/event-att/gtz_re_in_developing_countries.pdf

Baum L (2008). The energy industry in the Kyrgyz Republic: current state, problems, and reforms, *Central Asia and the Caucasus* 6(54): 101–112

CASA-1000 (2018) Homepage. Retrieved 10 March 2019, http://casa-1000.org

Doukas H, Marinakis V, Karakosta C, Psarras J (2012) Promoting renewables in the energy sector of Tajikistan, *Renewable Energy* 39(1): 411–418

Dukhovny VA (2010) *Current Problems in Irrigated Agriculture in Central Asia and Future Solutions.* Paper presented at international scientific symposium titled 'Water in Central Asia', held on 24–26 November 2010. SIC-ICWC, Tashkent, Uzbekistan

Dukhovny VA, de Schutter JL (2011) *Water in Central Asia: Past, Present, Future.* Taylor and Francis, London

Embassy of Uzbekistan to New Delhi' (2017, June 2) *Information Digest of Press of Uzbekistan # 108* [Government]. Retrieved 16 July 2018, http://www.uzbekembassy.in/information-digest-of-press-of-uzbekistan-108-june-2-2017/

Eshchanov BR, Grinwis Plaat Stultjes, Eshchanov M, Salaev SK (2013) Prospects of renewable energy penetration in Uzbekistan—perception of the Khorezmian people. *Renewable and Sustainable Energy Reviews* 21: 789–797

IUCN Evaluation Report – April (2013) *Tajikistan – Tajik National Park* (Mountains of the Pamirs), Geneva, Switzerland

Jigarev S (2008). *Water and Energy Issues and their Solutions in Central Asia.* Paper presented at the conference 'Problems of Aral: Impact on the Gene Pool of Population, Flora, Fauna and International Cooperation for Mitigating Consequences', 11–12 March 2008, Tashkent, Uzbekistan

Jorde K, Biegert A (2009) Country chapter: Republic of Uzbekistan. In: Deutsche Gesellschaft für technische Zusammenarbeit, GTZ (ed.), *Regional Reports on Renewable Energies. 30 Country Analyses on Potentials and Markets in West Africa (17), East Africa (5), Central Asia (8).* Deutsche Gesellschaft für technische Zusammenarbeit (GTZ) GmbH, Frankfurt/Eschborn, Germany, pp 171–191. Retrieved 3 March 2019, from www.ecreee.org/sites/default/files/event-att/gtz_re_in_developing_countries.pdf

Karatayev M, Clarke ML (2014). Current energy resources in Kazakhstan and the future potential of renewables: a review, *Energy Procedia* 59: 97–104

Karatayev M, Clarke ML (2016) A review of current energy systems and green energy potential in Kazakhstan, *Renewable and Sustainable Energy Reviews* 55: 491–504

Karimov KS, Akhmedov KM, Abid M, Petrov GN (2013) Effective management of combined renewable energy resources in Tajikistan, *Science of The Total Environment* 461–462: 835–838

Khamraev SP (2005) *The Tasks of Integrated Water Resources Management in the Amudarya Basin from the Standpoint of the Strategic Interests of Uzbekistan*. Jaihun" project INCO-CT-2005-516761, Risk Management of Interstate Water Resources: Towards a Sustainable Future for the Aral Basin. Retrieved 12 March 2019, www.cawater-info.net/library/rus/hamraev.pdf

Kochnakyan A, Khosla SK, Buranov I, Hofer K, Hankinson D, Finn J (2013) *Uzbekistan Energy/Power Sector Issues Note* (Report No. No. ACS4146). World Bank, Washington, DC, USA, p 108

Petrushkov M (2018). Nurek HHP: so that reality does not become a fairy tale. *The Rhythm of Eurasia*. Retrieved 10 March 2019, www.ritmeurasia.org/news--2018-01-12-nurekskaja-ges-chtoby-byl-ne-stala-

Rahimov S (2014) Tajikistan: Turn water into cooperation. *EP Today*. Retrieved 10 March 2019, http://eptoday.com/tajikistan-turn-water-cooperation

Rakhmatullaev S, Frédéric H, Philippe LC, Mikael M, Mashari M (2010) Facts and perspectives of water reservoirs in Central Asia: a special focus on Uzbekistan, *Water* 2: 307–320

Rizk J, Utemuratov B (2012) *Balancing the Use of Water Resources in the Amu Darya Basin*. Policy Brief. Retrieved 30 August 2018, www.amudaryabasin.net

Shenhav R, Xenarios S, Domullodzhanov D (2019) *The Role of Water User Associations in Improving the Water for Energy Nexus in Tajikistan*. OSCE Programme Office in Tajikistan. Retrieved 6 March 2019, www.osce.org/programme-office-in-dushanbe/413228

SNC-Lavalin (2011) *Central Asia–South Asia Electricity Transmission and Trade (CASA-1000) Project Feasibility Study Update*. Retrieved 19 October 2016, www.casa-1000.org/1)Techno-EconomicFeasbilityStudy_MainRep_English.pdf

Statistical Agency under President of the Republic of Tajikistan (2018) *Results of One-Off Sample Survey on "The State of the Energy Facilities and Efficiency of Use in 2016"* (Government of Tajikistan Statistics). Retrieved 4 January 2019, http://stat.ww.tj/publications/July2018/English.pdf

Temiraliev U (2015) *Energy Sector of Kyrgyz Republic Overview*. Presentation presented at the Workshop on the Investment Climate in Kyrgyzstan: Investing into Sustainable Energy Sources, Brussels, Belgium. Retrieved 4 January 2019, https://energycharter.org/fileadmin/DocumentsMedia/Events/KICW_Brussels_2015_P4_Temiraliev.pdf

UNIDO & ICSHP (2016) *The World Small Hydropower Development Report 2016*. United Nations Industrial Development Organization (UNIDO) and International Center on Small Hydro Power (ICSHP), Vienna and Hangzhou. Retrieved 3 February 2019, www.smallhydroworld.org/fileadmin/user_upload/pdf/2016/WSHPDR_2016_full_report.pdf

Valentini KL, Orolbayev EE, Abylgazieva AK (2004) *Water Problems in Central Asia*. MISI, Socinfromburo & Friedrich-Ebert-Stiftung, Bishkek, Kyrgyzstan

Wegerich K (2010) *Handing Over the Sunset: External Factors Influencing the Establishment of Water user Associations in Uzbekistan: Evidence from Khorezm Province*. Cuvillier Verlag, Göttingen, Germany

Weinthal E (2002) *State Making and Environmental Cooperation: Linking Domestic and International Politics in Central Asia*. MIT Press, Cambridge, MA, USA, pp 274

World Bank (2014) *Final Reports Related to the Proposed Rogun HPP*. Retrieved 10 March 2019, www.worldbank.org/en/country/tajikistan/brief/final-reports-related-to-the-proposed-rogun-hpp

World Bank (2018) *The CASA-1000 Project Crosses Another Important Milestone* [press release]. Retrieved 18 February 2019, www.worldbank.org/en/news/press-release/2018/09/24/the-casa1000-project-crosses-another-important-milestone

# 6

# Environmental degradation

*Alisher Mirzabaev, Jamal Annagylyjova and Iroda Amirova*

## Key messages

- Land and water degradation are among the major hindrances to sustainable development in the Aral Sea Basin (ASB). Land degradation alone is estimated to cost about US$3 billion of losses in ecosystem services annually. No information is available about the economic costs of water quality deterioration, but qualitative evidence suggests that they are substantial, especially when the detrimental impacts of water pollution on human health are considered.

- Causes of land and water degradation are numerous, but the major ones consistently highlighted in the literature are land tenure insecurity, lack of access to markets and advisory services, limitations with crop choices, overstocking of rangelands and constraints on herd mobility, inefficient irrigation methods, institutional gaps in water management and lack of investments into irrigation and drainage infrastructure maintenance.

- A number of technological options are available that can be used to avoid, reduce and reverse land and water degradation in the ASB. These are conservation agriculture practices, more efficient irrigation methods such as drip and sprinkler irrigation, crop diversification, afforestation, improving rotational grazing in rangelands and terracing in sloping areas. However, their adoption in the ASB remains low. Only about 26 per cent of agricultural households apply any of these sustainable land and water management practices in the Basin.

- Investments into the adoption of technological and policy options for sustainable land and water management are projected to yield considerable economic and social co-benefits, in addition to helping address land and water degradation. Approximate estimations assume the return investment ratio to be as high as 1:5—US\$1 will provide US\$5 in a period of 6 to 30 years.

- The success of measures for sustainable land and water management in the Aral Sea Basin requires increasing investments into environmental research and educational programs. There are currently very few educational programs in the region for training professionals in issues related to environmental sustainability. Collaboration with leading international universities could provide a good platform for expanding environmental education in the region.

## Introduction

Environmental degradation is among the major threats to sustainable development in the Aral Sea Basin (Glantz 1999; Gupta *et al.* 2009; Qadir *et al.* 2009). The ASB countries of Afghanistan, Kazakhstan, Kyrgyzstan, Tajikistan, Turkmenistan and Uzbekistan are faced with extensive scales of land degradation, growing water scarcity and deterioration of water quality (Mirzabaev *et al.* 2013; Mirzabaev *et al.* 2016; Ahmad and Wasiq 2004). All these facets of environmental degradation are reducing agricultural production and productivity in the region. Since the majority of the rural population depends on agriculture for employment, land and water degradation are resulting in lower agricultural incomes, with subsequent negative effects on poverty reduction and food security in many rural parts of the Basin (Mirzabaev *et al.* 2016; Ahmad and Wasiq 2004; Habib 2014).

The total area of ASB is about (1,760,000 km² (176 million hectares), of which about 1,500,000 km² (150 million hectares) are in Central Asia (CA)[1], the remaining being mostly in Afghanistan and a very small area in Iran (FAO 2012a). In the Central Asian part of the ASB approximately 52 million hectares are rangelands, 10.5 million hectares represent irrigated lands and about 1.5 million hectares are rainfed arable areas, with the remaining major extent of the area (86 million hectares) being covered by deserts, mountains and glaciers, and to some extent with settlements and forested areas (Mirzabaev 2013). The forested areas are mainly located in the mountainous parts of the Basin. The area falling under the ASB in Afghanistan represents only 12% of the country's territory but is home to 25% of the population and is considered to be the most agriculturally productive region within the country (Ahmad and Wasiq 2004). Of about 20–25 million hectares of ASB in northern Afghanistan, approximately 1.16 million hectares are irrigated, and 2.5 million hectares are rainfed arable lands, whereas the remaining areas fall under rangelands, mountains, and to a limited extent, settlements (Ahmad and Wasiq 2004). Unsustainable agricultural practices and water resources allocation are the main drivers of land degradation in the Basin. Irrigated agriculture consumes almost 96 per cent of the water discharge of the two main rivers in ASB—the Amu Darya and the Syr Darya (Abdullaev 2004).

Assessments of land degradation in Central Asia (Gupta *et al.* 2009; Mirzabaev *et al.* 2015) highlight secondary salinization as the major form of land degradation in the irrigated areas, whereas soil erosion (by wind and water) and deforestation are the key sources of land degradation in upstream mountainous areas. Soil erosion is also the major reason for the siltation of water courses and reservoirs. The rangelands are affected by loss of vegetation due to overgrazing and encroachment of unpalatable plant species (Gupta *et al.* 2009). Similarly, rangeland degradation and deforestation are also major forms of land degradation in northern Afghanistan (Ahmad and Wasiq 2004) (Table 6.1).

The aridity of the climate requires that land and water management are considered together, rather than separately, calling for integrated approaches of sustainable land and water management. This chapter takes stock of the extent and causes of land and water degradation in the ASB. A qualitative assessment in Table 6.1 indicates the major environmental issues in the ASB according to their location within the Basin.

The chapter also reviews and assesses various opportunities for sustainable natural resource management and environmental protection within the ASB's boundaries. The most critical environmental issue in the countries of CA is land and water degradation, which has been extensively studied from numerous angles.

## Land degradation

The studies on the extent of land degradation in the ASB and the broader CA region can be classified into three categories. First, qualitative expert evaluations on the extent of land degradation have been the most frequently applied approach (Gupta *et al.* 2009; Ji 2008).

*Table 6.1* Qualitative assessment of key environmental issues along the upstream–downstream transect in the Aral Sea Basin

| Environmental issues | Upstream | Midstream areas | Downstream |
|---|---|---|---|
| **Secondary salinization** | Small | High | Very high |
| **Sodic soils** | Small | High in specific locations (e.g., Arys Turkestan, Kazakhstan) | Small |
| **Soil erosion** | High in sloping areas | Small | Small |
| **Overgrazing** | Small in high mountain pastures; very high near population settlements | Medium to high | Medium to high |
| **Sand and dust storms** | Small | Small | High |
| **Water pollution** | Small | High | Very high |
| **Deforestation** | High | Small to medium | Small to medium |

*Note:* the qualifying descriptors from 'small' to 'very high' are used to signify the severity of the issue

This is not unique to the region. Even at the global level, until very recently, qualitative expert evaluations were the dominant methodology. The major advantage of qualitative expert evaluations is that they are not encumbered by the lack of observed data. The major shortcoming is in the accuracy of the obtained land degradation descriptions and the difficulty of replicating the results of these assessments with a different set of experts. The second approach relies on detailed case studies based on the collection of soil samples. Obviously, this approach provides a much more accurate representation of soil quality, but due to high costs it can be applied only on a case-study basis and not on large scales. The third approach makes use of now widely available remotely sensed satellite data. Remotely sensed data provide opportunities to assess the status of land degradation on a large scale, often with very high local granularity, specifically on the net primary productivity of soils, an important proxy for soil quality in dryland areas.

In the ASB, most of the earlier studies were based on qualitative expert opinions (e.g., CACILM 2016; Gupta *et al.* 2009). However, the number of studies using soil surveys and remotely sensed data has increased rapidly over the last decade (Dubovyk *et al.* 2013; Akramkhanov *et al.* 2011; de Beurs *et al.* 2015; Akramkhanov and Vlek 2012; Zhumanova *et al.* 2018). Many of these studies are concentrated on specific locations of the Basin, e.g., the Khorezm region in Uzbekistan (Dubovyk *et al.* 2013) or focus on CA (Mirzabaev *et al.* 2016).

The usual scales of the studies on land degradation may be at plot level, on specific districts and provinces, or at the regional level for the whole CA, rather than at the scale of the ASB. For this reason, the extent of land degradation is reviewed below from regional perspectives, highlighting Basin-specific aspects whenever possible.

Studies based on remotely sensed data show that the extent of land degradation varies in different parts of the CA region. To illustrate this, Le *et al.* (2016) showed that, between the 1980s and the 2000s, the extent of land degradation in CA varied between 8 per cent in Turkmenistan and Uzbekistan to 60 per cent in Kazakhstan (Figure 6.1). Cropland degradation was widespread across CA, ranging from about 25 per cent in Kyrgyzstan to slightly more than 50 per cent in Kazakhstan (Mirzabaev *et al.* 2016). Specific locations of land degradation hotspots in the ASB are given in Figure 6.1. For Afghanistan, Le *et al.* (2016) indicate that land degradation covers about 35 per cent of croplands, 51 per cent of forests and 27 per cent of rangelands. However, these figures are for the whole country, and the extent to which they could be also representative of the ASB part of Afghanistan is not clear.

Irrigation has been practiced in the lower reaches of the Amu Darya and the Syr Darya Rivers for several millennia. However, the irrigated land area in the whole Aral Sea Basin was not more than 3.6 million hectares until last century. This area increased so fast that in 1986 about 7.6 million hectares were irrigated (Lewis 1966; Micklin 1988). The rapid extension of irrigation systems without sufficient care for drainage led to rising groundwater levels and heavy salinization of soils. Because of these problems, large areas in the Hungry Steppe, the area between the Tashkent and Samarkand oases in Uzbekistan, and other similar places in the region can no longer be used for agriculture (Obertreis 2017). In post-Soviet Uzbekistan, for instance, at the end of the 1990s, more than half of the country's irrigated area was affected by salinization (Opp 2004: 49). In this regard, the remotely sensed studies corroborate an earlier, more qualitative assessment by Qadir *et al.* (2009), who estimated that about 40–60 per cent of irrigated lands in the region were affected by secondary salinization.

The problem of secondary salinization remains acute across all the irrigated areas in the Basin, but particularly in the downstream areas of Kazakhstan, Turkmenistan and Uzbekistan

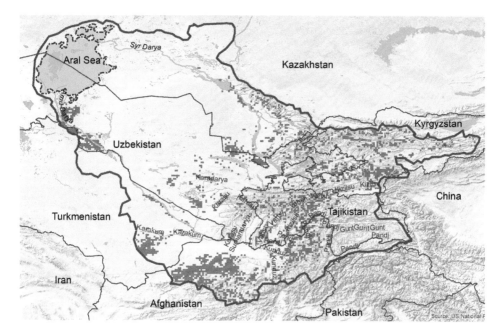

*Figure 6.1* Land degradation hotspots in the Aral Sea Basin (in red)
Source: Adapted from Le *et al.* (2016)

(CACILM 2016). Secondary salinization puts a heavy burden on scarce water resources as well, since it requires substantial quantities of additional water for leaching these saline soils to make it possible to plant crops. The return waters from the leaching process, in turn, often lead to the salinization and marginalization of freshwater resources and are often channeled to constantly growing drainage lakes. The information on land and water use in Afghanistan is sketchy due to the impacts of the wars (Ahmad and Wasiq 2004). However, what is known is that irrigation systems in northern Afghanistan were severely damaged by ongoing conflicts; on the other hand, secondary salinization is not widespread in northern Afghanistan due to natural drainage of the soils (Ahmad and Wasiq 2004).

The rangelands in the ASB are affected by large-scale vegetation losses, especially near settlements where the livestock grazing pressure is highest (Gintzburger *et al.* 2003; Kerven *et al.* 2008; Robinson *et al.* 2010). On the contrary, de Beurs *et al.* (2015) indicate that rangeland areas with minimal grazing pressure have an increased growth of vegetation. The upstream parts of the ASB—Kyrgyzstan and Tajikistan—are mostly located in mountainous areas. Mountain glaciers are the key source of freshwater. Mountains in the ASB are also rich in biodiversity, including globally important and rare animal breeds, plant species and plant genetic resources. However, the mountainous ecosystems suffer from unregulated grazing and reduction of forest and vegetation resources, which leads to increased water run-off, the formation of gullies and risk of landslides.

The desiccation of the Aral Sea has considerably contributed to dust and sand storm activity, with substantial negative impacts on the adjacent areas (Indoitu *et al.* 2015). Dust and sand storms are intensifying land degradation processes and further add to the deterioration of water

*Figure 6.2* An example of formerly irrigated lands abandoned due to secondary salinization in Uzbekistan
Source: Kristina Toderich

quality (Issanova *et al.* 2015). The causes of land degradation in the ecologically and socially diverse settings of CA are many and varied. The proximate drivers of secondary salinization are dilapidated irrigation and drainage systems in the region that have not received sufficient investments for their repair and maintenance over the last several decades (ADB 2007). Lack of water pricing, subsidization of water use in agriculture and continued application of water-de-manding furrow irrigation techniques are the major reasons for inefficient and high water use in agriculture (Bekchanov *et al.* 2016). Moreover, land tenure insecurity, lack of access to credit and lack of access to markets were found to be the major deterrents for the adoption of sustainable land and water management technologies (Mirzabaev *et al.* 2016). Ambiguity in rangeland tenure and rangeland use regulations and overgrazing of rangelands near villages due to limited livestock mobility serve as the major causes of rangeland degradation (Pender *et al.* 2009).

Sustainable land and water management in the Basin is hindered by insufficient access to agricultural advisory and extension services, resulting in lack of farmers' knowledge about soil and water conservation (Kienzler *et al.* 2012; Mirzabaev *et al.* 2016). Deforestation due to war and fuelwood harvesting is a major form of land degradation in northern Afghanistan (Ahmad and Wasiq 2004). Demand for fuelwood and timber and the expansion of crop-lands have also been driving high rates of deforestation in the mountainous areas of Tajikistan and Kyrgyzstan (Muslimshoeva *et al.* 2013; Novikova and Yatimov 2011; Bekberdieva 2011), with negative impacts on the rich biodiversity of forest ecosystems in these upstream areas (Novikova and Yatimov 2011; Bekberdieva 2011).

Land degradation has enormous socio-economic costs in the region. Sutton *et al.* (2007) estimated that secondary salinization in Uzbekistan results in annual losses of about US$1 billion, whereas in Kazakhstan, Saigal (2003) indicated the annual costs of land degradation at US$6.2 billion. For the whole of CA, the World Bank (1998) estimated that land degradation leads to agricultural losses of up to US$2 billion per annum. The most recent estimate of the costs of land degradation in CA was done by Mirzabaev *et al.* (2016), who calculated that the annual cost of land degradation in the region due to land use and cover change between 2001 and 2009 was US$6 billion, including rangeland degradation (US$5.4 billion), deforestation (US$0.3 billion) and abandonment of croplands (US$0.1 billion). About half of these costs, namely US$3 billion, are estimated to be incurred within the ASB, excluding northern Afghanistan.

Information on the costs of land degradation in northern Afghanistan is not available. Although there are slight differences in the amounts of these estimated costs between various studies due to methodological differences, overall, these estimates are broadly consistent with each other and point to the fact that the economic costs of land degradation in the Aral Sea Basin are high. Currently, there are land and water management technologies that can be economically sustainable and revert the land degradation to a high extent (Pender *et al.* 2009). Mirzabaev *et al.* (2016) estimated that each US$1 invested into land restoration in the ASB may yield about US$5 over a period of 6 to 30 years, implying that investments into land and water management in the Basin are economically profitable.

## Water quality deterioration

Water systems of Central Asia are threatened by human activities and are expected to be affected by anthropogenic climate change. Land cover change, intensive agriculture, irrigation and reservoir schemes, urbanization and industrialization are recognized as direct stressors affecting water quality and quantity (Vörösmarty *et al.* 2004). These stressors jointly have had far-reaching repercussions on the Aral Sea Basin as well. The countries of the Basin, like in any other region of the world, have been reaping economic benefits through trading off agricultural production against ecosystems and biodiversity for a long time. When planning for a major expansion of irrigation in the ASB, predictions were made that this expansion would reduce inflow to the Aral Sea and considerably decrease its size. However, some experts saw this as a sensible trade-off: a unit of water for irrigation was assumed to generate more tangible, short-term economic returns than the same unit of water delivered to the Aral Sea (Micklin 1988; Morrison 2008: 232).

Currently, the Basin is struggling with water scarcity, on the one hand, and with the inadequate provision of basic water services like clean drinking water and sanitation, on the other (Vörösmarty *et al.* 2010). In the 2000s, three and four out of ten of Tajikistan's and Turkmenistan's inhabitants, respectively, had no access to safe drinking water. In 1994, Uzbekistan documented 271,000 and in 2005 Kyrgyzstan recorded 122,000 cases of water-related illnesses (FAO 2018). As a result of water pollution and the spread of toxic substances from agriculture, the population's health, especially in the Karakalpakstan region of Uzbekistan, is at high risk (Obertreis 2017: 460). In 1998, water sample tests revealed that 65 per cent of drinking water in Karakalpakstan did not correspond to the required quality standards. However, during the last two decades there have been tangible improvements in the access to piped water and sanitation in Central Asia (Bekturganov *et al.* 2016). Currently, more than 90 per cent of the population in CA have access to basic sanitation, while the basic access to piped drinking water ranges from 75 to 94 per cent (WHO/UNICEF 2017).

Agriculture in the ASB is the major source of water pollution due to discharges of return waters into the rivers and natural lowlands. The major pollutant components originate from return waters of collector drainage systems especially during low-flow periods of the year and contain salts, residues of pesticides, herbicides, fertilizers and other chemicals that are applied in agricultural production (Abdullaev 2004; Crosa *et al.* 2006a; Qadir *et al.* 2009). In northern Afghanistan, the discharge of non-treated sewage is a major source of water contamination (Habib 2014).

In downstream sites of the Syr Darya River, high accumulation of water and soil salinity has been traced due to discharged drainage (Cai *et al.* 2002). The Amu Darya River also presents high levels of soluble contents, which in turn prevent further water use for drinking or other purposes (Crosa *et al.* 2006b). Similar to the Amu Darya and the Syr Darya, the Zerafshan River flow also contains high concentrations of ions, dissolved metals, salts, ammonium, nitrates, phosphates and pesticides. These pollutants are released mostly to the lower parts of the Zerafshan from municipal wastewater, manufacturing industries, mining, irrigation and agricultural cultivation (Toderich *et al.* 2002a; Kulmatov *et al.* 2013).

The water quality of these rivers has deteriorated gradually since the 1950s mainly due to large-scale irrigation systems introduced in the Soviet period. The salinity level and other parameters of river water quality such as major cations and anions, organic compounds, pH and pesticide levels were within safe limits (between 0.33 and 0.72 g $L^{-1}$) in the first half of the last century (Altiyev 2005). The twentieth century is recorded into the history of CA as the period of expanding cotton production through increases in irrigated areas (Allworth and Allworth 1994; Obertreis 2017). From the 1970s onwards, an increase in the discharge of return waters to the Amu Darya and Syr Darya River flows, mostly from irrigation schemes, dramatically increased salinity levels. In 1985, the salinity levels were lower in the upstream sections of the Amu Darya and Syr Darya Rivers (range of 1.2–1.4 g $L^{-1}$), but noticeably higher in downstream areas of both river basins, ranging from 1.4 to 2.3 g $L^{-1}$ (Altiyev 2005). The irrigation infrastructure, serving to provide cotton autonomy for the whole Soviet Union, often suffered from the consequences of haphazard construction. For example, the Karakum canal in the early 1950s brought water to Ashgabat in Turkmenistan and converted non-arable into arable lands. This achievement, however, came at an enormous price, because this construction was responsible to a considerable extent for the siltation of the Aral Sea and its related consequences (Obertreis 2017: 478–479).

There is a distinctive type of water quality deterioration in some zones of CA, which is characterized by irrigation water containing higher levels of magnesium (Mg2+) with the Mg2+ to calcium (Ca2+) ratio not lower than one. The increase of magnesium in water content beyond the safe levels induces the irrigated soil to behave like sodic soil (Qadir *et al.* 2018). In many downstream zones of Kazakhstan many such cases of magnesium-induced soil degradation are traced. Studies by Qadir *et al.* (2009) and Karimov *et al.* (2009) showed that significant areas of Central Asia suffer from similar magnesium-induced soil degradation, demonstrating the direct links between water and land quality in the Basin.

Moreover, in the downstream countries of the Basin, a part of the return waters with rich content of salt and fertilizer and pesticide residues feeds into drainage water lakes. One example is the Aydarkul Lake (Uzbekistan), with a capacity of 30 $km^3$. It was artificially created in 1969 with the return waters from the Hungry Steppe's irrigated land and the water released from the Chardarya reservoir. Similar depression lakes exist in Turkmenistan as well, namely the Sarykamysh and Golden Age (Altyn Asyr) (Spoor 1998: 417; Kamalov 2012: 84; Frenken 2013: 113). Because the levels of pesticide and herbicides are dangerously high in these depression lakes, there is little economic use of these water bodies.

CA countries use marginal quality water resources (e.g., saline water from agricultural drainage systems) to irrigate crops like cotton in summer, when freshwater is scarce. For example, cotton producers in the Syr Darya River in Uzbekistan use moderately saline water (total soluble salts, TSS~2.5 g$L^{-1}$; electrical conductivity, ECe ~ 4dS $m^{-1}$) in combination with freshwater. However, there are places where such moderately saline water is not available in sufficient quantities, so that farmers have no choice but to use water containing even higher levels of soluble salts (Qadir *et al.*

2007; Qadir *et al.*, 2009). In the Amu Darya catchment areas, return water is used for leaching and irrigating the surrounding land (Crosa *et al.* 2006a, 2006b). Use of water with salts and pollutants without appropriate management leads to negative impacts on crop productivity and, ultimately, on farming income (Karimov *et al.* 2009; Bezborodov *et al.* 2010; Toderich *et al.* 2002b).

In the 1980s, the environmental crisis resulting from mismanagement of water was no longer a secret. The decision-making echelons of the Soviet Union started to acknowledge the problem and began to ponder mitigation measures and remedies. The collapse of the Soviet Union, however, prevented the implementation of concrete coordinating actions within the boundaries of a single country (Obertreis 2017). After the collapse, five politically independent CA republics had to accept responsibility for the transboundary water streams and irrigation systems. The common resource, which was managed previously by a single decision-maker, now became a matter for international disputes and compromises (O'Hara 2000; Bernauer and Siegfried 2012). Currently, all five CA governments have similar systems for regulatory requirements regarding the quality of water resources. However, there are notable variations in the rates of development and implementation of legislative standards and norms among these countries. A major reason for these discrepancies is the limited investment and institutional capacity across countries. Consequently, these variations might hinder CA cooperation in water quality protection activities. This might leave the water quality improvement prospects dependent only on bilateral or at best multilateral agreements (UNECE and CAREC 2011).

## Options for sustainable land and water management

### *Technological options*

The World Overview of Conservation Approaches and Technologies (WOCAT) database has documented a total of 206 technologies and approaches for the countries of CA, with the majority of practices coming from Tajikistan (153) and the fewest from Turkmenistan (3). The sustainable land management options applied in the region cover a broad spectrum of technologies including enhancing soil fertility, improving irrigation water use efficiency, improving rangelands, soil erosion control, agroforestry, sustainable fuel consumption and energy efficiency. These technologies are applicable across the diverse landscape and land use types and, therefore, can be grouped according to the four main agro-ecosystems of the region: rainfed, irrigated areas, mountains and rangelands. Crop diversification is a management option that can replace conventional mono-cropping systems. For example, alternating wheat cultivation with legumes provides better incomes to farmers (Pender *et al.* 2009). Most promising for diversified crop rotations in the ASB are field pea, chickpea, lentil, safflower, mustard, buckwheat, millet and alfalfa (Gupta *et al.* 2009). In addition to improved soil fertility, crop diversification enhances fodder supply for livestock and reduces water use for irrigation (Suleimenov *et al.* 2005).

Mulching is an important soil management technique, which increases soil quality, water content and organic carbon stock in irrigated agriculture. In Uzbekistan, according to a three-year study, soils with mulching showed a 20-per cent increase in salinity compared to soils without mulching. Application of mulching could substantially increase crop productivity and water saving, reaching up to 500 m$^3$ of water savings for each ton of cotton produced (Bezborodov *et al.* 2010).

Various irrigation techniques were tested on croplands in the ASB, such as alternate furrow irrigation and drip and sprinkler irrigation. Both techniques require significant up-front investments. However, drip irrigation is gaining popularity in Tajikistan and Kyrgyzstan, where it is used predominantly by smallholder farmers for gardens and greenhouses.

(a)

(b)

*Figure 6.3* Various irrigation techniques successfully tested in the Aral Sea Basin: (a) drip irrigation, (b) use of plastic mulching in furrow irrigation, (c) alternate furrow irrigation, (d) sprinkler irrigation, (e) use of plastic chutes for irrigation in sloping areas

Source: Ikramov Rakhimdjan

Replacement of furrow irrigation with drip irrigation for open-field tomato cultivation in Kyrgyzstan has shown a 100-per cent increase in tomato yields and a proportional increase in farming income. Afforestation of degraded lands was shown to be economically viable in the downstream areas of the Basin, also contributing to carbon sequestration and improved ecosystem health (Khamzina *et al.* 2008; Djanibekov *et al.* 2013).

In the ASB, rangelands are spread across steppes, deserts and the mountains by occupying approximately one-third of the basin area (Gintzburger 2004). Livestock breeding has been an important traditional occupation in the region. With the collapse of the Soviet Union, the support for livestock production (veterinary services, maintenance of rangeland infrastructure, stable markets for meat and wool) was halted. Uneven grazing pressure has led to overgrazing of rangelands next to the settlements and undergrazing in remote areas. Numerous studies on rotational grazing—when animals are moved from one grazing sector to another allowing for natural rehabilitation of vegetation—have illustrated its effectiveness in the Basin. In the framework of CACILM II, rotational grazing was applied on 5,000 hectares of desert rangelands in Uzbekistan. Since the majority of livestock is privately owned, the adoption of grazing management requires

*Figure 6.3* continued

(c)

(d)

(e)

community mobilization. Development of communal grazing plans supported by adequate water sources increases the mobility of herds, which removes the pressure from the stressed rangelands around the settlements and water sources. In Turkmenistan, community-based grazing with the combination of moving sand fixation and further afforestation with native desert species demonstrated socio-economic and environmental effectiveness (CACILM 2016).

The most common sustainable land and water management (SLWM) technologies adopted in the mountainous regions are terracing, moisture-retaining trenches, drainage ditches and check-dams to reduce river velocity, prevent soil erosion, preserve soil fertility and maintain long-term sustainable production on sloping lands (Sanz *et al.* 2017). For example, in Turkmenistan, the technology of producing Juniper seedlings in nurseries for reforestation of sloping areas was adopted by local communities (CACILM 2016).

The dissemination of the SLWM technologies is an iterative process, and experts are looking for upscaling options in CA. Earlier studies were focused only on technical solutions, while considerations of economic profitability were added later to make SLWM more attractive to farmers. Currently, the carbon sequestration potential of SLWM technologies (Sanz *et al.* 2017) is also assessed for various technologies in CA in order to demonstrate climate adaptation and mitigation capacity that is required for climate-funded initiatives.

Despite the availability of locally tested SLWM technologies, the upscaling is still limited. It was found that only about 26 per cent of agricultural households apply any of these sustainable land and water management technologies (Mirzabaev 2013). The barriers and success factors influencing the adoption of new practices vary from external (political, institutional and socio-economic) to internal (value and behavioral systems, resources) (Djanibekov *et al.* 2016). Lack of knowledge on conservation agriculture and land tenure insecurity coupled with insufficient rural extension services are salient barriers for the adoption of innovative technologies. The household data by Mirzabaev (2013) further suggest that poor farmers who lack access to credits and resources are unlikely to switch to new practices that require high upfront costs and do not bring quick returns to investments. The role of the governments is to provide regulatory programs and effective incentives that will make the adoption of SLWM practices feasible, such as subsidies or water pricing (Gupta *et al.* 2009). Despite these constraints, the adoption of some technologies was successful in the region, such as planting of winter wheat among standing cotton, practiced widely in Uzbekistan (Mirzabaev 2016), or the application of phosphogypsum to sodic soils in southern Kazakhstan (Vyshpolsky *et al.* 2010). These cases indicate that the upscaling of SLWM technologies is viable when economic, institutional and policy-level enabling conditions are present. Given the diversity of external and internal factors influencing the adoption of SLWM technologies, tailoring them to the local context is important. In this regard, the role of national and regional partners to generate, apply and disseminate knowledge is essential.

Notably, the Central Asian Regional Environmental Center (CAREC), an independent inter-governmental agency established in 2001, has promoted the dialogue on regional cooperation among CA countries and Afghanistan in the field of environmental protection and sustainable development. CAREC develops and promotes innovative tools and practices in the field of ecosystems services, water initiatives, climate change and sustainable energy. The Central Asian Countries Initiative for Land Management (CACILM, www.fao.org/in-action/cacilm-2/en/), a multi-country program to address land degradation in CA, has tested and documented several proven land conservation technologies. The knowledge management component of the program reviewed and systematized numerous technologies on increasing soil fertility, improving sowing methods, agroforestry, reforestation, prevention of slope erosion, improving irrigation techniques, rangeland management and increasing the capacity of land users. The ongoing CACILM intervention aims at scaling up integrated natural resources management in drought-prone and salt-affected agricultural production landscapes in CA and Turkey. With the support of a broad community of technical and development partners, such as FAO, CGIAR, WOCAT, GIZ, CAREC and the University of Central Asia, this multi-country program will help in stabilizing and potentially reversing the trends of soil salinization, reduce erosion, improve water capture and retention, increase the sequestration of carbon and reduce the loss of agrobiodiversity in the Basin.

## *Policy options*

Numerous enabling policy opportunities for sustainable natural resource management and environmental protection exist in the region, such as promoting incentives for sustainable land management through land restoration and rehabilitation, including afforestation, as well as promoting institutional reforms in water management.

After dismantling the centralized and planned agricultural production of the Soviet era, the countries of CA made efforts toward market-oriented systems. This transition took various forms: the appearance of agro-holdings in Kazakhstan, decentralized natural resource management in Kyrgyzstan, the transition to smallholder farming in Tajikistan, moderate land reforms in Uzbekistan and still-prevailing governmental orders for cash crops in Turkmenistan. The Central Asian countries also took steps in diversifying the land tenure. Gradual decentralization of natural resources management occurred in all countries of CA. Kyrgyzstan is a remarkable example of land tenure reform where community-based resource management was institutionally introduced in the form of pasture committees and community forest management. The experience of Kyrgyzstan showed that the mechanism of leasing plots substantially increased the income of local families (FAO 2012b). However, cautious planning and detailed analysis are required for community-based resources management to avoid potential conflicts like unfair land distribution and lack of compliance with sustainability principles (BIOM 2008).

In regard to water management, the experience of water user associations (WUAs) represents another significant example of agricultural reforms. WUAs were introduced to empower farmers to take direct decisions and responsibility for water distribution and the maintenance of water infrastructures. WUA members and administrators were also trained in applying integrated water management principles in their operations. The innovative basin approach of water management was first adopted in Kyrgyzstan in 1997 and subsequently in Kazakhstan, Uzbekistan and Tajikistan, with respective legislative acts adopted in each country. Given the challenges of introducing such reforms, substantial financial and technical support has been provided by the World Bank and Asian Development Bank, and a number of bilateral development agencies. The application of WUAs in various national contexts of these four countries offers lessons and recommendations. The main barriers to developing the full potential of WUAs, such as low fees to achieve cost-recovery and top-down implementation approaches, need to be addressed to improve their effectiveness in water management (Abdullaev *et al.* 2009). The continuing conflicts in Afghanistan act as a major barrier for stepping up policy actions to address land and water degradation. Although there have been many international research and demonstration projects for improving the sustainability of land and water management in Afghanistan, the impact of these activities remains limited.

Inter-state water management in the ASB remains a challenge. Since most of the irrigation water comes from transboundary rivers, water management is addressed by international or inter-governmental agreements. Nevertheless, bilateral agreements for water distribution are still more common (Abdullaev 2004). However, during recent years the CA governments have stepped up their efforts to mitigate the Aral Sea impacts and improve water allocation within the ASB. An example is the recently established (2018) Multi-Partner Human Security Trust-Fund for the Aral Sea Region in an initiative taken by the government of Uzbekistan. Common threats for the region such as changing climate, more frequent droughts and increasing water scarcity require further concerted efforts from the CA countries.

Recognizing the urgency of the action to mitigate the Aral Sea crisis, the Summit of the Heads of the Founding States of the International Fund for Saving the Aral Sea (IFAS) took place in August 2018 in Turkmenistan. The Summit highlighted the fact that the Aral Sea crisis went far beyond the regional scale, and the joint efforts of the governments, as well as of the international community, are required to counterbalance the adverse negative effect of the disaster. As a result, the Joint Communiqué has been adopted, which calls, among other things, for the creation of a sustainable regional mechanism for the integrated use of water and energy resources in CA.

## *Land degradation neutrality as an integrated framework for scaling SLWM options*

The land–water nexus is fundamental to the most important global policy developments concerning land degradation in recent years. At international level, an innovative framework for sustainable natural resource management is proposed through the Land Degradation Neutrality (LDN) concept. The United Nations Convention to Combat Desertification (UNCCD) defines LDN as 'a state whereby the amount and quality of land resources, necessary to support ecosystem functions and services and enhance food security, remains stable or increases within specified temporal and spatial scales and ecosystems' (decision3/COP12, paragraph 2, UNCCD 2015).

LDN is a policy tool that encourages counterbalancing the anticipated loss of land with measures to regain healthy land through avoiding the risk of land degradation and restoring already-degraded areas (Orr *et al.* 2017). As the counterbalancing mechanism should be best embedded in the integrated land use planning and will likely be managed within a biophysical (e.g., catchment) or administrative (e.g., province) spatial domain, the decisions on land intervention are closely interlinked with water management. LDN is a part of the Sustainable Development Goals agenda and is a 'strong vehicle' to implement the United Nations Convention to Combat Desertification (decision3/COP12, paragraph 4, UNCCD 2015). The LDN framework underpins the critical role of national policy-makers in establishing LDN targets and measures, as well as integrating the LDN agenda in the national development strategies and country commitments, such as the Nationally Determined Contribution under Paris Agreement targets.

The CA countries and Afghanistan are Parties of the United Nations Convention to Combat Desertification (UNCCD), and three of them (Kazakhstan, Kyrgyzstan and Uzbekistan) took part in the LDN Target Setting Program, which guided countries in the definition of national baselines, targets and associated measures. Multi-stakeholder national working groups have been leading consultation processes in three countries to assess initial conditions of land use, based on the proposed LDN indicators (trends in land cover, trends in primary productivity, trends in soil organic carbon) and establish national voluntary and time-bound LDN targets.

By embracing LDN, the countries of CA are now working under a common global policy approach that has been designed to maintain or improve the sustainable delivery of ecosystem services and land productivity. The key goal is to enhance food security, increase resilience of the land and populations dependent on the land, seek synergies with other social, economic and environmental objectives and reinforce responsible and inclusive governance of land (UNCCD 2017).

The role of land as an accelerator to some SDGs, including those that address poverty, food, water and energy security, human health, migration and biodiversity, was politically recognized (Akhtar-Schuster *et al.* 2017). Most of the mentioned sustainable development challenges are critical for Central Asia. UNCCD has recently promoted two policy options highly relevant for CA through the decisions of the Conference of the Parties at its 13th Session in China in 2017. Namely, the Policy Advocacy on Drought and Policy Advocacy Framework to Combat Sand and Dust Storms (decision 29/COP13, UNCCD 2017 and decision 31/COP13, UNCCD 2017). The same Conference of the Parties endorsed the Drought Initiative that aims at building drought resilience of the countries. The UNCCD assists in developing a comprehensive national drought policy plan, as well as a toolbox to provide stakeholders with easy access to tools, case studies and other resources to support the design of the plan. The second policy option promoted by UNCCD is mitigation of impacts of sand and dust storms and anthropogenic dust sources by advocating the following three-pillars approach—early warning systems, preparedness and resilience, and anthropogenic source mitigation. With the anticipated increasing pressures on water and with more intense and severe droughts and sand and dust storms, the new holistic approach to land and water management needs to inform policy-level, infrastructural, local and, increasingly, digital interventions.

Both phenomena—drought, and sand and dust storms—are widespread in CA, with increasing frequency and intensity. CA countries should take an active part in nationalizing the proposed policy options to move from reactive and crisis-based management towards a proactive and risk mitigation-based approach.

## Conclusions

Land and water degradation are among the key barriers for the attainment of the Sustainable Development Goals in the ASB area. Environmentally sound and economically viable technological solutions for addressing land and water degradation are available in the region. However, their adoption remains limited. To support the adoption of SLWM practices, a number of financial, institutional and policy constraints need to be removed. Key policy measures for facilitating SWLM adoption in Central Asia include improving access to markets and extension services, securing land tenure and providing access to credit to cover the upfront investment costs of SLWM adoptions, as well as continuing investments for the upgrade and maintenance of irrigation and drainage networks. A wide adoption of SWLM practices without returning to peace and stability would be challenging in the northern Afghanistan.

Transborder collaboration and knowledge exchange is also crucial for accelerating SWLM infiltration for improving transboundary water management in the Basin. Various regional and international organizations and programs, such as CAREC, IFAS, CGIAR centers, UN agencies and others, could play an active role to facilitate these exchanges. The existing and widening networks built by the regional and international actors suggest optimism for innovative solutions and learning opportunities. Involvement of stakeholders from Afghanistan in these networks could build knowledge exchange and experience between the countries.

The success of measures for sustainable land and water management in the Aral Sea Basin also requires increasing investments into environmental research and educational programs. Knowledge gaps exist in such areas as the actual costs of water quality degradation, the health impacts of land and water degradation and interactions between land and water degradation under climate change, as well as optimization methods on SLWM policies. There are currently very few educational programs in the region for training professionals in issues related to environmental sustainability. Collaboration with leading international universities could provide a good platform for expanding environmental education in the region.

## Note

1 According to the United Nations classification, Central Asia includes five countries: Kazakhstan, Kyrgyzstan, Tajikistan, Turkmenistan and Uzbekistan (Source: https://unstats.un.org/unsd/methodology/m49/). Afghanistan is part of South Asia.

## References

Abdullaev I (2004) *Water Management Policies of Central Asian Countries: Integration or Disintegration?* Paper presented at USDS organized conference 'Celebrating 10 years of Contemporary Issues Fellowships Conference on Water Issues in Central Asia', Tashkent, Uzbekistan, 25 September

Abdullaev I, Kazbekov J, Manthrithilake H, Jumaboev K (2009) Water user groups in Central Asia: emerging form of collective action in irrigation water management, *Water Resources Management* 24: 1029–1043

Kerven C, Shanbaev K, Alimaev I, Smailov A, Smailov K (2008) Livestock mobility and degradation in Kazakhstan's semi-arid rangelands. In: Behnke R (ed.), *The Socio-Economic Causes and Consequences of Desertification in Central Asia*. NATO Science for Peace and Security Series (Series C: Environmental Security). Springer, Dordrecht, the Netherlands

Ahmad M, Wasiq M (2004) *Water Resource Development in Northern Afghanistan and Its Implications for Amu Darya Basin*. World Bank Working Paper Series No. 36. World Bank, Washington, DC, USA

Akramkhanov A, Martius C, Park SJ, Hendrickx JMH (2011) Environmental factors of spatial distribution of soil salinity on flat irrigated terrain, *Geoderma* 163(1–2): 55–62

Akramkhanov A, Vlek PLG (2012) The assessment of spatial distribution of soil salinity risk using neural network, *Environmental Monitoring and Assessment* 184(4): 2475–2485

Akhtar-Schuster M, Stringer LC, Erlewein A, Metternicht G, Minelli S, Safriel U, Sommer S (2017) Unpacking the concept of land degradation neutrality and addressing its operation through the Rio Conventions, *Journal of Environmental Management* 195(1): 4–15

Allworth EA, Allworth E (eds) (1994) *Central Asia, 130 Years of Russian Dominance: A Historical Overview*. Duke University Press, Durham, NC, USA

Altiyev T (2005) *Water for Development in Central Asia*. Paper presented at the sub-regional workshop on the Implementation of Integrated Water Resources Management for Well-Being and Future Development in Central Asia, 31 October–2 November 2005, Almaty, Kazakhstan

Bekberdieva T (2011) Role of floodplain forests of Talas river basin in biodiversity conservation. In: Laletin A, Parrotta J, Domashov I (eds), *Traditional Forest Related Knowledge, Biodiversity Conservation and Sustainable Forest Management in Eastern Europe, Northern and Central Asia*. International Union of Forest Research Organizations (IUFRO) World Series Volume 26, Vienna, Austria

Bekchanov M, Ringler C, Bhaduri A, Jeuland M (2016) Optimizing irrigation efficiency improvements in the Aral Sea Basin, *Water Resources Economics* 13: 30–45

Bekturganov Z, Tussupova K, Berndtsson R, Sharapatova N, Aryngazin K, M Zhanasova (2016) Water-related health problems in Central Asia—a review, *Water* 8(219). DOI: https://doi.org/10.3390/w8060219

Bernauer T, Siegfried T (2012) Climate change and international water conflict in Central Asia, *Journal of Peace Research* 49(1): 227–239

Bezborodov GA, Shadmanov DK, Mirhashimov RT, Yuldashev T, Qureshi AS, Noble AD, Qadir M (2010) Mulching and water quality effects on soil salinity and sodicity dynamics and cotton productivity in Central Asia, *Agriculture, Ecosystems & Environment* 138(1–2): 95–102

BIOM (2008) *Implementation of the Forestry Policy of Kyrgyzstan through the Informational Campaign on Involvement of the Communities to Joint Forest Management*. BIOM Ecological Movement, Bishkek, Kyrgyzstan

Cai X, McKinney DC, Lasdon LS (2002). A framework for sustainability analysis in water resources management and application to the Syr Darya Basin, *Water Resources Research* 38(6): n.p.

Central Asian Countries Initiative for Land Management (CACILM) (2016). *Phase II. Technologies and approaches on Sustainable Land Management in Central Asia.* Composed with support of the Project 'Knowledge Management' in CACILM phase II, International Center for Agricultural Research in the Dry Areas (ICARDA) in Central Asia and the Caucasus

Crosa G, Froebrich J, Nikolayenko V, Stefani F, Galli P, Calamari D (2006a) Spatial and seasonal variations in the water quality of the Amu Darya River (Central Asia), *Water Research* 40(11): 2237–2245

Crosa G, Stefani F, Bianchi C, Fumagalli A (2006b) Water security in Uzbekistan: implication of return waters on the Amu Darya Water Quality, *Environmental Science and Pollution Research* 13(1): 37–42

de Beurs KM, Henebry GM, Owsley BC, Sokolik I (2015) Using multiple remote sensing perspectives to identify and attribute land surface dynamics in Central Asia, 2001–2013, *Remote Sensing of Environment* 170: 48–61

Djanibekov U, Djanibekov N, Khamzina A, Bhaduri A, Lamers JPA, Berg E (2013) Impacts of innovative forestry land use on rural livelihood in a bimodal agricultural system in irrigated drylands, *Land Use Policy* 35: 95–106

Djanibekov U, Villamor GB, Dzhakypbekova K, Chamberlain J, Xu J (2016) Adoption of sustainable land uses in post-Soviet Central Asia: the case for agroforestry, *Sustainability*. DOI: 10.3390/su101030

Dubovyk O, Menz G, Conrad C, Kan E, Machwitz M, Khamzina A (2013). Spatio-temporal analyses of cropland degradation in the irrigated lowlands of Uzbekistan using remote-sensing and logistic regression modeling, *Environmental Monitoring and Assessment*, 185(6): 4775–4790

FAO (2010) *Forest Tenure in West and Central Asia, the Caucasus and the Russian Federation.* Forestry Policy and Institutions Working Paper www.fao.org/docrep/012/k7544e/k7544e00.pdf

FAO (2012a) *Irrigation in Central Asia in figures – AQUASTAT Survey.* Retrieved 18 December 2018, www.fao.org/nr/water/aquastat/basins/aral-sea/aral.sea-CP_eng.pdf

FAO (2012b) *Regional Office for Europe and Central Asia.* Policy Studies on Rural Transition

FAO (2018) *AQUASTAT*. Retrieved 29 September 2018, www.fao.org/nr/water/aquastat/data/query/index.html?lang=en

Frenken K (ed) (2013) *Irrigation in Central Asia in Figures.* FAO Water Reports. Food and Agriculture Organization of the United Nations, Rome, Italy

Glantz M (ed) (1999) *Creeping Environmental Problems and Sustainable Development in the Aral Sea Basin.* Cambridge University Press, Cambridge, UK

Gintzburger G, Toderich K, Mardonov B, Mahmudov MM (2003) *Rangelands of the Arid and Semi-Arid Zones in Uzbekistan.* CIRAD, ICARDA, Montpellier, France

Gintzburger G (2004) Agriculture and rangelands in Middle Asian Countries. In: Ryan J, Vlek P, Paroda R (eds), *Agriculture in Central Asia: Research for Development.* International Center for Agricultural Research in the Dry Areas (ICARDA), Syria, pp 154–175

Gupta R, Kienzler K, Martius C, Mirzabaev A, Oweis T, de Pauw E, Qadir M, Shideed K, Sommer R, Thomas R, Sayre K, Carli C, Saparov A, Bekenov M, Sanginov S, Nepesov M, Ikramov R (2009) *Research Prospectus: A Vision for Sustainable Land Management Research in Central Asia.* ICARDA Central Asia and Caucasus Program. Sustainable Agriculture in Central Asia and the Caucasus Series No.1. CGIAR-PFU, Tashkent, Uzbekistan

Habib H (2014) Water-related problems in Afghanistan, *International Journal of Educational Studies* (Vol. 1). Retrieved 10 March 2019, www.escijournals.net/IJES

Indoitu R, Kozhoridze G, Batyrbaeva M, Vitkovskaya I, Orlovsky N, Blumberg D, Orlovsky L (2015) Dust emission and environmental changes in the dried bottom of the Aral Sea, *Aeolian Research* 17: 101–115

Issanova G, Abuduwaili J, Galayeva O, Semenov O, Bazarbayeva T (2015). Aeolian transportation of sand and dust in the Aral Sea region, *International Journal of Environmental Science and Technology* 12(10): 3213–3224

Ji CY (2008) Land degradation in central asia. ADB TA 6356-REG: Central Asian Countries Initiative for Land Management Multicountry Partnership Framework Support Project. Retrieved 10 March 2019, www.adb.org/sites/default/files/project-document/66804/38464-reg-tar.pdf

Kamalov YS (2012) A last movement for a lost sea. In Edelstein M, Cerny A, Gadaev A (eds), *Disaster by Design: The Aral Sea and its Lessons for Sustainability.* Research in Social Problems and Public Policy Vol. 20. Emerald Group Publishing Limited, Vienna, Austria, pp 77–88

Karimov A, Qadir M, Noble A, Vyshpolsky F, Anzelm K (2009) Development of magnesium-dominant soils under irrigated agriculture in southern Kazakhstan, *Pedosphere* 19(3): 331–343

Khamzina A, Lamers JPA, Vlek PLG (2008) Tree establishment under deficit irrigation on degraded agricultural land in the lower Amu Darya River region, Aral Sea Basin, *Forest Ecology and Management* 255(1): 168–178

Kienzler KM, Lamers J, Mcdonald A, Mirzabaev A, Ibragimov N, Egamberdiev O, Ruzibaev E, Akramkhanov A (2012) Conservation agriculture in Central Asia–what do we know and where do we go from here? *Field Crops Research* 132: 95–105

Kulmatov R, Opp C, Groll M, Kulmatova D (2013) Assessment of water quality of the trans-boundary Zarafshan River in the territory of Uzbekistan, *Journal of Water Resource and Protection* 5(1): 17

Le QB, Nkonya E, Mirzabaev A (2016) *Biomass Productivity-Based Mapping Of Global Land Degradation Hotspots. Economics of Land Degradation and Improvement – A Global Assessment for Sustainable Development.* DOI: https://doi.org/10.1007/978-3-319-19168-3_4

Lewis, R (1966) Early irrigation in west Turkestan, *Annals of the Association of American Geographers* 56(3): 467–491

Micklin P (1988) Desiccation of the Aral Sea: a water management disaster in the Soviet Union, *Science* 241(4870): 1170–1176

Mirzabaev A (2013) *Climate Volatility and Change in Central Asia: Economic Impacts and Adaptation.* Doctoral thesis at Agricultural Faculty, University of Bonn, Germany

Mirzabaev A (2016) Land degradation and sustainable land management innovations in Central Asia. In: Gatzweiler F, von Braun J (eds), *Technological and Institutional Innovations for Marginalized Smallholders in Agricultural Development.* Springer, Cham, Switzerland

Mirzabaev A, Goedecke J, Dubovyk O, Djanibekov U, Le Q, Aw-Hassan, A (2015) *Economics of Land Degradation in Central Asia.* ZEF Policy Briefs, 1–6

Morrison A (2008) *Russian Rule in Samarkand 1868–1910: A Comparison with British India.* No. 2012-1 European and Central Asian Agriculture Towards 2030 and 2050. Oxford University Press, Oxford, UK

Muslimshoeva B, Samimi C, Kirchhoff J, Koellner T (2013) Analysis of costs and people's willingness to enroll in forest rehabilitation in Gorno Badakhshan, Tajikistan, *Forest Policy and Economics* 37: 75–83

Novikova T, Yatimov O (2011) Biodiversity of mountain forest ecosystems supporting the socio-economic wellbeing of local communities in south-eastern Tajikistan. In: Laletin A, Parrotta J, Domashov I (eds), *Traditional Forest Related Knowledge, Biodiversity Conservation and Sustainable Forest Management in Eastern Europe, Northern and Central Asia.* IUFRO World Series Volume 26. Vienna, Austria

Obertreis J (2017) *Imperial Desert Dreams: Cotton Growing and Irrigation in Central Asia, 1860–1991.* Vandenhoeck & Ruprecht, Göttingen, Germany

Opp, C (2004) Desertifikation in Usbekistan: Ursachen, Wirkung und Verbreitung *Geographische Rundschau* 56(10): 44–51

Orr BJ, Cowie AL, Castillo Sanchez VM, Chasek P, Crossman ND, Erlewein A, Louwagie G, Maron M, Metternicht GI, Minelli S, Tengberg AE, Walter S, Welton S (2017) Scientific conceptual framework for land degradation neutrality. *A Report of the Science-Policy Interface. United Nations Convention to Combat Desertification (UNCCD),* Bonn, Germany

Pender J, Mirzabaev A, Kato E (2009) *Economic Analysis of Sustainable Land Management Options in Central Asia.* Consultant Report to the Asian Development Bank. IFPRI, Washington, DC, USA

Qadir M, Noble AD, Qureshi AS, Gupta RK, Yuldashev T, Karimov A (2009) Salt-induced land and water degradation in the Aral Sea Basin: a challenge to sustainable agriculture in Central Asia, *Natural Resources Forum* 33(2): 134–149

Qadir M, Schubert S, Oster JD, Sposito G, Minhas PS, Cheraghi SAM, Saqib M (2018) High-magnesium waters and soils: emerging environmental and food security constraints, *Science of The Total Environment* 642: 1108–1117

Qadir M, Sharma BR, Bruggeman A, Choukr-Allah R, Karajeh F (2007) Non-conventional water resources and opportunities for water augmentation to achieve food security in water scarce countries, *Agricultural Water Management* 87(1): 2–22

Robinson S, Whitton M, Biber-Klemm S, Muzofirshoev N (2010) The impact of land-reform legislation on pasture tenure in Gorno-Badakhshan: from common resource to private property? *Mountain Research and Development* 30(1): 4–13

Saigal S (2003) Kazakhstan: Issues and Approaches to Combat Desertification. TA-5941-REG. The Asian Development Bank and the Global Mechanism. Retrieved 10 March 2019, www.droughtmanagement. info/literature/ADB_kazakhstan_combat_desertification_2003.pdf

Sanz MJ, de Vente J, Chotte J-L, Bernoux M, Kust G, Ruiz I, Almagro M, Alloza J-A, Vallejo R, Castillo V, Hebel A, Akhtar-Schuster M (2017) Sustainable land management contribution to successful land-based climate change adaptation and mitigation. *A Report of the Science-Policy Interface. United Nations Convention to Combat Desertification (UNCCD),* Bonn, Germany

Spoor M (1998) The Aral Sea Basin crisis: transition and environment in former Soviet Central Asia, *Development and Change* 29(3): 409–435

Suleimenov M, Akhmetov K, Kaskarbayev Z, Kireyev A, Martynova L, Medeubayev R (2005) Role of wheat in diversified cropping systems in dryland agriculture of Central Asia, *Turkish Journal of Agriculture and Forestry* 29: 143–150

Sutton W, Whitford P, Stephens EM, Galinato SP, Nevel B, Plonka B, Karamete E (2007) *Integrating Environment into Agriculture and Forestry: Progress and Prospects in Eastern Europe and Central Asia*. World Bank, Prishtina, Kosovo

Toderich KN, Tsukatani T, Mardonov BK, Gintzburger G, Zemtsova OY, Tsukervanik ES, Shuyskaya EV (2002a) *Water Quality, Cropping and Small Rumminants: Future Agriculture in Dry Areas of Uzbekistan*. Retrieved 10 March 2019, https://repository.kulib.kyoto-u.ac.jp/dspace/bitstream/2433/129510/1/DP553.pdf

Toderich KN, Tsukatani T, Goldshtein RI, Aparin VB, Ashurmetov AA (2002b) Ecological conservation and reclamation of arid/saline lands under agricultural system development in Kyzylkum Deserts of Uzbekistan. In Ahmad R, Malik KA (eds), *Prospects for Saline Agriculture*. Tasks for Vegetation Science Vol. 37. Springer, Dordrecht, the Netherlands, pp 19–28

UNCCD (2015) Report of the Conference of the Parties on its twelfth session, held in Ankara from 12 to 23 October 2015. *Part Two: Actions Taken by the Conference of the Parties at its Twelfth Session*. ICCD/COP(12)/20/Add. United Nations Convention to Combat Desertification, Bonn, Germany

UNCCD (2017) Report of the Conference of the Parties on its thirteenth session, held in Ordos, China. *Part Two: Actions Taken by the Conference of the Parties at its Thirteenth Session*. ICCD/COP(13)/21/Add.1. United Nations Convention to Combat Desertification, Bonn, Germany

UNECE and CAREC (2011) *Development of Regional Cooperation to Ensure Water Quality in Central Asia*. Diagnostic Report and Cooperation Plan. Retrieved 10 March 2019, http://wedocs.unep.org/handle/20.500.11822/7533

Vörösmarty C, Lettenmaier D, Leveque C, Meybeck M, Pahl-Wostl C, Alcamo J, Cosgrove W, Grassl H, Hoff H, Kabat P, Lansigan F (2004) Humans transforming the global water system. *Eos, Transactions American Geophysical Union* 85(48): 509–514

Vörösmarty CJ, McIntyre PB, Gessner MO, Dudgeon D, Prusevich A, Green P, Glidden S, Bunn SE, Liermann RC, Davies PM (2010). Rivers in crisis: global water insecurity for humans and biodiversity, *Nature* 467(7315): 555–561

Vyshpolsky F, Mukhamedjanov K, Bekbaev U, Ibatullin S, Yuldashev T, Noble AD, Qadir M et al. (2010) Optimizing the rate and timing of phosphogypsum application to magnesium-affected soils for crop yield and water productivity enhancement, *Agricultural Water Management* 97(9): 1277–1286

WHO/UNICEF (2017) *A Snapshot of Drinking Water, Sanitation and Hygiene in the Central Asia and Southern Asia Region*. Retrieved 24 February 2019, https://washdata.org/file/588/download

World Bank (1998) *Aral Sea Basin Program (Kazakhstan, Kyrgyz Republic, Tajikistan, Turkmenistan and Uzbekistan): Water and Environmental Management Project*. Project Document. Global Environment Division, World Bank, Washington, DC, USA, pp 55

World Bank (2006) *Sustainable Land Management: Challenges, Opportunities, and Trade-offs*. Agriculture and Rural Development. World Bank, Washington, DC, USA

Zhumanova M, Moennig C, Hergarten C, Darr D, Wrage-Moennig N (2018) Assessment of vegetation degradation in mountainous pastures of the Western Tien-Shan, Kyrgyzstan, using eMODIS NDVI, *Ecological Indicators* 95: 527–543

# 7

# Water for agriculture and other economic sectors

*Asel Murzakulova, Dietrich Schmidt-Vogt, Dagmar Balla,*
*Dietrich Darr, Ahmad Hamidov, Ulan Kasymov,*
*Roman Mendelevitch and Serik Orazgaliyev*

## Key messages

- Expansion of intensive and wasteful irrigated farming during the Soviet period was the main cause of the drying-up of the Aral Sea. Intensive irrigation continues as 90 per cent of the total water withdrawal in Central Asia (CA) is for irrigated agriculture. Using irrigation water in agriculture plays a key role in the economy of the five CA countries as the agriculture sector contributes from 10 to 45 per cent of GDP and employs 20 to 50 per cent of the rural population.
- Annual water use for irrigation of the five CA states is 124.6 km$^3$. Water shortages are not due to water scarcity, but rather to inadequate or dilapidated infrastructure and wasteful practices coupled with intra-regional competition between upstream and

downstream countries, and inter-sectoral competition for water, mainly between the agriculture and energy sectors.

• Water resources allocation and water use by agriculture and other sectors lacks coordination because of the emergence of nation-states and the transition from collective to individual farming systems. The newly independent states of CA pursued, at least initially, a policy of prioritizing their own national economies at the expense of regional cooperation, which manifested itself in managing water resources primarily for national interests. The transition from collective to individual farming systems left farmers isolated. Establishment of institutions for decentralized management of natural resources, such as management of water resources through water user associations, is slow and inadequate.

• Sustainable water management requires intra-regional cooperation, especially between upstream and downstream countries, and integrated approaches to rural development.

## Introduction

Water as the core factor of economic prosperity and sustainable livelihoods is a theme that runs through the history of Central Asian states. The Soviet preoccupation with water as the key to socio-economic development was adopted after the collapse of the Soviet Union by the newly independent states of CA and continues to impact the management of water resources in the region. Since ancient times, but especially since the beginning of Russian dominance in the Aral Sea region, water has been primarily used for agriculture. Today, more than 90 per cent of the total water withdrawn in CA is used for irrigated agriculture (Abdullaev *et al.* 2006).

Despite the predominantly dry climate, water resources are sufficient in CA—mainly because of vast reserves stored in the glaciers and snowfields of the Tien Shan and Pamir Mountains and feeding from there into the headwaters of the Amu Darya and Syr Darya Rivers (Xenarios *et al.* 2019). Under Soviet Union rule, the local management of water was replaced by central management of water for agriculture, but increasingly also for energy and manufacturing. After the collapse of the Soviet Union in 1991, central management was replaced by competition between the new CA countries for water resources allocation in irrigated agriculture, hydropower generation and the mining and manufacturing sectors. Competition for water between countries intersects with competition between sectors. Most of the water resources in the region are transboundary, with the main rivers flowing through several countries.

Hydropower generation potential is upstream, where key reservoirs regulate river flows, while the irrigated land is located downstream. Differences between the interests of upstream and downstream countries are complicating water management in the region (Stadelbauer 2007). Rivalry and conflicts between sectors and countries are expected to increase further with more conflicting water demands and resource scarcity caused by climate change. Precarious transboundary water supply between upstream and downstream countries, amplified by climate change as well as by poor land and water management, may seriously deteriorate the socio-economic and environmental situation in the region with the danger of translating into political conflicts. Conflict over water is an issue that plays out not only on the level of national governments, but also on the local level between users living along the borders of the CA states (Murzakulova and Mèstre 2018).

Agriculture plays a lead role in this field of conflicting forces. Unsustainable extraction of water for irrigation has been a prime cause of water scarcity and environmental degradation since the onset of the Soviet era in the region, leading up to the well-documented Aral Sea catastrophe. Excessive overuse of freshwater resources in the Aral Sea Basin caused the gradual desiccation of the Aral Sea and salt-induced land degradation in irrigated areas (Bekchanov *et al.* 2015). Agriculture remains an important pillar of the economies of CA countries. Despite attempts to reduce its extent and impact, irrigated farming continues to place a heavy demand on water resources, and the resulting consequences for the environment in turn put a heavy burden on the economies of the CA states.

In the relatively new and dynamic political landscape of CA, water continues to play a pivotal role in the socio-economic development of the region with the potential to enhance human welfare (Qadir *et al.* 2009). Water as a contested resource remains to be the key not only for the economic development of the individual states, but also for the livelihoods of their people, for food and water security and finally for political stability.

While political stability has been threatened since the collapse of the Soviet Union by conflicts over the region's water resources, there are encouraging developments that indicate growing awareness of water as a limited resource that must be managed on a regional scale through cooperation of the national governments and their constituents. In the end, water as a shared resource may prove to be the most potent medium for overcoming divisiveness and achieving regional cohesion.

## Water perception in a context of historical transformation

The perception of the role of water in the economic and social development of the countries situated in the Aral Sea Basin has changed over the course of the nineteenth and twentieth centuries until today. We will illustrate this shift in the case of the CA states that share the Soviet past. It must be noted that the USSR did not include the interests of Afghanistan and Iran in its water basin policy (Wegerich 2008) and that the CA states follow the same.

The economic basis of the Aral Sea Basin historically was irrigated agriculture (Lipovsky 1995) and animal husbandry (Zhdanko 1966). The demand of water was significantly larger for irrigated agriculture than for animal husbandry, even though animal husbandry in its nomadic form was the mainstay of rural livelihoods in regions that today belong to Kyrgyzstan and Kazakhstan. Before the Russian colonization of Central Asia, irrigation systems were managed locally.

After the conquest of Turkestan, which started around 1865, the administration of the Russian Empire continued to manage the water resources of the region on a small scale, but in a largely decentralized fashion. However, the stepping-up of cotton production in the Aral Sea Basin, in response to cotton shortages caused by the American Civil War, was a first indication of the massive changes that were to take place under Soviet Union rule. In a context of developing a centrally planned economy, water was perceived as a common good to be managed for goals determined by the central government. After the Second World War, the economic development of the Central Asian Soviet Republics focused on agriculture, with the aim of developing what was considered an underutilized region to produce agricultural commodities on an industrial scale. Exploitation and management of the region's water resources for agricultural production was a central element in the scheme. Large-scale irrigation systems were put into place as part of the 'virgin-land' project announced by Nikita Khrushchev in 1953 for improving agricultural productivity. In the Aral Sea Basin, soviet agrarian policy led to rapid and intensified cotton production between 1965 and 1983 (Glantz *et al.* 1993; Rumer 1989). As a consequence of the intensification of cotton production, the total irrigated area in Central Asia increased from 4.5 million ha in 1965 to 7 million ha in 1991 (Wegerich 2008).

During the Soviet period, CA states produced 95 per cent of the raw cotton, about 40 per cent of the rice, 25 per cent of the vegetables and melons and 32 per cent of the fruits and grapes of the total production in the former Soviet Union (Kurbanbaev and Artykov 2010). In addition, dozens of large hydropower plants for both electricity generation and irrigation were built on the Syr Darya and Amu Darya, the main rivers of the Aral Sea Basin. Water was thus an important element in the policy of creating a centralized planned economy in CA. The perception of water as an *unlimited* resource played a key role in the Soviet modernization plan and translated into the ideological imperative of the conquest of nature. This perception led not only to unrestrained, but also to inefficient water management resulting in massive water losses through seepage and evaporation along irrigation channels, followed by waterlogging and salinization (Zhupankhan *et al.* 2018).

From the 1980s, *perestroika* and *glasnost* created an intellectual environment that was conducive to a reassessment of the Aral Sea catastrophe and an acknowledgment of the overexploitation and mismanagement of the water resources of the Aral Sea Basin as root causes of this disaster. This understanding paved the way for a revised perception of water as a limited resource that requires careful and conservative management. This new perception lost some of its force after independence, when the new countries of the region were confronted with the task of building market-based national economies and reverted, to some extent, to the Soviet perception of water as the key agent of economic development based on agriculture and energy. The perception of water as a resource to be exploited for economic gains still prevails over a view of water as a limited resource to be used sustainably for livelihoods and socio-economic development. Overuse of water has continued, exacerbated by the competition of countries across national borders and across the irrigation and energy divide.

## Water in the agriculture, energy and mining sectors

The perceived role of water in the region has been shaped by increased water demand and consumption (Bekchanov *et al.* 2016) as the population has grown by well above 1 per cent in the last decade (Worldometers.info 2018). It is also characterized by a great potential of intra-regional and sectorial cooperation on the one hand, and risk of competition and resource overuse on the other.

The challenges posed by intra-regional and sectoral cooperation or competition for water are well illustrated by the food–water–energy–mining interdependence in CA. Using irrigation water in agriculture plays a key role in the economy of the five CA countries as the agriculture sector contributes from 10 to 45 per cent of the GDP and employs 20 to 50 per cent of the rural population (Qushimov *et al.* 2007). There is an estimated 7.84 million ha of land suitable for irrigation in CA (0.77 million ha in Kazakhstan; 0.42 in Kyrgyzstan; 0.72 in Tajikistan; 1.73 in Turkmenistan; and 4.2 in Uzbekistan) with annual water use reported at 124.6 km$^3$ (Hamidov 2015; Russel 2018). Figure 7.1 illustrates the contribution of agriculture to national GDP, share of rural population and agricultural employment as well as water withdrawal from the major rivers. Figure 7.2 provides information on the proportional representation of total agricultural land, irrigated land and crop land.

For instance, in Uzbekistan (a downstream country), up to 90 per cent of the available water resources are currently used for irrigated agriculture, producing food and export crops

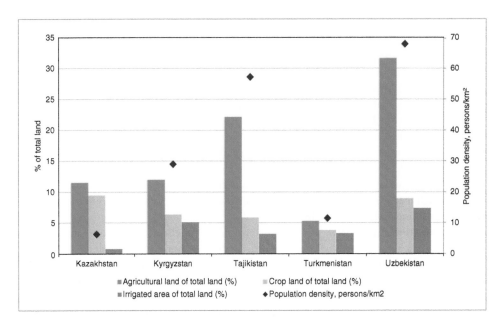

*Figure 7.1* Agriculture in Central Asia

Source: AGRIWANET project database (2017)

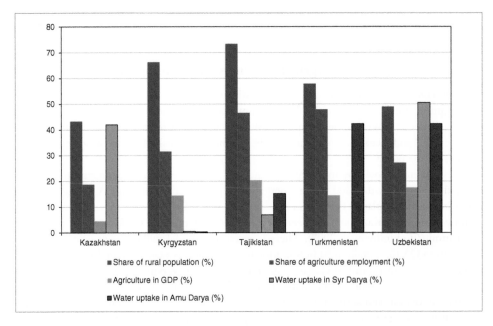

*Figure 7.2* Proportional representation of agricultural land, crop land and irrigated land in the five
countries of Central Asia

Source: AGRIWANET project database (2017)

(Swatuk *et al.* 2015). The establishment of an export-oriented agricultural sector that largely depends on the production of irrigated cash crops, such as cotton, has increased Uzbekistan's water withdrawals to levels that substantially exceed the internally renewable and total renewable water resources (World Bank 2014 cited in Swatuk *et al.* 2015). Also, other CA countries export agricultural commodities produced from significant amounts of water (Porkka *et al.* 2012; Rudenko *et al.* 2013, cited in Karthe *et al.* 2015), even though the absolute quantities remain very difficult to assess. Due to its dependence on irrigated food production, the region is highly vulnerable to food insecurity. Even though the prevailing rate of undernourishment has come down from its peak in the early 2000s, it has recently slightly increased again to 6.2 per cent of the total population (FAO 2018; World Bank 2018).

Estimates reveal that the CA countries have plenty of water relative to their populations but have a very low economic return on water compared with other parts of the world (Varis 2014, cited in Hamidov *et al.* 2016). For instance, Turkmenistan uses almost three times the amount of water that is used in India to produce US$1 of equivalent product (Varis 2014; World Bank 2013a). This is mainly related to physical loss of water as well as cultivation of water-intensive crops. It is worth noting that less-water-intensive crops may not necessarily recover the cost of the very large and expensive irrigation and drainage system in CA, and alternative crops may be integrated into the crop rotation program of these countries.

The agriculture sector in CA went through a significant reorganization process. The new Land and Water Codes that were implemented by CA states after independence initiated a trend toward liberalization. Farmers received the rights for land that was state-owned during the Soviet time. Land reforms led to the emergence of a new type of land user—individual farmers. This shift, however, created new challenges for farmers and pastoralists, such as: How to negotiate divergent interests and demands for the water with a multiple set of users? How to increase agriculture productivity and how to deal with land degradation? How to manage canals in a context of overlapping hydraulic and administrative boundaries? As a response to these challenges, new institutions for decentralized natural resource management were created at the community level—water user associations and pasture user committees. These institutions are still weak and are often imbued with the Soviet legacy (Isaeva and Shigaeva 2017).

In the course of post-Soviet-era reforms, previously functioning collective and state farms were reorganized into three farm types: farm enterprises, family/private farms and household farms (Table 7.1). Among these types, the family/private farms cover the largest share of the total agricultural area. Furthermore, according to Suleimenov (cited in Hamidov *et al.* 2016), the agricultural sectors have also been changed. For instance, during the time of the Soviet Union, each country specialized in certain agricultural productions, with Kazakhstan in grain production, Kyrgyzstan in sheep production, alfalfa and maize and Tajikistan, Turkmenistan and Uzbekistan producing as much irrigated cotton and karakul sheep for pelts as possible. This has now changed: in Kazakhstan, the cropland area was reduced significantly, and monoculture wheat production followed, with producers recently beginning to include food legumes such as dry peas and chickpeas. Kyrgyzstan significantly increased wheat production and dry beans under irrigation. Tajikistan's wheat area has doubled but has a low yield level. Turkmenistan and Uzbekistan have increased bread wheat production in recent years (Figure 7.3). Thus, the area under cotton has been reduced in CA, and alfalfa and other forage crops have been removed from irrigated land while livestock and forage production has decreased in most CA countries. However, the demand for food, forage and water is increasing. Poorly managed rangelands lead to a lack of fodder, land degradation, loss of plant biodiversity and expanding desertification (Gintzburger *et al.* 2003; IPBES 2018; Zhumanova *et al.* 2018).

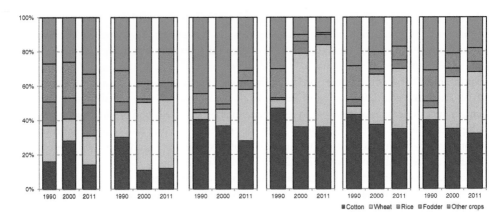

*Figure 7.3* Cropping patterns in irrigated areas of the five CA countries and the entire region

Source: AGRIWANET project database (2017)

*Table 7.1* Farm types in Central Asia

| Country/farm types | Farm enterprises | Family and private farms | Households | Notes |
|---|---|---|---|---|
| Kazakhstan | Agricultural enterprises | Average size = 309 ha | Household (HH) plots | 64 per cent of farms are <50 ha, operate 2 per cent of land |
| Kyrgyzstan | State and collective farms | Average size = 2.8 ha | HH plots | Arable land is mostly distributed equally to rural families |
| Tajikistan | Agricultural enterprises (e.g., collectives, state farms and cooperatives) | Average size = 55 ha | HH plots | 80 per cent of farms are <10 ha |
| Turkmenistan | Peasant associations | Average size = 10 ha | HH plots | Farmers' associations operate 94 per cent of arable land; family farms about 1 per cent of land |
| Uzbekistan | Agricultural enterprises | Average size = 80 ha | HH plots | 54 per cent of farms in cotton-wheat (average size = 106 ha), 31 per cent in vegetable and fruit (average size = 15 ha) |

Source: Adapted from AGRIWANET project database (2014), Djanibekov and Wolz (2015), Djanibekov *et al.* (2015).

Another challenge for agriculture and effective water management is rural outmigration. Because of the declining rural livelihoods after independence and persisting risks posed by environmental challenges and economic shocks, a large number of young people have left the farms for labor in cities in Russia and Kazakhstan. In this process, communities suffer from being drained of their most active members and from inadequate management of natural resources (Thieme 2014). Consequently, the remittances that could potentially be invested in modernizing farming are mostly spent on consumer goods (Sagynbekova 2017).

The energy sector is vital for the industrial development of the CA region. The economic downturn in the 1990s had a strong impact on the energy sector as the intra-regional and sectoral cooperation collapsed. Independent CA countries gradually enforced the national energy supply systems and prioritized their own interests. Energy demand has increased strongly in recent years encouraging countries to invest in the sector to meet their own needs, but also to export to neighboring countries.

The environmental impact of mining, which consumes large quantities of water and energy, makes the situation complex. An illustrative example is the gold mining sector in Kyrgyzstan where the treatment of water quality and destruction of river ecosystems is of concern. For instance, the country's largest gold mine, Kumtor, is located within the Issyk-Kul Biosphere Territory and close to the Sarychat-Ertash District Nature Reserve. In general, experts argue that unclear responsibilities of environmental, forest, geological and water authorities for the ecosystems of rivers are to blame for ineffective control of mining activities. Furthermore, local environmental and land registration authorities are often excluded from land allocation for mining purposes (Simonett *et al.* 2012).

Climate change may even worsen the situation and lead to changes in precipitation patterns, intensities and, potentially, water shortages in the future. Currently available scenarios of climate change predict decreasing physical availability of water for CA because of increasing temperatures and evaporation, shifting temporal patterns of precipitation with an increasing frequency of droughts (Barros *et al.* 2014; Xenarios *et al.* 2019). For instance, in Uzbekistan, the projections suggest that by 2050 a temperature increase is expected of between 1.7 and 4.7°C (Milanova *et al.* 2018). Even though runoff from melting glaciers may increase in the short-term (Cruz *et al.* 2007, cited in Aleksandrova *et al.* 2014), water availability will likely decline in the long term. The water discharges may potentially decrease by 2–5 per cent in the Syr Darya River Basin and by 10–15 per cent in the Amu Darya Basin by 2050, aggravating water scarcity (World Bank 2013b).

Climate change is likely to cause yield reductions in the order of 20–50 per cent by 2050 for nearly all crops, thus threatening food security and rural livelihoods (World Bank 2013b). There are, however, differences in climate change impacts across the region: some positive income gains in large-scale commercial farms in the northern regions of Kazakhstan and negative impact on small-scale farms in arid zones of Tajikistan, and short-term increase coupled with a long-term decrease of agricultural incomes in Uzbekistan (Bobojonov and Aw-Hassan 2014).

Lack of coordination and cooperation between sectors and countries in water use resulted in environmental degradation and deterioration of water quality. Diminishing water flows from the Pamir and Tien Shan Mountains lead to increasing water salinity (Siderius and Schoumans 2009, cited in Aleksandrova *et al.* 2014). Surface water bodies are often affected by the influx of fine sediments, nutrients and agrochemicals (Karthe *et al.* 2015). Agricultural and industrial pollution affect downstream areas, where heavy metal concentrations often exceed maximum permissible values (Bekturganov *et al.* 2016). Soil salinization critically affects large areas of previously fertile agricultural land (Abdullaev 2004; Qadir *et al.* 2009).

There is growing scientific interest to explore the food–water–energy–mining nexus in CA, particularly in the large transboundary Amu Darya and Syr Darya Rivers (Wegerich *et al.* 2012; Bekchanov and Lamers 2016; Jalilov *et al.* 2016; Keskinen *et al.* 2016). These studies highlight that a water–energy–food nexus approach in the context of climate change in CA that includes conflicting interests of different sectors and countries can enhance understanding of the water use interconnectedness.

*Figure 7.4* Field irrigation in the Ferghana Valley of Central Asia

Source: IWMI archive

## Competition for water resources and challenges
## for regional water cooperation

The availability of water for agriculture and other sectors is determined not only by supply, but more prominently by increasing demand and competition between countries and water users. In the upstream countries (Tajikistan and Kyrgyzstan), surface water resources are of significant economic importance for hydroelectric power generation. With a total installed capacity of 8,200 MW, more than 90 per cent of the electricity demand in Tajikistan and Kyrgyzstan is currently produced through hydropower (IHA 2018).

With the collapse of the Soviet Union, the barter system of fossil fuel and irrigation water between the upstream and the downstream countries collapsed (Bekchanov *et al.* 2015). Consequently, upstream countries typically intensify their hydro-energy production during the winter season to cater for their energy needs. To respond to the rising energy demand along with economic growth, and to further reduce their dependence on the import of fossil fuels, the expansion of their hydropower capacity in the upstream countries is a preferred strategy by these countries. Downstream countries (Uzbekistan, Kazakhstan and Turkmenistan) are concerned that with the construction of dams there will be a shortage of water for agricultural needs. In downstream countries, water is in demand for irrigation purposes during spring and summer months, whereas for upstream states there is a greater necessity to release water during winter time, when there is increased demand for electricity.

One example of a dispute over sharing water resources in the Aral Sea Basin is the Rogun Hydropower Project (HPP). The project generated prolonged controversy between two

Central Asian states, namely Tajikistan and Uzbekistan, while the other two downstream countries Kazakhstan and Turkmenistan also expressed their concerns about the potential negative impacts of the project. With the change of leadership in Uzbekistan, the relations between Tajikistan and Uzbekistan improved. Uzbekistan's president, Shavkat Mirziyoyev, made a state visit to Tajikistan in March 2018, following which the parties came to consensus on the long-standing dispute. In November 2018, the first turbine of the Rogun HPP was launched in Tajikistan.

The Rogun HPP is not the only disputed transboundary water management case in Central Asia. Another major river dam, Kambarata 1, is planned for construction in Kyrgyzstan on the Naryn River. This $3-billion project received approval of its feasibility study by the Kyrgyz government, but Uzbekistan expressed its opposition to the project (Hashimova 2014). The dispute was resolved in 2017 when the governments of Kyrgyzstan and Uzbekistan came to a consensus on the construction of the project.

Transboundary conflicts over water exist also on a local scale, as was shown by Murzakulova and Mestre (2018) for the case of the Kyrgyz–Tajik border in the Ferghana Valley (see Figure 1.1). Resource-sharing mechanisms that had been developed during Soviet times were disrupted by the transformation of administrative borders into international borders and continue to be obstructed by dissymmetry and uncertainty of institutions on both sides of the border.

CA countries have attempted to cooperate to jointly solve water management problems in the Aral Sea Basin. The Interstate Commission for Water Coordination in Central Asia (ICWC) was established in 1992 and the agreement on transboundary water management was signed by the neighboring states in the region. Another agreement was signed in March 1992, when heads of state met in Kyzylorda, Kazakhstan. Governments of all five Central Asian states established the International Fund to Save the Aral Sea (IFAS) in 1993. To achieve coordinated transboundary water governance bilateral agreements were concluded between Uzbekistan and Turkmenistan (1996), Kyrgyzstan and Uzbekistan (1998) and Kazakhstan and Kyrgyzstan (2000).

However, like in many multilateral agreements, most of the provisions of the bilateral agreements have declaratory character and none of the agreements offers a comprehensive mechanism for joint water management between the states. The joint efforts resulted in limited coordination of actions by the member states. The governments also attempted to support less-water-intensive agricultural activities, but so far with limited success. Nevertheless, having achieved consensus as in the case of the Rogun HPP signals an increasing willingness for cooperation.

## Conclusions

Water plays a key role in the economic and social development of the countries of the Aral Sea Basin, and especially so in the sustainable development of their agricultural potential. Though water is still available in sufficient quantities, this may change when melting of glaciers due to climate change will reduce the natural storage capacity for water. Irrigated agriculture would temporarily benefit from more water released by accelerated glacier melting but would lose out in the long term. Scarcity of water, as experienced today by local users in the CA states, is mainly due to competition between upstream and downstream countries and between the agriculture and energy sectors, to water lost in derelict infrastructure and water wasted on water-intensive crops, to weak institutions and to still inadequate regional and cross-border collaboration.

Though closely intertwined with the energy sector, and to a lesser extent with other sectors, agriculture, because of its comparatively much larger demand for irrigation water, lies at the core of water management in the Aral Sea Basin. Managing water more sustainably must therefore focus on the agricultural sector.

The critical issues that need consideration include the need to stimulate cooperation between agricultural units—farm enterprises, family and private farms, households—for sustainable use of water. This also requires the strengthening of new institutions, i.e., water user associations; investment in the repair of derelict irrigation infrastructure and the development of more cost-effective irrigation systems; a switch to less-water-intensive agricultural activities and to less-water-demanding crops; enhancement of coordination among agriculture and other sectors for water resources allocation and water use; application of a comprehensive policy for rural development to slow down migration outflows from rural areas and to provide incentives for investing remittances in improving agricultural activities.

Furthermore, the food–water–energy–mining nexus approach that incorporates conflicting interests of different sectors and countries can enhance understanding of the water use interconnectedness and provide a basis for transforming governance in these sectors (e.g., designing the complementary institutional arrangements in the sectors that allow for progressive distribution of economic benefits and fair burden-sharing of environmental costs), as well as introducing innovative technological solutions in order to use synergies and strengthen cooperation among water users, sectors and countries in the region.

# References

Abdullaev I (2004) *How to Live with Salinity in Central Asia: Review of the Situation and Strategy for Remediation.* Paper presented at the meeting on Research Program Development for Land Degradation, Drainage and Waste-Water Reuse in Central Asia. 11– 17 May 2004, Tashkent, Uzbekistan

Abdullaev I, Manthrithilake H, Kazbekov J (2006) Water security in Central Asia: troubled future or pragmatic partnership? In: *Paper 11, International Conference 'The Last Drop?' Water, Security and Sustainable Development in Central Eurasia,* 1–2 December 2006. Institute of Social Studies (ISS), The Hague, the Netherlands

AGRIWANET project database (2014) *Provincial-Level Data on Agricultural Development Indicators for Central Asia.* IAMO, Halle (Saale), Germany

AGRIWANET project database (2017) *Provincial-Level Data on Agricultural Development Indicators for Central Asia.* IAMO, Halle (Saale), Germany

Aleksandrova M, Lamers JPA, Martius C, Tischbein B (2014) Rural vulnerability to environmental change in the irrigated lowlands of Central Asia and options for policy-makers: a review, *Environmental Science & Policy* 41: 77–88

Barros VR, Mach KJ, Mastrandrea MD, van Aalst M, Adger WN, Arent DJ, Barnett J, Betts R, Bilir TE, Birkmann J, Carmin J, Chadee DD, Challinor AJ, Chatterjee M, Cramer W, Davidson DJ, Estrada YO, Gattuso J-P, Hijioka Y, Hoegh-Guldberg O, Huang HQ, Insarov GE, Jones RN, Kovats RS, Romero-Lankao P, Larsen JN, Losada IJ, Marengo JA, McLean RF, Mearns LO, Mechler R, Morton JF, Niang I, Oki T, Olwoch JM, Opondo M, Poloczanska ES, Pörtner H-O, Redsteer MH, Reisinger A, Revi A, Schmidt DN, Shaw MR, Solecki W, Stone DA, Stone JMR, Strzepek KM, Suarez AG, Tschakert P, Valentini R, Vicuña S, Villamizar A, Vincent KE, Warren R, White LL, Wilbanks TJ, Wong PP, and Yohe GW (2014) Technical summary. In: Field CB, Barros VR, Dokken DJ, Mach KJ, Mastrandrea MD, Bilir TE, Chatterjee M, Ebi KJ, Estrada YO, Genova RC, Girma B, Kissel ES, Levy AN, MacCracken S, Mastrandrea PR, White LL (eds), *Climate Change 2014: Impacts, Adaptation, and Vulnerability. Part A: Global and Sectoral Aspects.* Contribution of Working Group II to the Fifth Assessment Report of the Intergovernmental Panel on Climate Change. Cambridge University Press, Cambridge, UK and New York, USA, pp 35–94

Bekchanov M, Lamers JPA (2016) The effect of energy constraints on water allocation decisions: the elaboration and application of a system-wide economic-water-energy model (SEWEM), *Water* 8(6): 253

Bekchanov M, Ringler C, Bhaduri A, Jeuland M (2015) How would the Rogun Dam affect water and energy scarcity in Central Asia? *Water International* 40(5–6): 856–876

Bekturganov Z, Tussupova K, Berndtsson R, Sharapatova N, Aryngazin K, Zhanasova M (2016) Water-related health problems in Central Asia—a review, *Water* 8(6): 219

Bobojonov I, Aw-Hassan A (2014) Impacts of climate change on farm income security in Central Asia: an integrated modeling approach, *Agriculture, Ecosystems & Environment* 188: 245–255

Cruz RV, Harasawa H, Lal M, Wu S, Anokhin Y, Punsalmaa B, Honda Y, Jafari M, Li C, Huu NN (2007) Climate change 2007: impacts, adaptation and vulnerability. In: Parry ML, Canziani OF, Palutikof JP, van der Linden PJ, Hanson CE (eds), *Contribution of Working Group II to the Fourth Assessment Report of the Intergovernmental Panel on Climate Change.* Cambridge University Press, Cambridge, UK, pp 469–506

Djanibekov N, Djanibekov U, Sommer R, Petrick M (2015) Cooperative agricultural production to exploit individual heterogeneity under a delivery target: the case of cotton in Uzbekistan, *Agricultural Systems* 141: 1–13

Djanibekov N, Wolz A (2015) *Development Problems in Central Asia: Example of Agricultural Services Cooperatives.* IAMO Jahreszahl 17, Halle (Saale), Germany, pp 65–73

FAO (2018) *Publication Regional Overview of Food Security and Nutrition in Europe and Central Asia 2018.* Budapest, Hungary

Gintzburger G, Toderich K, Mardonov B, Mahmudov M (2003) *Rangelands of the Arid and Semi-Arid Zones in Uzbekistan.* ICARDA International Center for Agricultural Research in the Dry Areas, Tashkent, Uzbekistan

Glantz MH, Rubinstein AZ, Zonn I (1993) Tragedy in the Aral Sea Basin: looking back to plan ahead? *Global Environmental Change* 3(2): 174–198

Hamidov A (2015) Institutional design in transformation: a comparative study of local irrigation governance in Uzbekistan, *Environmental Science & Policy* 53: 175–191

Hamidov A, Helming K, Balla D (2016) Impact of agricultural land use in Central Asia: a review, *Agronomy for Sustainable Development* 36(1): 6

Hashimova U (2014) Rogun Dam studies set the scene for further disputes among central Asian countries, *Eurasia Daily Monitor* 11: 1

International Hydropower Association (2018) *Hydropower Status Report: Sector Trends and Insights.* Retrieved 9 May 2019, www.hydropower.org/publications/2018-hydropower-status-report

IPBES (2018) Summary for policymakers of the regional assessment report on biodiversity and ecosystem services for Europe and Central Asia of the Intergovernmental Science-Policy Platform on Biodiversity and Ecosystem Services. In: Fischer M, Rounsevell M, Torre-Marin Rando A, Mader A, Church A, Elbakidze M, Elias V, Hahn T, Harrison PA, Huack J, Martín-López B, Ring I, Sandström C, Sousa Pinto I, Visconti P, Zimmerman NE, Christie M (eds), Secretariat of the Intergovernmental Science-Policy Platform on Biodiversity and Ecosystem Services, Bonn, Germany

Isaeva A, Shigaeva J (2017) Soviet legacy in the operation of pasture governance institutions in present-day Kyrgyzstan, *Journal of Alpine Research* 105–1

Jalilov SM, Keskinen M, Varis O, Amer S, Ward FA (2016) Managing the water–energy–food nexus: gains and losses from new water development in Amu Darya River Basin, *Journal of Hydrology* 539: 648–661

Karthe D, Chalov S, Borchardt D (2015) Water resources and their management in Central Asia in the early twenty first century: status, challenges and future prospects, *Environmental Earth Sciences* 73(2): 487–499

Keskinen M, Guillaume JHA, Kattelus M, Porkka M, Rasanen TA, Varis O (2016) The water-energy-food nexus and the transboundary context: insights from large Asian rivers, *Water* 8(5): 193

Kurbanbaev E, Artykov OKS (2010) *Aralskoe more i vodohozyastvennaya politika v respublikai Centralnoi Azii.* Karakalpak branch CANIIRI. Retrieved 10 May 2019, www.cawater-info.net/library/rus/aral-nukus.pdf

Lipovsky I (1995) The Central Asian cotton epic, *Central Asian Survey* 14(4): 529–542

Milanova E, Nikanorova A, Kirilenko A, Dronin N (2018) Water deficit estimation under climate change and irrigation conditions in the Ferghana Valley, Central Asia. In: Mal S *et al.* (eds), *Climate Change, Extreme Events and Disaster Risk Reduction.* Sustainable Development Goals Series. Springer International Publishing AG, Berlin, Germany

Murzakulova A, Mèstre I (2018) Border change and conflict in Central Asia: the case of agro-pastoral communities in cross-border areas of the Ferghana Valley. In: Zurayk R, Woertz E, Bahn R (eds), *Crisis and Conflict in Agriculture*. CAB International, Wallingford, UK, pp 176–189

Porkka M, Kummu M, Siebert S, Flörke M (2012) The role of virtual water flows in physical water scarcity: the case of Central Asia, *International Journal of Water Resources Development* 28(3): 453–474

Qadir M, Noble AD, Qureshi AS, Gupta RK, Yuldashev T, Karimov A (2009) Salt-induced land and water degradation in the Aral Sea Basin: a challenge to sustainable agriculture in Central Asia, *Natural Resources Forum* 33: 134–149

Qushimov B, Ganiev I, Rustamova I, Haitov B, Islam K (2007) Land degradation by agricultural activities in Central Asia. In: Doraiswamy P, Lal R, Suleimenov M, Stweart BA, Hansen DO (eds), *Climate Change and Terrestrial Carbon Sequestration in Central Asia*. Taylor & Francis, London, UK, pp 137–146

Rudenko I, Bekchanov M, Djanibekov U, Lamers JPA (2013) The added value of a water footprint approach: micro- and macroeconomic analysis of cotton production, processing and export in water bound Uzbekistan, *Global and Planetary Change* 110: 143–151

Rumer BZ (1989) *Soviet Central Asia. A Tragic Experiment*. Unwin Hyman, Boston, MA, USA

Russel M (2018) *Water in Central Asia: An Increasingly Scarce Resource*. Briefing, European Parliamentary Research Service. Retrieved 10 May 2019, www.europarl.europa.eu/RegData/etudes/BRIE/2018/625181/EPRS_BRI(2018)625181_EN.pdf

Sagynbekova L (2017) Environment, rural livelihoods, and labour migration: a case study in Central Kyrgyzstan, *Mountain Research and Development* 37(4): 456–463

Siderius C, Schoumans OF (2009) *Baseline Assessment Amudarya: On Water Quantity, Quality and Ecosystem Issues*. Report of the NeWater project – New Approaches to Adaptive Water Management under Uncertainty. Retrieved 10 May 2019 ,www.newater.info

Simonett O, Stuhlberger Ch, Sairinen R, Tiainen H, Honkonen T, Rinne P, Halonen M, Tommila P, Isabaev K, Soronkulov G, Tursunbaev O, Mendibaev N, Ibraev K, Mezgin I, Pak N, Ashirkulov K, Dzhumabaev N, Khakdodov M, Illarionova F, Pechenyuk O (2012) *Mining, Development and Environment in Central Asia: Toolkit Companion with Case Studies*. Zoï Environment Network, University of Eastern Finland, Gaia Group Oy, Finland

Stadelbauer J (2007), Wasser und energie in Zentralasien: Politisch-geographische aspekte von Ressourcenverfügbarkeit und -nutzung. In: Mächtle B, Nüsser M, Schmid H, Siegmund A (eds), *Inszenierte Landschaften und Städte*. GFZ Potsdam, Heidelberg, Germany, pp 65–94

Swatuk L, McMorris M, Leung C, Zu Y (2015) Seeing 'invisible water': challenging conceptions of water for agriculture, food and human security, *Canadian Journal of Development Studies* 36(1): 24–37

Thieme S (2014) Coming home? Patterns and characteristics of return migration in Kyrgyzstan, *International Migration* 52(5): 127–143

Varis O (2014) Resources: curb vast water use in Central Asia, *Nature* 514: 27–29

Wegerich K (2008) Hydro-hegemony in the Amu Darya Basin, *Water Policy* 10(2): 71–88

Wegerich K (2011) Water resources in Central Asia: Regional stability or patchy make-up? *Central Asian Survey*. DOI: https://doi.org/10.1080/02634937.2011.565231

Wegerich K, Kazbekov J, Kabilov F, Mukhamedova N (2012) Meso-level cooperation on transboundary tributaries and infrastructure in the Ferghana Valley, *International Journal of Water Resources Development* 28(3): 525–543

World Bank (2013a) *World Development Indicators*. World Bank, Washington, DC, USA. Retrieved 30 June 2015, http://databank.worldbank.org/data/download/WDI-2013-ebook.pdf

World Bank (2013b) *Uzbekistan: Overview of Climate Change Activities*. Retrieved 10 May 2019, http://documents.worldbank.org/curated/en/777011468308642720/text/855660WP0Uzbek0Box382161B00PUBLIC0.txt

World Bank (2014) *Key Issues for Consideration on the Proposed Rogun Hydropower Project*. World Bank, Washington, DC, USA

World Bank (2018) Prevalence of undernourishment 2000–2016 (% of population). Retrieved 14 December 2018, https://data.worldbank.org/indicator/SN.ITK.DEFC.ZS?locations=UZ-KZ-KG-TJ-TM

Worldometers.info (2018) *Based on data of UN DESA (2017): World Population Prospects: The 2017 Revision, Key Findings and Advance Tables*. United Nations Department of Economic and Social Affairs, Population Division, ESA/P/WP/248. Retrieved 13 December 2018, https://population.un.org/wpp/Publications/Files/WPP2017_KeyFindings.pdf

Xenarios S, Gafurov A, Schmidt-Vogt D, Sehring J, Manandhar S, Hergarten C, Shigaeva J, Foggin M (2019) Climate change and adaptation of mountain societies in Central Asia: knowledge gaps and data constraints, *Regional Environmental Change* 19: 1339

Zhdanko T (1966) Sedentarisation of the nomads of Central Asia, including Kazakhstan, under the Soviet regime, *International Labour Review* 93. Retrieved 14 May 2019, https://heinonline.org/HOL/Page?handle=hein.journals/intlr93&id=614&div=&collection=

Zhumanova M, Moenning C, Hergarten C, Darr D, Wrage-Moenning N (2018) Assessment of vegetation degradation in mountainous pastures of the Western Tien-Shan, Kyrgyzstan, using eMODIS NDVI, *Ecological Indicators* 95: 527–543

Zhupankhan A, Tussupova K, Berndtsson R (2018) Water in Kazakhstan, a key in Central Asian water management, *Hydrological Science Journal* 63(5): 752–762

# 8

# The status and role of the alpine cryosphere in Central Asia

*Martin Hoelzle, Martina Barandun, Tobias Bolch, Joel Fiddes,
Abror Gafurov, Veruska Muccione, Tomas Saks and
Maria Shahgedanova*

## Key messages

- The alpine cryosphere, including snow, glaciers and permafrost, is critical to water management in the Aral Sea Basin (ASB) and larger Central Asia (CA) under the changing climate, as it stores large amounts of water in its solid forms. Most cryospheric components in the Aral Sea Basin are close to melting point, and hence very vulnerable to a slight increase in air temperature with significant consequences to long-term water availability and to water resources variability and extremes.

- Current knowledge about different components of the cryosphere and their connection to climate in the Basin and in the entire Central Asia region varies. While it is advanced in the topics of snow and glaciers, knowledge on permafrost is rather limited.
- Observed trends in runoff point in the direction of increasing water availability in July and August at least until mid-century and increasing possibility for water storage in reservoirs and aquifers. However, eventually this will change as glaciers waste away. Future runoff may change considerably after mid-century and start to decline if not compensated by increasing precipitation.
- Cryosphere monitoring systems are the basis for sound estimates of water availability and water-related hazards associated with snow, glaciers and permafrost. They require a well-distributed observational network for all cryospheric variables. Such systems need to be re-established in the Basin after the breakup of the Soviet Union in the early 1990s. This process is slowly emerging in the region. Collaboration between local operational hydro-meteorological services and the academic sector, and with international research networks, may improve the observational capabilities in high-mountain regions of CA in general and in the ASB in particular.

## Introduction

Water resources in arid continental regions like Central Asia (CA) depend strongly on cryosphere components: snow, glaciers and permafrost (Figure 8.1). Two of the world's largest mountain systems, the Tien Shan (also: Tian Shan) and Pamir Mountains, serve as water towers for CA, of which the Aral Sea Basin (ASB) is a major part. The cryosphere components of these mountain systems store large amounts of water in a solid form and play a key role for current and future water availability and management under a changing climate. Several recent studies (Hagg et al. 2007; Hagg et al. 2013; Huss and Hock 2018; Kaser et al. 2010) indicate that a) in arid regions like CA, water release by snow and glaciers is fundamental to maintaining sufficient run-off during the dry summer months and b) by the end of this century the water contribution of glaciers will be drastically reduced and some catchments may completely dry out. This may pose significant challenges to water resources management, energy production, the environment, disaster risk reduction, security and public health. High-quality baseline data on cryosphere components may help develop sound climate-based scenarios of future water availability.

*Snow* is temporally and spatially the most variable component of the cryosphere. It has a strong influence on the climate system but is also strongly controlled by climate. Recent observed trends derived from remote sensing (Adnan et al. 2017; Immerzeel et al. 2009; Peters et al. 2015), and ground measurements (e.g., Marty 2008; Serquet et al. 2013) indicate a seasonally reduced duration of snow cover, as well as reduced extent of snow cover, especially at lower elevations. Such changes are important as snow has strong feedback processes, particularly over its strong albedo difference in comparison to surfaces like water, vegetation, bedrock or sediments. While glaciers release most of their melt water during the hot summer months, water from snowmelt mostly comes in spring (April, May, June). Therefore, snow storage can be seen as a temporary water reservoir accumulating snow in winter and releasing water to rivers and streams in spring and early summer. The melt water from seasonal snow cover is vital for the environment, economic development and social security of CA (e.g., (Aizen et al. 1995; Sorg et al. 2012; Unger-Shayesteh et al. 2013). Many ground-based stations measuring snow were abandoned after the breakup of the Soviet Union in the early 1990s, and a new observational network is only slowly evolving in the region.

*Figure 8.1* Abramov glacier: one of the reference glaciers in Central Asia, where measurements were re-established in 2011 after a gap of 12 years

Source: Martin Hoelzle, August 2011

*Glaciers* today are the most well-known symbol of changing climate as they are at many places close to melting point and therefore highly sensitive to changes in air temperatures. Their response to warming trends is well-manifested by fast-retreating glacier tongues or even whole glaciers collapsing and decaying over very short time periods (Kääb *et al.* 2018; Paul and Mölg 2014; Zemp *et al.* 2006). These images have become icons of climate change (Haeberli 2008). CA is especially vulnerable to glacier changes, because runoff during the dry summer months is mainly dependent on the vast glacierized areas in Tien Shan and Pamir. In the coming decades, enhanced melting may lead to an increased runoff in spring and summer (Kaser *et al.* 2010) and cause glacier lake outburst floods (GLOF), debris flow and landslides, which can be very damaging to settlements and agriculture (Bolch *et al.* 2011; Erokhin *et al.* 2018; Kapitsa *et al.* 2017; Petrov *et al.* 2017; Stoffel and Huggel 2012). Towards the end of this century, runoff during the dry summer months is likely to continuously decrease due to reduced ice volumes (Hagg *et al.* 2007; Hagg *et al.* 2013; Huss and Hock 2018; Kure *et al.* 2013). Similar to *in situ* measurements of snow, long-term monitoring programs in CA and the ASB collapsed after the break-up of the Soviet Union and are currently being re-established with great national and international efforts.

*Permafrost* is defined by ground temperatures, which are continuously below 0°C over a period of at least one year. It mainly occurs in continental areas at high latitudes, but also in high mountains. Climate-induced changes in permafrost can lead to strong feedbacks with the climate system, for example, by releasing bound methane or carbon dioxide through thawing of permafrost soils, which in turn reinforces the greenhouse effect (Koven *et al.* 2011). In addition, ice melting in permafrost soils causes land subsidence due to the loss of its volume. Furthermore, on inclined topography and with an increasing active layer,[1] erosion can intensify. Infrastructure can directly heat the permafrost in the ground, leading to local destabilization with associated prevention and maintenance costs. In high-mountain regions, recent increases in air temperature cause rock falls, landslides, debris flows and increased creep rates in rock glaciers (Delaloye *et al.* 2010; Sorg *et al.* 2015), and increased runoff from permafrost zones with high ice contents (Bolch and Marchenko 2009; Mateo and Daniels 2018). CA and the Tibetan Plateau host the largest permafrost area outside the polar regions (Gruber 2012). It covers around $3.5 \times 10^6$ km$^2$, which corresponds to about 15 per cent of the total areal extent of permafrost in the Northern Hemisphere. Permafrost research in high mountains only began in the late twentieth century, but since then considerable progress has been made in understanding mountain permafrost processes (Haeberli et al., 2010).

The following sections examine each of the three cryosphere components in the ASB and larger CA in more detail.

## Alpine snow cover

The ASB river flow during the vegetation period in summer months is dominated by snowmelt, followed by glacier melt in late summer. Thus, high mountains of the ASB can be considered natural water towers of the region where winter precipitation is stored in the form of snow and melts during a warm period. The meltwater from seasonal snow cover is vital for the economic development and social security (e.g., Aizen *et al.* 1995; Sorg *et al.* 2012; Unger-Shayesteh *et al.* 2013). Also, through the snow-albedo feedback, changes in seasonal snow cover may affect local and regional climate and reinforce surface warming in ASB headwaters (e.g., Aizen *et al.* 2000).

Despite the importance of snow, the snow-related data (snow depth, snow cover, snow water equivalent, snow density) are limited in the ASB. Snow depth and snow water equivalent data are mainly available from the meteorological stations of national hydrometeorology services. In addition to station data, field survey and airborne snow data were collected during the Soviet era (Krenke 1998)[2]. Snow measurement surveys of this type were regularly carried out in CA, predominantly in Kyrgyzstan, Tajikistan and Uzbekistan, to assess snow storage and corresponding water availability during the vegetation period. Snow density and snow water equivalent were also recorded during these surveys. Unfortunately, the frequency of field and airborne snow surveys decreased dramatically after the breakup of the Soviet Union in 1991 (Unger-Shayesteh *et al.* 2013). Uzbekistan and Tajikistan continued a few airborne surveys but there were no such surveys in Kyrgyzstan (Glazirin 2007).

There are few studies of the long-term trends and shorter-term variability in snow cover. Aizen *et al.* (1997) analyzed snow data from 110 stations for the time period from 1940 to 1991 and found a decrease of mean annual snow depth of 8–14 cm at elevations below 2000 masl (meters above sea level) and of 6–19 cm at higher elevations. They also concluded that the number of days with snow cover decreased by nine days during this period. Glazirin (2006) analyzed snow cover duration at the Oigaing and Tashkent stations from the 1930s and reported slight negative trends, which were, however, not statistically significant. Tsarev (2006) analyzed snow depth, precipitation and temperature data to estimate how climate change impacts on the maximum snow storage in the mountains of Central Asia based on a temperature-precipitation approach. According to his results, scenarios of a temperature increase of 2°C and a precipitation decrease of 30 per cent would lead to about 30 per cent less snow storage in March when snow accumulation peaks. Merkushkin and Tsarev (2007) introduced an empirical relationship between snow parameters and elevation that can be used to estimate snow parameters for any basin and assess impacts of climate change scenarios.

Remote sensing products have become an important source of snow data in mountainous areas of the ASB in the past three decades. Moderate-resolution imaging spectroradiometer (MODIS) and advanced very-high-resolution radiometer (AVHRR) data are among those that are widely used in the ASB to assess water availability. Yakovlev (2005) used the end-of-March AVHRR snow cover data for runoff modeling in the Pyanj Basin in Tajikistan. Gafurov et al. (2013) assessed the quality of MODIS snow cover data against manual observations from stations in the ASB and reported about 93 per cent accuracy. However, cloud cover prevents the efficient use of optical remote sensing in hydrological studies. Gafurov and Bàrdossy (2009) developed a methodology for cloud removal and applied it in the Kokcha River Basin of the ASB. Zhou et al. (2013) used cloud-removed AVHRR snow cover data to understand snow characteristics in the ASB and reported that in the plain areas maximum snow coverage can reach up to 32 per cent of the total area and in the mountainous areas this value can exceed 80 per cent. Dietz et al. (2014) used combined AVHRR and MODIS snow cover data to understand the elevation-dependent snow cover characteristics and reported a snow cover duration rise of ~4 days per 100 m elevation. Gafurov et al. (2016) developed an all-in-one package MODSNOW-Tool, which includes all processing steps of raw MODIS data including cloud removal. This tool allows operational and automated monitoring of snow coverage at pre-defined river basins using the MODIS data. The MODSNOW-Tool is officially implemented in all five Central Asian countries (Kazakhstan, Kyrgyzstan, Tajikistan, Turkmenistan and Uzbekistan) and is currently used to improve seasonal river flow forecasts based on snow cover information obtained from remote areas (Apel et al. 2018; Kalashnikova and Gafurov 2017). Besides MODIS and AVHRR optical remote sensing snow cover maps, LANDSAT snow cover maps with high spatial resolution (30 m) were used to reconstruct glacier mass balance in selected glaciers in the ASB (Barandun et al. 2018; Kronenberg et al. 2016). Satellite radar systems, e.g., Sentinel-1, are being tested for their capability to detect snow water equivalent (Conde et al. 2019). Overall, snow cover maps, obtained from remote sensing, can significantly improve understanding of the hydrological processes in the ASB. Maps of mean monthly snow-covered area and snowmelt for the periods 1961–1990 and 2001–2010 for six major river basins in Asia, including the Amu Darya and the Syr Darya, are available at http://waterdata. iwmi.org/Applications/Glacier_Snow_Asia/.

## Alpine glaciers

The Tien Shan Mountains host almost 15,000 glaciers, covering a surface area of about 12,400 km$^2$ (RGI Consortium 2017; Sorg et al. 2012), while both reported glacier coverage and glacier number in the Pamir (incl. Pamir Alay) Mountains are slightly higher according to the most recent inventory (~13,800 km$^2$, No. ~17,000 (Mölg et al. 2018)). The mountain ranges are conventionally divided into Western/Northern Tien Shan (glacierized area: ~2,265 km$^2$), Eastern Tien Shan (~2,210 km$^2$), Central Tien Shan (~7,270 km$^2$) and Jetisu (Dzhungarsky Alatau) (~520 km$^2$), Pamir-Alay (~2,080 km$^2$), Western Pamir (~9,470 km$^2$) and Eastern Pamir (~2280 km$^2$) (Aizen et al. 1995; Bolch et al. 2019; RGI Consortium 2017; Mölg et al. 2018). A significant part of this glacierized area is situated in the ASB (Figure 8.2). The median glacier

*Figure 8.2* Overview map of the study region. The ASTER-derived geodetic mass change for each
subregion of the Tien Shan and the Pamir derived in Barandun *et al.* (2019) are shown.
The locations of all glaciers with continuous, long-term mass balance series are indicated red.

Source: Constructed by authors

elevation is highest in the Eastern Pamir (>5000 masl), slightly less than 4900 masl in the
Western Pamir, approximately 4200 masl in the Central Tien Shan, and lowest in the Western
and Northern Tien Shan (3700 to 3900 masl) (Mölg *et al.* 2018; Sakai *et al.* 2015). Glaciers
in the west receive more precipitation during winter, whereas summer accumulation regimes
become predominant towards the east (Dyurgerov *et al.* 1994; Sakai *et al.* 2015) where a combination of low temperature and summer precipitation maximum is more common (Kutuzov
and Shahgedanova 2009).

## Glacier mass balance

In the 1950s, an extensive system of cryospheric monitoring in Central Asia was launched
under the auspices of the USSR Committee for the International Hydrological Decade and the
measurements intensified during the following years (Dyurgerov 2002; Kuzmichenok 2006;
WGMS 2017). Monitoring included extensive glacier mass balance measurements on several glaciers (e.g., Central Tuyuksu, Golubin, Karabatkak, Abramov in the Soviet Union and
Urumchi Glacier No. 1 in China, (WGMS 2017)). The majority of these *in situ* monitoring
programs stopped in the early 1990s after the breakup of the USSR. Regular mass balance
measurements continued only on Central Tuyuksu (Shahgedanova *et al.* 2018) and Urumchi
Glacier No. 1 in China (WGMS 2017), neither of which, however, is in the ASB. Efforts to
re-establish *in situ* glacier monitoring of the formerly monitored glaciers located in the ASB and
nearby catchments have started since around 2010 through intensive international and national
collaboration projects (Hoelzle *et al.* 2017). At present, mass balance is monitored on more
than ten glaciers with longer measurement series. These glaciers are distributed throughout the
mountain ranges of Central Asia, and Abramov, Barkrak, Batysh Sook and Glacier No. 354 are
located in the ASB.

Despite this success, relative to the large number of glaciers located in this area, data remain sparse. To fill the gap in direct glaciological observations and to cover larger mountain areas, glacier volume changes were assessed on different catchment scales from local (e.g., Aizen *et al.* 2007; Bolch *et al.* 2011; Goerlich *et al.* 2017; Holzer *et al.* 2015; Li *et al.* 2017; Pieczonka and Bolch 2015) to regional studies (e.g., Brun *et al.* 2017; Gardelle *et al.* 2013; Gardner *et al.* 2013; Lin *et al.* 2017; Wang *et al.* 2017) using remote sensing. Furthermore, several mass balance time series are available from modeling studies (Farinotti *et al.* 2015; Li *et al.* 2011; Liu and Liu 2016). However, detailed region-wide mass balance time series with a high temporal resolution are still lacking.

*In the Tien Shan*, region-wide geodetic mass balance assessments agree on a glacier mass loss during the past two decades. Results of mass change estimates range from about −0.3 m w.e. a$^{-1}$ to −0.7 m w.e. a$^{-1}$ (Brun *et al.*, 2017; Farinotti *et al.*, 2015; Gardner *et al.*, 2013). Accelerated mass loss since the 1970s was reported for most regions (e.g., Farinotti *et al.*, 2015; Pieczonka *et al.* 2013; Sorg *et al.*, 2012). No significant acceleration of glacier mass loss could be identified since the onset of the century through snowline-constrained mass balance modeling (Barandun 2019). However, an increase in inter-annual variability was observed, pointing toward a change in the mass balance regime from a more continental to a more maritime setting, as described by Dyurgerov and Dwyer (2001). The highest geodetic mass loss was observed in the Eastern Tien Shan (not in the ASB). The lowest rates of mass loss were found in the Central Tien Shan (Figure 8.2). Comparison of the different geodetic assessments showed a good agreement for the Inner and Central Tien Shan (Brun *et al.* 2017; Farinotti *et al.* 2015; Gardner *et al.* 2013) and for the Northern/Western Tien Shan (Bolch 2015; Brun *et al.* 2017; Gardner *et al.* 2013). For the two aforementioned regions, an ASTER-derived average mass loss of approximately −0.5 m w.e. a$^{-1}$ and −0.4 m w.e. a$^{-1}$, respectively, was calculated from 2004 to 2012.

Published mass change assessments for *the Pamir* show quite large divergence and range from a close to balanced budget (+0.14 to −0.13 m w.e. a$^{-1}$; Brun *et al.* 2017; Gardelle *et al.* 2013; Gardner *et al.* 2013; Kääb *et al.* 2015) to strongly negative mass balances (−0.48 to −0.52 m w.e. a$^{-1}$; Kääb *et al.* 2015; Pohl *et al.* 2017). There are still discrepancies between the assessments, leading to debates over the ambiguous mass balance regime and its change. Important methodological differences, input data quality, inconsistent study periods and spatial divisions can explain the differences to some extent. A compilation and reassessment of the different published geodetic estimates revealed, on average, a mass loss of 0.26 m w.e. a$^{-1}$ for Western Pamir and balanced budgets for the Eastern Pamir (−0.02 m w.e. a$^{-1}$) for the period after 2000 (Bolch *et al.* 2019). This is in line with recent geodetic estimates by Barandun (2019), who reports −0.37 ± 0.42 m w.e. for the Western Pamir and +0.19 ± 1.47 m w.e. for the Eastern Pamir.

The reanalyzed historical glaciological measurements, reconstructed mass balance data and re-initiated *in situ* measurements provide a comprehensive and complete mass balance time series for Abramov, Golubin, Batysh Sook and Glacier No. 354 (Barandun *et al.* 2018; Barandun *et al.* 2015; Kenzhebaev *et al.* 2017; Kronenberg *et al.* 2016). Modern direct measurements revealed mass losses ranging from −0.25 to −0.51 m w.e. a$^{-1}$ (2011–2016) and reconstructed mass balances confirmed the negative signal for the last decades for these glaciers (−0.30 to −0.43 m w.e. a$^{-1}$ from 2000 to 2014; Barandun *et al.* 2018; Barandun *et al.* 2015; Hoelzle *et al.* 2017; Kenzhebaev *et al.* 2017; Kronenberg *et al.* 2016). The first mass balance calculations reported moderate mass loss of −0.10 to −0.25 m w.e. a$^{-1}$ for Barkrak Middle Glacier in Uzbekistan for 2017–2018 (unpublished results). For Abramov, geodetic mass change of −0.36 ±0.07 m w.e. a$^{-1}$ derived from aerial photographs and high-resolution optical satellite data for 1975 to 2015 also agreed with reanalyzed and reconstructed time series (Denzinger 2018). Unfortunately, meaningful comparison between the different studies was not always straightforward due to a very limited number of studies assessing the mass change of individual glaciers in the region and due to differences in study periods. New developments try to implement tools allowing for an optimal reconstruction of annual mass balance time series of a large amount of unmeasured and remote glaciers on mountain range scales based on a sophisticated combination of *in situ* and remote measurements coupled with different models. The resulting mass balance time series permit the determination of annual mass balance variability for a large number of glaciers in the Tien Shan and Pamir, with minimal cost and labor effort (Barandun 2019).

## Glacier area changes

Several studies employed remote sensing techniques to map glacier area fluctuations (e.g., Bolch 2007; Khromova *et al.* 2003; Khromova *et al.* 2006; Narama *et al.* 2010; Ozmonov *et al.* 2013; Shangguan *et al.* 2006). Glacier retreat and reduction in glacier area are observed throughout the Tien Shan but rates of retreat vary (Sorg *et al.* 2012; Unger-Shayesteh *et al.* 2013). In the Pamir, glacier change follows a more heterogeneous pattern (Knoche *et al.* 2017). Generally, the highest loss of glacierized area is observed in the outer regions and at lower elevations, while in the inner regions, in the inter-mountain basins and in the higher-elevation regions of the Pamir and Tien Shan the observed glacier shrinkage is slower (Aizen *et al.* 2014; Narama *et al.* 2010; Sorg *et al.* 2012; Unger-Shayesteh *et al.* 2013). A comparison of glacier area change in the Pskem (the western part of the ASB, with a mean elevation of the glacierized area of 3000 masl), At Bashy and the southeastern Ferghana ranges (inner ranges in the southeastern part of the ASB with a mean elevation of 3500 masl) show that in the Pskem Basin glaciers lost 19 per cent and 5 per cent in the 1970–2000 and 2000–2007 periods, while in the At Bashy and Ferghana ranges they lost 12 per cent and 4 per cent, and 9 per cent and 0 per cent in the same periods, respectively (Narama *et al.* 2010). For the Naryn Basin, Kriegel *et al.* (2013) reported a glacier area reduction of 23 per cent for the 1970s–mid-2000s period. In most basins (e.g., Pskem, At Bashy, Naryn) an acceleration in glacier shrinkage is reported in the twenty-first century, particularly in the catchments where small glaciers (which also tend to be located at lower elevations) prevail.

## Climate considerations

The sensitivity of glaciers to increasing summer temperatures is assumed to be responsible for the long-term retreat (Glazirin *et al.* 2002); however, changes in precipitation should not be disregarded. In particular, a well-documented step reduction in precipitation observed in Central Asia in the 1970s, driven by changes in atmospheric circulation, led to a decrease in annual mass balance due to a reduction in accumulation (Cao 1998; Shahgedanova *et al.* 2018). More recent changes in precipitation are non-uniform across the region. Many studies report no statistically significant long-term trends in annual or seasonal precipitation while stressing strong inter-annual variability (Chevallier *et al.* 2014; Finaev 2006; Kriegel *et al.* 2013; Narama *et al.* 2010; Shahgedanova *et al.* 2018). An increasing number of positive-degree days in the high-elevation regions suggests a growing frequency of days with liquid precipitation (Kriegel *et al.*, 2013) which results in lower accumulation and further enhancement of glacier melt through suppressing surface albedo. Some of the earlier studies, considering changes in precipitation in the 1970s–2000s, reported increasing trends (Unger-Shayesteh *et al.* 2013). Considering the negative mass budget of the glaciers in the ASB, these did not compensate for the increase in air temperature (Khromova *et al.* 2006).

## Alpine permafrost

### *Tien Shan and Pamir permafrost distribution*

Continuous permafrost exists above 3600 masl in the central-northern Tien Shan Mountains. The discontinuous zone extends from 3200 to 3600 masl, while the sporadic zone is present from 2700 to 3200 masl (Figure 8.3). Within these zones, permafrost spread may be strongly influenced by local topography and other ground conditions.

Since the end of the Little Ice Age, permafrost in the Tien Shan Mountains experienced a continuous warming until present (Marchenko *et al.* 2007). The first systematic permafrost temperature measurements in the Northern Tien Shan began in 1973 (Gorbunov and Nemov 1978). Initial geothermal observations (1974–1977) in boreholes in the Northern Tien Shan showed that permafrost temperatures within loose deposits and bedrock at an altitude of 3300

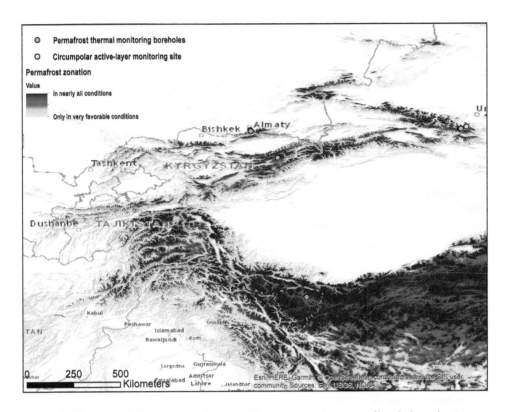

*Figure 8.3* World Terrain Base map showing permafrost zonations, locations of boreholes and active-layer monitoring sites in Central Asia (after Gruber 2012)

Source: Constructed by authors

masl vary from −0.3°C to −0.8°C (Gorbunov and Nemov 1978). Thickness of permafrost in this area varied from 15 to 90 m and the maximum active-layer thickness reached 3.5–4.0 m (Gorbunov and Nemov 1978). Permafrost investigations in the Inner Tien Shan were performed between 1985 and 1992. The results of these investigations included permafrost temperature records, active-layer thickness measurements, descriptions of the cryogenic structures of frozen ground maps, and charts of the distribution of permafrost, ground ice, and periglacial landforms. Ground temperature measurements were carried out in 20 boreholes in the Akshiirak massif (42°N, between 4000 and 4200 masl), and in more than 25 boreholes in the Kumtor Valley (between 3560 and 3790 masl). In the Akshiirak Mountain Range, at elevations of 4100–4200 masl, the lowest measured ground temperature was −5°C in the bedrock (Paleozoic schist) and −6.7°C in the ice-rich Late Pleistocene moraines. The corresponding thickness of permafrost was 350–370 m and 250–270 m, respectively (Gorbunov et al. 1996). Thickness of the active layer on the western slope of the Akshiirak massif decreased from 2.5–3.5 to 0.5–0.7 m within 3200–4000 masl. In the southwestern part of the Tien Shan (Chatyr-Kol and Aksai depressions, 40°30'N), at an elevation of 3500–3600 masl, the thickness of permafrost in loose deposits was 60–90 m and its temperatures were between −1.2 and −1.6°C. The geothermal gradient in the Tien Shan changes from 0.01°C m$^{-1}$ at the mountain ridges and up to 0.02–0.03°C m$^{-1}$ at the bottom of the valleys and within the mountain depressions (Schwarzman 1985).

It is important to note that the 3D topography of the mountains strongly controls the heat flow direction. In steep mountainous topography the heat flow in the area of mountain peaks in the Northern Hemisphere is in general not vertical but more horizontal from the warm southern side to the cold northern side (Magnin et al. 2015; Magnin et al. 2017; Noetzli and Gruber 2009; Noetzli et al. 2007). This is particularly important when interpreting temperature measurements in deep boreholes as the temperature profiles are heavily impacted by the topography (Gruber et al. 2004). Relict Pleistocene permafrost was found in the Aksai depression (40°55'N, 76°25'E) at an elevation of 3160 masl. A 400-meter-deep borehole revealed a two-layered permafrost structure with a lower layer of frozen clay at a depth between 214 and 252 m (Aubekerov and Gorbunov 1999). The thickness of the modern upper layer of permafrost is 90–110 m. This is the one single observation of relic permafrost in the Tien Shan Mountains.

Permafrost temperature observations during 1974–1977 and 1990–2009 indicate that the ground has warmed in the Kazakh part of Tien Shan Mountains over the past 35 years. The increase from 1974 to 2009 varies from 0.38°C to 0.68°C at depths of 14–25 m. Based on interpolation of borehole temperature data, the active layer increased in thickness from 3.2 to 3.4 m in the 1970s to a maximum of 5.2 m in 1992 and to 5.0 m in 2001 and 2004. The average active-layer thickness for all measured sites increased by 23 per cent in comparison to the early 1970s (Marchenko et al. 2007).

In the Pamir, knowledge on permafrost distribution, properties and impacts is limited. Müllebner (2010), using datasets provided by the Tajik hydro-meteorological service, derived an elevation of 3300 masl for the 0°C isotherm. This elevation was interpreted as the approximate lower boundary (without considering surface offsets) for permafrost in Tajikistan. That constitutes approximately 44.3 per cent of the total area of Tajikistan as potential permafrost area and up to 84.1 per cent of the eastern Pamir (Gruber and Mergili 2013; Mergili et al. 2012).

The few thickness measurements in the Karakul lake area suggest relatively thin and unevenly distributed permafrost. Reported thicknesses range from 21–22 m near the eastern coast of Lake Karakul (elevation approximately 3900 masl), while approximately 1 km to the east of this site a permafrost thickness of 120–140 m has been reported from drill sites (Gorbunov *et al.* 1996). Further south, in the Rangul depression, the permafrost thickness is just 15 m at an altitude of 3800 masl, while at the elevation range from 4050–4150 masl it varies from 25 to 110 m (Gorbunov 1978).

## *Permafrost and changing climate*

Overall, very little progress has been made so far in quantifying changes of permafrost under future climate in CA. Mean annual air temperature is the primary climate control on permafrost extent, further modified by surface–atmosphere interactions (mainly snow cover thickness and duration) controlled by topography and surface cover. Clear increasing trends in mean temperature have been observed throughout the CA region (Figure 8.4) with the possible exception of the central Pamir region, which may be affected by the so-called *Karakoram anomaly* (Forsythe *et al.* 2017). These trends are expected to continue in the future, as projected by all scenarios from the most recent Coupled Model Intercomparison Project- Phase 5 (CMIP5) and hence widespread permafrost degradation in CA may be expected.

As mountain permafrost slopes warm, they tend to destabilize, primarily through reduced mechanical strength, potentially leading to various types of mass movements such as debris flows, rock avalanches, or, in the case of ice-cored moraine dams, glacial lake outburst floods (GLOFS). Mass movements are complex phenomena and while climate-induced permafrost degradation (observed at GTNP[3] sites in Tien Shan, e.g., Marchenko *et al.* 2007) can be a key driver of such events, it is not straightforward to disentangle the climate signal from normal erosional processes in mountain regions. However, there is increasing evidence that increased incidence of thermally induced slope instabilities should be expected as high-mountain regions warm. Due to the high potential energy inherent in steep environments and the possibility of compound events that entrain moisture sources (glacier ice, snow or water), the consequences of mass movements can be far-reaching and affect communities many kilometers downstream.

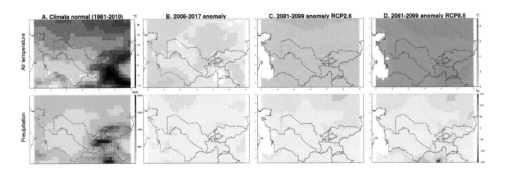

*Figure 8.4* Observed and projected climate change in Central Asia as reported by ERA-Interim reanalysis for climate normals: (A) and current anomaly (B) and GCM multimodal means (Hempel *et al.* 2013) for RCP2.6 (C) and RCP8.5 (D) (projected changes 2081–2099). Cooling of up to 1°C is shown in white in panel (B) and corresponds to the so-called 'Karakoram anomaly'. Note all temperature scales are in °C, precipitation normal is in mm whereas precipitation anomalies are in % change

Source: Constructed by authors

## *Permafrost as a potential water resource*

Central Asia is a relatively arid region where permafrost ice could be a significant contributor to the hydrological cycle as compared to more humid regions, e.g., the European Alps, particularly in the latter half of the twenty-first century, when glacial contributions are projected to be sharply reduced (Huss and Hock 2018). According to Gorbunov *et al.* (1996) the total volume of ground ice in the Northern Tien Shan is 56 km$^3$, which equals 62 per cent of the surface ice volume for the same territory, underscoring its potential value as a water resource (Bolch and Marchenko 2009). Furthermore, it has been estimated that the ratio of rock glacier to surface ice in CA is several times higher than in the European Alps. Permafrost responds more slowly to climate due to the insulating effect of the overlying active layer together with ventilation effects, mainly caused by coarse blocky surface material. Permafrost-based water resources are therefore likely to be available long after surface ice is heavily depleted.

## *Improving permafrost monitoring*

While much research activity has focused on permafrost of the Northern Tien Shan (Bolch and Gorbunov 2014; Bolch and Marchenko 2009; Gorbunov and Titkov 1992; Marchenko *et al.* 2007; Sorg *et al.* 2015), to date there is a paucity of information throughout the Pamir region. The region has a total of two permafrost boreholes listed in the GTNP database.[4] Continuous *in situ* measurements and monitoring in remote mountain areas of CA are challenging due to difficult access, complex topography, financial and logistic constraints, political instability, as well as lack of appropriate infrastructure (Hoelzle *et al.* 2017; Unger-Shayesteh *et al.* 2013). There are very few datasets above 3000 masl and virtually none above 5000 masl. If meteorological stations are present at all, they are usually located at lower elevations where most of the population lives. Remote sensing data as well as model-assimilated observations (from reanalysis data) are used to fill the observational gap. However, the relatively short time series and coarse resolution do not allow for robust assessments of changes in mountain areas with complex topography (Prein *et al.* 2015). This makes the case for denser observational networks in remote mountain areas ever more urgent.

## Runoff trends and water-related hazards in headwater catchments

While a decline in the streamflow of the Amu Darya and Syr Darya are well documented (Micklin 2007), changes in discharge of their tributaries, particularly in the headwater catchments, have received less attention (Chevallier *et al.* 2014). Existing studies suggest that long-term (50–60 years) trends in streamflow in the unmodified headwater catchments (where cryosphere components are present) annually or during the melt season are either insignificant or inconsistent. Contrasting runoff trends were reported by Khan and Holko (2009), who analyzed the reconstructed natural flow time series for the gauges Chinaz (close to the border between Uzbekistan and Kazakhstan, taken as representative for the upper Syr Darya Basin) and Kerki (near the border between Afghanistan and Turkmenistan, representing the upper Amu Darya Basin). Kriegel *et al.* (2013) examined the long-term time series of monthly streamflow in the headwaters of the Big Naryn and the Small Naryn, concluding that August is a month in which glacier melt is expected to make the most prominent contribution to discharge. According to these observations, streamflow declined by 20 per cent and increased by 21 per cent in the Small Naryn and Big Naryn respectively, but there were no significant trends in summer streamflow of either river between 1960 and 2007.

Studies analyzing trends in streamflow over shorter time periods show an increase in streamflow; for example, Finaev (2006) reported positive trends at several gauges in the Pamir for the 1990–2005 period but acknowledges that poor data quality may have affected the results. Although few, the existing analyses of discharge in the headwater catchments of the ASB and some neighboring basins (Shahgedanova *et al.* 2018; Duethmann *et al.* 2015; Krysanova *et al.* 2015; Kundzewicz *et al.* 2015) show that, to date, summer streamflow has not declined. However, the relative contributions of glacier, snow and permafrost may need more insights.

The current observed trends in streamflow and runoff point in the direction of increased water availability in July and August at least until mid-century and increased possibility for water storage in reservoirs and aquifers. However, eventually this will change as glaciers waste away; runoff patterns may change considerably after mid-century (Sorg *et al.* 2014). In a global effort to model runoff changes of more than 50 glacierized basins worldwide, Huss and Hock (2018) found that annual glacier runoff from about half of the basins will continue to increase until *peak water* is reached, and will start to decline afterwards (Figure 8.5).

Peak water will occur later in areas with a higher glacierized fraction and a higher altitudinal distribution of the ice masses. Seasonal runoff is expected to increase in early summer but decrease in late summer. Depending on the climate scenarios employed, peak water will occur in 2030 ± 2 (RCP2.6, mitigation scenario), and 2044 ± 15 (RCP8.5, business-as-usual scenario) in the ASB. Correspondingly, their model predicts an annual runoff increase for the ASB until peak water of 37 per cent (RCP2.6) and 53 per cent (RCP8.5). The increasing rate of evapotranspiration in a warmer climate combined with reduced water flow once peak water has been reached might further intensify challenges for water availability and management (Cretaux *et al.* 2013). However, studies focusing on predicting peak water on a smaller scale (individual catchments) are still lacking for the region and they are necessary in order to better assess the vulnerability of the local populations in the coming decades.

Changes in water resources are to be understood also in terms of changes in the frequency and distribution of water-related hazards, such as river floods, glacier lake outburst floods (GLOFs) and rain-on-snow events (rapid melting of snow combined with intense precipitation). Recent studies have reported that the number and extent of glacier lakes is increasing around the world as a consequence of increased temperatures in high-mountain areas (Carrivick and Tweed 2013; Tweed and Carrivick 2015; Wang and Zhang 2013) and that this trend will continue into the future (Kapitsa *et al.* 2017). The formation of ice- and moraine-dammed lakes from increased glacier melt has the potential to generate glacier lake outburst floods (GLOFs) hazards. The Dasht event in 2002 in the Shakhdara Valley (southwestern Tajik Pamir) is a stark reminder of the destructive power of such events. During the event 250,000 $m^3$ of water was released. The flow of water and entrained debris travelled about 10 km downstream to Dasht village. Large sections of the village were destroyed and the event claimed the life of more than 20 people (Komatsu and Watanabe 2014). Another more recent GLOF event in the Teztor Valley (Ala-Archa River catchment in the Tien Shan) in northern Kyrgyzstan in 2012 entrained a debris flow that caused minor disruption to the capital Bishkek (Erokhin *et al.* 2018). The event was preceded by intense precipitation and rapid rise in temperature, which are believed to have been the cause (Erokhin *et al.* 2018).

More than 1500 lakes extending across Tajikistan, Kyrgyzstan and Afghanistan have been identified through satellite remote sensing (Mergili *et al.* 2013). Kyrgyzstan alone counts about 2000 glacial lakes in its territory, of which 20 per cent are considered at potential risk of outburst due to unstable dams and frequent overflows (Janský *et al.* 2008). A new glacier lake inventory for the Uzbekistan territory (Petrov *et al.* 2017) identified 242 in the four Uzbekistan

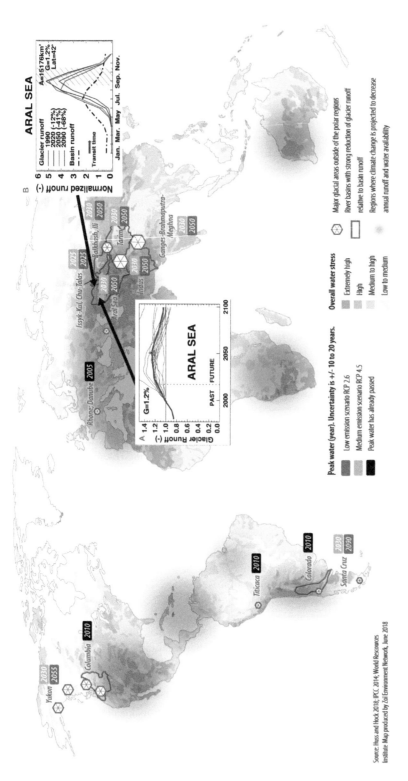

*Figure 8.5* Simulated timing of peak water for individual macro-scale glacierized drainage basins (Huss and Hock 2018). Results refer to runoff from the initially glacierized area and are based on the multi-model mean of 14 GCMs and the RCP4.5 emission scenario. Inset (A) shows calculated annual glacier runoff normalized by average runoff in 1990–2010 for the ASB. The red triangle depicts peak water and the thin lines show results for individual GCMs. Inset (B) shows monthly mean basin runoff and calculated mean glacier runoff (shifted by water transit time) for four 20-year periods (1980–2000, 2010–2030, 2040–2060, 2080–2100). All series are normalized by mean monthly basin/glacier runoff in 1980–2000. The modeled change in glacier area for 2020, 2050, and 2090 relative to model initialization (around 1990) is stated in brackets, and present glacier area (A), glacierization (G) and geographic latitude (Lat.) of the basin's center are given. Shaded areas on the Inset indicate a monthly share of glacier runoff relative to total basin runoff of >30 per cent and hatched areas denote months. Basin-specific transit times are indicated with the blue arrow

Source: Constructed by authors

mountain regions, with more than half (131) located in the Tashkent region. Forty per cent of these (97 lakes) were classified to have a high outburst potential.

The increase in number and extent of glacier lakes does not appear to be accompanied by an increase in glacial floods (Harrison *et al.* 2018). Conversely, some studies have suggested a global decline in glacier floods since the 1990s, possibly associated on the one side with delayed response of glacier flood activities with glacier retreat (Harrison *et al.* 2018), and on the other side with the capacity of successive floods to give rise to river channels, which are better suited to accommodate subsequent flood events (Carrivick and Tweed 2016).

Global and regional climate models point in the direction of future increase in the intensity and frequency of extreme precipitation (Seneviratne *et al.* 2016). At high altitude and/ or latitude, extreme precipitation and higher temperature might exacerbate the frequency of rain-on-snow flood events, which are responsible for the most damaging floods in mountain areas (Würzer *et al.* 2016). Furthermore, precipitation extremes combined with reduced snow and ice might be responsible for increased sediment transport and subsequent deterioration of water quality, infiltration in hydropower reservoirs and damage to infrastructure and agriculture (Huss *et al.* 2017). A wide knowledge gap exists on the future evolution of extreme events and their impacts on people, infrastructure and livelihoods in Central Asia (Unger-Shayesteh *et al.* 2013; Xenarios *et al.* 2019).

## Conclusions

With its two large mountain ranges, Tien Shan and Pamir, Central Asia contains a large part of the worldwide alpine cryosphere, which has a fundamental function as a water storage reservoir with very different temporal scales, as described in this chapter. The state of current knowledge about the different cryospheric components, their processes, and their connection to climate in the Central Asian mountains and in the ASB catchment in particular varies. In the areas of snow and glacier research, knowledge in CA is quite advanced, whereas in permafrost research it is still marginal. Considerable work has been undertaken since the 1990s to address existing research gaps. One basic prerequisite for sound future estimates of cryospheric changes in the Central Asian mountains is the re-establishment of high-quality monitoring sites and capacity building, i.e., the education of young local scientists being able to continue the existing as well as new monitoring programs and to independently build up local research capacities. Currently such capacities and innovations are strongly supported through international projects such as the Central Asia Water project (CaWA[5]), an international consortium of German and Central Asian institutions, the Capacity Building and Twinning Climate Observing System and Cryospheric Climate Services for Improved Adaptation of the Swiss Agency for Development and Cooperation (CATCOS[6] and CICADA[7]), and the Central Asia Hydrometeorology Modernization Project (CAHMP[8]) of the World Bank, and many projects supported by the UK Newton Fund and Global Challenges Research Fund.

The modern alpine cryospheric network of Switzerland can serve as a role model for the mountainous countries in Central Asia. Swiss cryospheric monitoring networks (such as PERMOS, GLAMOS, or the snow monitoring network) are mainly funded by the Swiss government and are implemented primarily by researchers of different universities and research institutions. This is a reliable model that has extended operational monitoring capabilities beyond regions traditionally serviced by standard operational centers. A focus on collaboration between traditional operational hydro-meteorological services with academic sectors and the international networks they leverage could also be a model for improving the observing capabilities in high-mountain regions of CA in general, and the ASB in particular.

## Notes

1 Active layer corresponds to the layer of ground that is subject to annual thawing and freezing in areas underlain by permafrost.
2 Data from National Snow and Ice Data Center, Central Asian snow cover from hydrometeorological surveys, 1932–1990, https://nsidc.org/data/g01171.
3 Global Terrestrial Network for Permafrost mandated by Global Climate Observing System/WMO.
4 http://gtnpdatabase.org/boreholes.
5 http://cawater-info.net/.
6 www.meteoschweiz.admin.ch/home/forschung-und-zusammenarbeit/projekte.subpage.html/de/data/projects/2011/catcos.html.
7 www.unifr.ch/geoscience/geographie/en/research/integrated-themes/international-cooperation-capacity-building.
8 http://projects.worldbank.org/P120788/central-asia-hydrometeorology-modernization-project?lang=en.

## References

Adnan, M., Nabi, G., Poomee, M. S., and Ashraf, A., 2017, Snowmelt runoff prediction under changing climate in the Himalayan cryosphere: the case of the Gilgit River Basin: *Geoscience Frontiers*, v. 8, pp. 941–949.

Aizen, E. M., Aizen, V. B., and Melack, J. M., 2000, Heat exchange during snow ablation in plains and mountains of Eurasia: *Journal of Geophysical Research*, v. 105, pp. 13–22.

Aizen, V. B., Aizen, E. M., 2014, The Central Asia climate and cryosphere/water resources changes. In CAIAG (ed.), Materials of the International Conference 'Remote- and Ground-based Earth Observations in Central Asia'. CAIAG, Bishkek, Kyrgyzstan, 8–9 September 2014

Aizen, V. B., Aizen, E. M., and Kuzmichenok, V., 2007, Glaciers and hydrological changes in the Tien Shan: simulation and prediction: *Environmental Research Letters*, v. 2, no. 045019, doi:10.1088/1748-9326/1082/1084/045019.

Aizen, V. B., Aizen, E. M., and Melack, J. M., 1995, Climate, snow cover, glaciers, and runoff in the Tien Shan, Central Asia: *Water Resources*, v. 31, no. 6, pp. 1113–1129.

Aizen, V. B., Aizen, E. M., Melack, J. M., and Dozier, J., 1997, Climatic and hydrologic changes in the Tien Shan, Central Asia: *Journal of Climate*, v. 10, pp. 1393–1404.

Apel, H., Abdykerimova, Z., Agalhanova, M., Baimaganbetov, A., Gavrilenko, N., Gerlitz, L., Kalashnikova, O., Unger-Shayesteh, K., Vorogushyn, S., and Gafurov, A., 2018, Statistical forecast of seasonal discharge in Central Asia using observational records: development of a generic linear modelling tool for operational water resource management: *Hydrology and Earth System Science*, v. 22, pp. 2225–2254.

Aubekerov, B. and Gorbunov, A., 1999, Quaternary permafrost and mountain glaciation in Kazakhstan: *Permafrost and Periglacial Processes*, v. 10, pp. 65–80.

Barandun, M., 2019, A novel approach to estimate glacier mass balance in the Tien Shan and Pamir based on transient snowline observations [Ph.D]. University of Fribourg, Switzerland.

Barandun, M., Huss, M., Berthier, E., Kääb, A., Azisov, E., Bolch, T., Usulbajev, R., and Hoelzle, M., 2018, Multi-decadal mass balance series of three Kyrgyz glaciers inferred from transient snowline observations: *The Cryosphere*, v. 12, pp. 1899–1919.

Barandun, M., Huss, M., Sold, L., Farinotti, D., Azisov, E., Salzmann, N., Usulbaliev, R., Merkushkin, A., and Hoelzle, M., 2015, Re-analysis of seasonal mass balance at Abramov glacier 1968–2014: *Journal of Glaciology*, v. 61, no. 230, pp. 1103–1117.

Bolch, T., 2004, *Using ASTER and SRTM DEMs for Studying Glaciers and Rockglaciers in Northern Tien Shan: Theoretical and Applied Problems of Geography on a Boundary of Centuries* [Teoretičeskije i Prikladnyje Problemy geografii na Rubešje Stoletij], Kazakh National University, Almaty, Kazakhstan, pp. 254–258

Bolch, T., 2007, Climate change and glacier retreat in northern Tien Shan (Kzakhstan/Kygyzstan) using remote sensing data: *Global and Planetary Change*, v. 56, pp. 1–12.

Bolch, T., 2015, Glacier area and mass changes since 1964 in the Ala Archa Valley, Kyrgyz Ala-Too, Northern Tien Shan: *Led i Sneg (Ice and Snow)*, v. 55, no. 1, pp. 28–39.

Bolch, T. and Marchenko, S., 2009, Significance of glaciers, rockglaciers and ice-rich permafrost in the Northern Tien Shan as water towers under climate change conditions. In Braun, L. N., Hagg, W., Severskiy, I. V., and Young, G. (eds), *Proceedings Assessment of Snow, Glacier and Water Resources in Asia,*

*Almaty, Kazakhstan, 2006, Volume 8*. UNESCO-IHP and German IHP/HWRP National Committee, Koblenz, Germany, pp. 132–144.

Bolch, T. and Gorbunov, A.P. 2014, Characteristics and origin of the rock glaciers in northern Tien Shan (Kazakhstan/Kyrgyzstan): *Permafrost and Periglacial Processes*, v. 25, no. 4, pp. 320–332.

Bolch, T., Peters, J., Yegorov, A., Pradhan, B., Buchroithner, M., and Blagoveshchensky, V., 2011, Identification of potentially dangerous glacial lakes in the northern Tien Shan: *Natural Hazards*, v. 59, no. 3, pp. 1691–1714.

Bolch, T., Shea, J. M., Shiyin, L., Azam, M. F., Gao, Y., Gruber, S., Immerzeel, W. W., Kulkarni, A., Li, H., Tahir, A. A., Zhang, G., and Zhang, Y., 2019, Status and change of the cryosphere in the extended Hindu Kush Himalaya region. In Wester, P., Mishra, A., Mukherji, A., and Shrestha, A. B. (eds), *The Hindu Kush Himalaya Assessment*. Springer, Cham, Switzerland, pp. 209–255.

Brun, F., Berthier, E., Wagnon, P., Kääb, A., and Treichler, D., 2017, A spatially resolved estimate of high mountain asia glacier mass balances, 2000–2016: *Nature Geoscience*, v. 10, no. 9, pp. 668–673.

Cao, M. S., 1998, Detection of abrupt changes in glacier mass balance in the Tien Shan Mountains: *Journal of Glaciology*, v. 44, no. 147, pp. 352–358.

Carrivick, J. L. and Tweed, F. S., 2013, Proglacial lakes: character, behaviour and geological importance: *Quaternary Science Reviews*, v. 78, pp. 34–52.

Carrivick, J. L., and Tweed, F. S., 2016, A global assessment of the societal impacts of glacier outburst floods: Global and Planetary Change, v. 144, p. 1–16.

Chen, Y., Li, W., Deng, H., Fang, G., and Li, Z., 2016, Changes in Central Asia's water tower: past, present and future: *Scientific Reports*, v. 6, pp. 35458.

Chevallier, P., Pouyaud, B., Mojaisky, M., Bolgov, M., Olsson, O., Bauer, M., and Froebrich, J., 2014, River flow regime and snow cover of the Pamir Alay (Central Asia) in a changing climate: *Hydrogeological Sciences Journal*, v. 59, no. 8, pp. 1491–1506.

Conde, V., Nico, G., Mateus, P., Catalao, J., Kontu, A., and Gritsevich, M., 2019, On the estimation of temporal changes of snow water equivalent by spaceborne SAR interferometry: a new application for the Sentinel-1 mission: *Journal of Hydrology and Hydromechanics*, v. 67, no. 1, p. 93–100.

Cretaux, J.-F., Letolle, R., and Bergé-Nguyen, M., 2013, History of Aral Sea level variability and current scientific debates: *Global and Planetary Change*, v. 110, pp. 99–113.

Delaloye, R., Lambiel, C., and Gärtner-Roer, I., 2010, Overview of rock glacier kinematics research in the Swiss Alps: *Swiss Journal of Geography*, v. 65, no. 2, pp. 135–145.

Denzinger, F., 2018, Structure from motion using historical aerial images to analyse changes in surface elevation and volume on Abramov Glacier, Kyrgyzstan [MSc thesis]. University of Zurich, Switzerland.

Dietz, A. J., Conrad, C., Kuenzer, C., Gesell, G., and Dech, S., 2014, Identifying changing snow cover characteristics in Central Asia between 1986 and 2014 from remote sensing data: *Remote Sensing*, v. 6, pp. 12752–12775.

Duethmann, D., Bolch, T., Farinotti, D., Kriegel, D., Vorogushyn, S., Merz, B., Pieczonka, T., Jiang, T., Su, B., and Güntner, A., 2015, Attribution of streamflow trends in snow and glacier melt-dominated catchments of the Tarim River, Central Asia: *Water Resources Research*, v. 51, pp. 4727–4750.

Dyurgerov, M. B., 2002, *Glacier Mass Balance and Regime: Data of Measurements and Analysis*. Occasional Paper No. 55. Institute of Arctic and Alpine Research, University of Colorado, Boulder, Colorado, USA

Dyurgerov, M. B. and Dwyer, J. D., 2001, The steepening of glacier mass balance gradients with northern hemisphere warming: *Zeitschrift für Gletscherkunde und Glazialgeologie*, v. 36, pp. 107–118.

Dyurgerov, M. B., Mikhalenko, V. N., Kunakhovitch, M. G., Ushnurtsev, S. N., Liu, C., and Xie, Z., 1994, On the cause of glacier mass balance variations in the Tien Shan Mountains: *GeoJournal*, v. 33, no. 2–3, pp. 311–317.

Erokhin, S. A., Zaginaev, V. V., Meleshko, A. A., Ruiz-Villanueva, V., Petrakov, D., Chernomorets, S. S., Viskhadzhieva, S. K., Tutubalina, O., and Stoffel, M., 2018, Debris flows triggered from non-stationary glacier lake outbursts: the case of the Teztor Lake complex (Northern Tian Shan, Kyrgyzstan): *Landslides*, v. 15, no. 1, pp. 83–98.

Farinotti, D., Longuevergne, L., Moholt, G., Duethmann, D., Mölg, T., Bolch, T., Vorogushyn, S., and Güntner, A., 2015, Substantial glacier mass loss in the Tien Shan over the past 50 years: *Nature Geoscience*, v. 8, pp. 716–722.

Finaev, A., 2009, Review of hydrometeorological observations in Tajikistan for the period of 1990–2005. In Braun, L. N., Hagg, W., Severskiy, I. V., and Young, G. (eds), *Proceedings Assessment of Snow, Glacier and Water Resources in Asia, Almaty, Kazakhstan, 2006, Volume 8*. UNESCO-IHP and German IHP/HWRP National Committee, Koblenz, Germany, pp. 55–64.

Forsythe, N., Fowler, H. J., Li, X.-F., Blenkinsop, S., and Pritchard, D., 2017, Karakoram temperature and glacial melt driven by regional atmospheric circulation variability: *Nature Climate Change*, v. 7, pp. 664–670.

Gafurov, A. and Bàrdossy, A., 2009, Cloud removal methodology from MODIS snow cover product: *Hydrology and Earth System Sciences*, v. 13, no. 7, pp. 1361–1373.

Gafurov, A., Kriegel, D., Vorogushyn, S., and Merz, B., 2013, Evaluation of remotely sensed snow cover product in Central Asia: *Hydrology Research*, v. 44, no. 3, pp. 506–522.

Gafurov, A., Ludtke, S., Unger-Shayesteh, K., Vorogushyn, S., Schöne, T., Schmidt, S., Kalashnikova, O., and Merz, B., 2016, MODSNOW-Tool: an operational tool for daily snow cover monitoring using MODIS data: *Environmental Earth Science*, v. 75, no. 1078, pp. 1–15.

Gardelle, J., Berthier, E., Arnaud, Y., and Kääb, A., 2013, Region-wide glacier mass balance over the Pamir-Karakoram-Himalaya during 1999–2011: *The Cryosphere*, v. 7, pp. 1263–1286.

Gardner, A. S., Moholdt, G., Cogley, J. G., Wouters, B., Arendt, A. A., Wahr, J., Berthier, E., Hock, R., Pfeffer, W. T., Kaser, G., Ligtenberg, S. R. M., Bolch, T., Sharp, M. J., Hagen, J. O., van den Broeke, M. R., and Paul, F., 2013, A reconciled estimate of glacier contributions to sea level rise: 2003 to 2009: *Science*, v. 340, pp. 852–857.

Glazirin, G. E., 2009, Hydrometeorological monitoring system in Uzbekistan. In Braun, L. N., Hagg, W., Severskiy, I. V., and Young, G. (eds), *Proceedings Assessment of Snow, Glacier and Water Resources in Asia, Almaty, Kazakhstan, 2006, Volume 8.* UNESCO-IHP and German IHP/HWRP National Committee, Koblenz, Germany, pp. 65–83.

Glazirin, G. E., Braun, L., and Shchetinnikov, A. S., 2002, Sensitivity of mountain glacierization to climatic changes in Centra Asia: *Zeitschrift für Gletscherkunde und Glazialgeologie*, v. 38, no. 1, pp. 71–76.

Goerlich, F., Bolch, T., Mukherjee, K., and Pieczonka, T., 2017, Glacier mass loss during the 1960s and 1970s in the Ak-Shirak range (Kyrgyzstan) from multiple stereoscopic corona and hexagon imagery: *Remote Sensing*, v. 9, no. 3, p. 275.

Gorbunov, A. and Nemov, A. E., 1978, Kissledovaniu temperatur rykhlooblomochnyh tolscsh vysokogornogo Tyan-Shanya (On temperature research of loose deposits in the Tien Shan high mountains). *Cryogenic Phenomena of High Mountains.* Nauka, Novosibirsk, Russia, pp. 92–99 [in Russian].

Gorbunov, A., Severskiy, E., and Titkov, S. N., 1996, Geocriologicheskie Usloviya Tyan-Shanya i Pamira (Geocryological Conditions of the Tien Shan and Pamir). Permafrost Institute, Yakutsk, Russia [in Russian].

Gorbunov, A. P., 1978, Permafrost investigations in high-mountain regions: *Arctic and Alpine Research*, v. 10, no. 2, pp. 283–294.

Gorbunov, A. P. and Titkov, S. N., 1992, Dynamics of rock glaciers of the northern Tien Shan and the Djungar Ala Tau, Kazakhstan: *Permafrost and Periglacial Processes*, v. 3, pp. 29–39.

Gruber, F. E. and Mergili, M., 2013, Regional-scale analysis of high-mountain multi-hazard and risk indicators in the Pamir (Tajikistan) with GRASS GIS: *Natural Hazards and Earth System Sciences*, v. 13, pp. 2779–2796.

Gruber, S., 2012, Derivation and analysis of a high-resolution estimate of global permafrost zonation: *The Cryosphere*, v. 6, pp. 221–233.

Gruber, S., King, L., Kohl, T., Herz, T., Haeberli, W., and Hoelzle, M., 2004, Interpretation of geothermal profiles perturbed by topography: The Alpine permafrost boreholes at Stockhorn Plateau, Switzerland: *Permafrost and Periglacial Processes*, v. 15, no. 4, pp. 349–357.

Haeberli, W., 2008, Changing views of changing glaciers. In Orlove, B., Wiegandt, E., and Luckman, B. H. (eds), The Darkening Peaks: Glacial Retreat in Scientific and Social Context, University of California Press, Oakland, California, USA, pp. 23–32.

Haeberli, W., Noetzli, J., Arenson, L., Delaloye, R., Gärtner-Roer, I., Gruber, S., Isaksen, K., Kneisel, C., Krautblatter, M., and Phillips, M., 2010, Mountain permafrost: development and challenges of a young research field: *Journal of Glaciology*, v. 56, no. 200, pp. 1040–1058.

Hagg, W., Braun, L., Kuhn, M., and Nesgaard, T. I., 2007, Modelling of hydrological response to climate change in glacierized Central Asian catchments: *Journal of Hydrology*, v. 332, pp. 40–53.

Hagg, W., Hoelzle, M., Wagner, S., Mayr, E., and Klose, Z., 2013, Glacier and runoff changes in the Rukhk catchment, upper Amu-Darya basin until 2050: *Global and Planetary Change*, v. 110, pp. 62–73.

Harrison, S., Kargel, J. S., Huggel, C., Reynolds, J., Shugar, D. H., Betts, R. A., Emmer, A., Glasser, N., Haritashya, U. K., Klimes, J., Reinhardt, L., Schaub, Y., Wiltshire, A., Regmi, D., and Vilimek, V., 2018, Climate change and the global pattern of moraine-dammed glacial lake outburst floods: *The Cryosphere*, v. 12, no. 4, pp. 1195–1209.

Hauck, C., 2013, New concepts in geophysical surveying and data interpretation for permafrost: *Permafrost and Periglacial Processes*, v. 24, no. 2, pp. 131–137.

Hoelzle, M., Azisov, E., Barandun, M., Huss, M., Farinotti, D., Gafurov, A., Hagg, W., Kenzhebaev, R., Kronenberg, M., Machguth, H., Merkushkin, A., Moldobekov, B., Petrov, M., Saks, T., Salzmann, N., Schöne, T., Tarasov, Y., Usubaliev, R., Vorogushyn, S., Yakovlev, A., and Zemp, M., 2017, Re-establishing glacier monitoring in Kyrgyzstan and Uzbekistan, Central Asia: *Geoscientific Instrumentation Method and Data Systems*, v. 6, pp. 397–418.

Holzer, N., Vijay, S., Yao, T., Xu, B., Buchroithner, M., and Bolch, T., 2015, Four decades of glacier variations at muztagh ata (eastern pamir): a multi-sensor study including hexagon kh-9 and pléiades data: *The Cryosphere*, v. 9, no. 6, pp. 2071–2088.

Huss, M., Bookhagen, B., Huggel, C., Jacobsen, D., Bradley, R. S., Clague, J. J., Vuille, M., Buytaert, W., Cayan, D. R., Greenwood, G., Mark, B. G., Milner, A. M., Weingartner, R., and Winder, M., 2017, Toward mountains without permanent snow and ice: *Earth's Future*, v. 5, no. 5, pp. 418–435.

Huss, M. and Hock, R., 2018, Global-scale hydrological response to future glacier mass loss: *Nature Climate Change*, v. 8, pp. 135–140.

Immerzeel, W. W., Droogers, P., de Jong, S. M., and Bierkens, M. F. P., 2009, Large-scale monitoring of snow cover and runoff simulation in Himalayan river basins using remote sensing: *Remote Sensing of Environment*, v. 113, pp. 40–49.

Janský, B., Šobr, M., Engel, Z., and Yerokhin, S., 2008, High-altitude lake outburst: Tien-Shan case study. In Dostál, P. (ed.), *Evolution of Geographical Systems and Risk Processes in the Global Context*. P3K Publishers, Prague, pp. 113–127.

Kääb, A., Leinss, S., Gilbert, A., Bühler, Y., Gascoin, S., Evans, S., Bartelt, P., Berthier, E., Brun, F., Chao, W.-A., Farinotti, D., Gimbert, F., Guo, W. Q., Huggel, C., Kargel, J. S., Leonard, G. J., Tian, L., Treichler, D., and Yao, T., 2018, Massive collapse of two glaciers in western Tibet in 2016 after surge-like instability: *Nature Geoscience*, v. 11, pp. 114–120.

Kääb, A., Treichler, D., Nuth, C., and Berthier, E., 2015, Contending estimates of 2003–2008 glacier mass balance over the Pamir–Karakoram–Himalaya: *The Cryosphere*, v. 9, pp. 557–564.

Kalashnikova, O. and Gafurov, A., 2017, Water availability forecasting for Naryn River using ground-based and satellite snow cover data: *Ice and Snow*, v. 57, no. 4, pp. 507–517.

Kapitsa, V., Shahgedanova, M., Machguth, H., Severskiy, E., and Medeu, A., 2017, Assessment of evolution and risks of glacier lake outbursts in the Djungarskiy Alatau, Central Asia, using Landsat imagery and glacier bed topography modelling: *Natural Hazards and Earth System Sciences*, v. 17, pp. 1837–1856.

Kaser, G., Grosshauser, M., and Marzeion, B., 2010, Contribution potential of glaciers to water availability in different climate regimes: *PNAS*, v. 107, no. 47, pp. 20223–20227.

Kenzhebaev, R., Barandun, M., Kronenberg, M., Yaning, C., Usubaliev, R., and Hoelzle, M., 2017, Mass balance observations and reconstruction for Batysh Sook Glacier, Tien Shan, from 2004 to 2016: *Cold Regions Science and Technology*, v. 135, pp. 76–89.

Khan, V. and Holko, L., 2009, Snow cover characteristics in the Aral Sea Basin from different data sources and their relation with river runoff: *Journal of Marine Systems*, v. 76, no. 3, pp. 254–262.

Khromova, T. E., Dyurgerov, M. B., and Barry, R. G., 2003, Late-twentieth century changes in glacier extent in the Ak-Shirak Range, Central Asia, determined from historical data and ASTER imagery: *Geophysical Research Letters*, v. 30, no. 16. doi:10.1029/2003GL017233.

Khromova, T. E., Osipova, G. B., Tsvetkov, D. G., Dyurgerov, M. B., and Barry, R. G., 2006, Changes in glacier extent in the eastern Pamir, Central Asia, determined from historical data and ASTER imagery: *Remote Sensing of Environment*, v. 102, pp. 24–32.

Knoche, M., Merz, R., Lindner, M., and Weise, S. M., 2017, Bridging glaciological and hydrological trends in the Pamir Mountains, Central Asia: *Water*, v. 9, no. 6, pp. 422.

Komatsu, T. and Watanabe, T., 2014, Glacier-related hazards and their assessment in the Tajik Pamir: a short review: *Geographical Studies*, v. 88, no. 2, pp. 117–131.

Koven, C. D., Ringeval, B., Friedlingstein, P., Ciais, P., Cadule, P., Khvorostyanov, D., Krinner, G., and Tarnocai, C., 2011, Permafrost carbon-climate feedbacks accelerate global warming: *PNAS*, v. 108, no. 36, pp. 14769–14774.

Krenke, A. N., 1998, Former Soviet Union hydrological snow surveys. In N. S. a. I. D. C. a. W. D. C. f. (ed.), *Glaciology*. National Snow and Ice Data Center and World Data Center for Glaciology, Boulder, CO, USA.

Kriegel, D., Mayer, C., Hagg, W., Vorogushyn, S., Duethmann, D., Gafurov, A., and Farinotti, D., 2013, Changes in glacierisation, climate and runoff in the second half of the 20th century in the Naryn basin, Central Asia: *Global and Planetary Change*, v. 110, pp. 51–61.

Kronenberg, M., Barandun, M., Hoelzle, M., Huss, M., Farinotti, D., Azisov, E., Usulbajev, R., Gafurov, A., Petrakov, D., and Kääb, A., 2016, Mass balance reconstruction for Glacier 354, Tien Shan, from 2003–2014: *Annals of Glaciology*, v. 57, no. 71, pp. 92–102.

Krysanova, V., Hattermann, F., Huang, S., Hesse, C., Vetter, T., and Liersch, S., 2015, Modelling climate and land-use change impacts with SWIM: lessons learnt from multiple applications: *Hydrogeological Sciences Journal*, v. 60, no. 4, pp. 606–635.

Kundzewicz, Z. W., Merz, B., Vorogushyn, S., Hartmann, G., Duethmann, D., Wortmann, M., Huang, S., Su, B., and Jiang, T., 2015, Analysis of changes in climate and river discharge with focus on seasonal runoff predictability in the Aksu River Basin: *Environmental Earth Sciences*, v. 73, no. 2, pp. 501–516.

Kure, S., Jang, S., Ohara, N., Kavvas, M. L., Chen, Z. Q., 2013, Hydrologic impact of regional climate change for the snowfed and glacierfed river basins in the Republic of Tajikistan: hydrological response of flow to climate change: *Hydrological Processes*, v. 27, no. 26, pp. 4057–4070.

Kutuzov, S. and Shahgedanova, M., 2009, Changes in the extent of glaciers in the Eastern Terskey Alatoo, the Central Tien-Shan, in response to climatic fluctuations in between the end of 19th and the beginning of the 21st century: *Global and Planetary Change*, v. 69, pp. 59–70.

Kuzmichenok, V., 2009, Monitoring of water, snow and glacial resources of Kyrgyzstan. In Braun, L. N., Hagg, W., Severskiy, I. V., and Young, G. (eds), *Proceedings Assessment of Snow, Glacier and Water Resources in Asia, Almaty, Kazakhstan, 2006, Volume 8*. UNESCO-IHP and German IHP/HWRP National Committee, Koblenz, Germany, pp. 84–98.

Li, J., Li, Z., Zhu, J., Li, X., Xu, B., Wang, Q., Huang, C., and Hu, J., 2017, Early 21st century glacier thickness changes in the Central Tien Shan: *Remote Sensing of Environment*, v. 192, pp. 12–29.

Li, Z. Q., Li, H. L., and Chen, Y. N., 2011, Mechanisms and simulation of accelerated shrinkage of continental glaciers: a case study of Urumqi Glacier No. 1 in Eastern Tianshan, Central Asia: *Journal of Earth Science*, v. 22, no. 4, pp. 423–430.

Lin, H., Li, G., Cuo, L., Hooper, A., and Ye, Q., 2017, A decreasing glacier mass balance gradient from the edge of the upper Tarim basin to the Karakoram during 2000–2014: *Scientific Reports*, v. 7, no. 1, p. 6712.

Liu, Q. and Liu, S., 2016, Response of glacier mass balance to climate change in the Tianshan mountains during the second half of the twentieth century: *Climate Dynamics*, v. 46, no. 1–2, pp. 303–303§306.

Magnin, F., Deline, P., Ravanel, L., Noetzli, J., and Pogliotti, P., 2015, Thermal characteristics of permafrost in the steep alpine rock walls of the Aiguille du Midi (Mont Blanc Massif, 3842m a.s.l): *The Cryosphere*, v. 9, pp. 109–121.

Magnin, F., Josnin, J.-Y., Ravanel, L., Pergaud, J., Pohl, B., and Deline, P., 2017, Modelling rock wall permafrost degradation in the Mont Blanc massif from the LIA to the end of the 21st century: *The Cryosphere*, v. 11, pp. 1813–1834.

Marchenko, S., Gorbunov, A., and Romanovsky, V. E., 2007, Permafrost warming in the Tien Shan Mountains, Central Asia: *Global and Planetary Change*, v. 56, pp. 311–327.

Marty, C., 2008, Regime shift of snow days in Switzerland: *Geophysical Research Letters*, v. 35, p. L12501.

Mateo, E. I. and Daniels, J. M., 2018, Surface hydrological processes of rock glaciated basins in the San Juan Mountains, Colorado: *Physical Geography*, p. 1–19.

Mergili, M., Kopf, C., Müllebner, B., and Schneider, J. F., 2012, Changes of the cryosphere and related geohazards in the high-mountain areas of Tajikistan and Austria: a comparison: *Geografiska Annaler Series a-Physical Geography*, v. 93, pp. 79–96.

Mergili, M., Müller, J. P., and Schneider, J. F., 2013, Spatio-temporal development of high-mountain lakes in the headwaters of the Amu Darya River (Central Asia): *Global and Planetary Change*, v. 107, pp. 13–24.

Merkushkin, A. and Tsarev, B., 2007, Regime of snow accumulation and snow melting in river basin Sangardak: *Proceedings of NGIMI*, v. 8, no. 253, pp. 63–70.

Micklin, P., 2007, The Aral Sea disaster: *Annual Review of Earth and Planetary Sciences*, v. 35, pp. 47–72.

Mölg, N., Bolch, T., Rastner, P., Strozzi, T., and Paul, F., 2018, A consistent glacier inventory for Karakoram and Pamir derived from Landsat data: distribution of debris cover and mapping challenges: *Earth System Science Data*, v. 10, pp. 1807–1827.

Müllebner, B., 2010, Modelling of potential permafrost areas in the Pamir and Alai Mountains (Tajikistan) using remote sensing and GIS techniques. M.Sc. thesis, BOKU University, Vienna, Austria. Retrieved 12 May 2019, https://abstracts.boku.ac.at/oe_list.php?paID=3&paSID=7913&paSF=-1& paCF=0&paLIST=0&language_id=DE.

Narama, C., Kääb, A., Duishonakunov, M., and Abdrakhmatov, K., 2010, Spatial variability of recent glacier changes in the Tien Shan Mountains, Central Asia Corona(~1970), Landsat (~2000), and ALOS (~2007) satellite data: *Global and Planetary Change*, v. 71, pp. 42–54.

Noetzli, J. and Gruber, S., 2009, Transient thermal effects in Alpine permafrost: *The Cryosphere*, v. 3, pp. 85–99.

Noetzli, J., Gruber, S., Kohl, T., Salzmann, N., and Haeberli, W., 2007, Three-dimensional distribution and evolution of permafrost temperatures in idealized high-mountain topography: *Journal of Geophysical Research*, v. 112, no. F02S13. doi:10.1029/2006JF000545.

Ozmonov, A., Bolch, T., Xi, C., Wei, J., Kurban, A., and Guo, W. Q., 2013, Glacier characteristics and changes in the Sary-Jaz river basin (Central Tien Shan) 1990–2010: *Remote Sensing Letters*, v. 4, no. 8, pp. 725–734.

Paul, F. and Mölg, N., 2014, Hasty retreat of glaciers in northern Patagonia from 1985 to 2011: *Journal of Glaciology*, 60, pp. 1033–1043.

Peters, J., Bolch, T., Gafurov, A., and Prechtel, N. 2015, Snow cover distribution in the Aksu catchment (Central Tien Shan) 1986–2013 based on AVHRR and MODIS data: *IEEE Journal of Selected Topics of Applied Earth Observation*, v. 8, no. 11, pp. 5361–5375.

Petrov, M. A., Sabitov, T. Y., Tomashevskaya, I. G., Glazirin, G. E., Chernomorets, S. S., Savernyuk, E. A., Tutubalina, O. V., Petrakov, D. A., Sokolov, L. S., Dokukin, M. D., Mountrakis, G., Ruiz-Villanueva, V., and Stoffel, M., 2017, Glacial lake inventory and lake outburst potential in Uzbekistan: *Science of the Total Environment*, v. 592, pp. 228–242.

Pieczonka, T., Bolch, T., Wei, J., and Liu, S. (2013) Heterogeneous mass loss of glaciers in the Aksu-Tarim Catchment (Central Tien Shan) revealed by 1976 KH-9 Hexagon and 2009 SPOT-5 stereo imagery: *Remote Sensing of the Environment*, 130, pp. 233–244.

Pieczonka, T. and Bolch, T., 2015, Region-wide glacier mass budgets and area changes for the Central Tien Shan between ~1975 and 1999 using Hexagon KH-9 imagery: *Global and Planetary Change*, v. 128, pp. 1–13.

Pohl, E., Gloaguen, R., Andermann, C., and Knoche, M., 2017, Glacier melt buffers river runoff in the Pamir Mountains: *Water Resources Research*, v. 53, pp. 2467–2489.

Prein, A. F., Langhans, W., Fosser, G., Ferrone, A., Ban, N., Goergen, K., Keller, M., Tölle, M., Gutjahr, O., Feser, F., Brisson, E., Kollet, S., Schmidli, J., van Lipzig, N. P. M., and Leung, R., 2015, A review on regional convection-permitting climate modeling: Demonstrations, prospects, and challenges: *Reviews of Geophysics*, v. 53, no. 2, pp. 323–361.

RGI Consortium, R., 2017, Randolph Glacier Inventory (RGI) – A Dataset of Global Glacier Outlines. In G. L. I. M. f. (ed.), *Space*. Global Land Ice Measurements from Space, Boulder, CO, USA.

Sakai, A., Nuimura, T., Fujita, K., Takenaka, S., Nagai, H., and Lamsal, D., 2015, Climate regime of Asian glaciers revealed by GAMDAM glacier inventory: *The Cryosphere*, v. 9, no. 3, pp. 865–880.

Schwarzman, U. G., 1985, Geotermicheskii rejim, dinamika litosfery I perspectivy ispolzovaniya geotermich-eskoi energii Tien-Shanya (in Russian) (Geothermic regime, dynamics of lithosphere, and perspectives of using geothermal energy of the Tien Shan). *Geothermic Investigations in Middle Asia and Kazakhstan*. Nauka, Moscow, pp. 236–250.

Seneviratne, S. I., Donat, M. G., Pitman, A. J., Knutti, R., and Wilby, R. L., 2016, Allowable CO2 emissions based on regional and impact-related climate targets: *Nature*, v. 529, no. 7587, pp. 477–483.

Serquet, G., Marty, C., and Rebetez, M., 2013, Monthly trends and the corresponding altitudinal shift in the snowfall/precipitation day ratio: *Theoretical and Applied Climatology*, v. 114, pp. 437–444.

Shahgedanova, M., Afzal, M., Severskiy, E., Usmanova, Z., Saidaliyeva, Z., Kapitsa, V., Kasatkin, N., and Dolgikh, S., 2018, Changes in the mountain river discharge in the northern Tien Shan since the mid-20th Century: Results from the analysis of a homogeneous daily streamflow data set from seven catchments: *Journal of Hydrology*, v. 564, pp. 1133–1152.

Shangguan, D., Liu, S., Ding, Y., Ding, L., Xiong, L., Cai, D., Li, G., Lu, A., Zhang, S., and Zhang, Y., 2006, Monitoring the glacier changes in the Muztag Ata and Konggur mountains, East Pamirs, based on Chinese glacier inventory and recent satellite imagery: *Annals of Glaciology*, v. 43, pp. 79–85.

Sorg, A., Bolch, T., Stoffel, M., Solomina, O., and Beniston, M., 2012, Climate change impacts on glaciers and runoff in Central Asia: *Nature Climate Change*. doi: 10.1038/nclimate1592.

Sorg, A., Huss, M., Stoffel, M., and Rohrer, M., 2014, The days of plenty might soon be over in glacierized Central Asian catchments: *Environmental Research Letters*, v. 9, p. 104018.

Sorg, A., Kääb, A., Roesch, A., Bigler, C., and Stoffel, M., 2015, Contrasting responses of Central Asian rock glaciers to global warming: *Scientific Reports*, v. 5, p. 8228.

Stoffel, M. and Huggel, C., 2012, Effects of climate change on mass movements in mountain environments: *Progress in Physical Geography*, v. 36, pp. 421–439.

Tsarev, B., 2006, Assessment of climate change impact on maximum snow storage in Central Asian mountains: *Proceedings of NGIMI*, v. 4, no. 249, pp. 15–32.

Tweed, F. S. and Carrivick, J. L., 2015, Deglaciation and proglacial lakes: *Geology Today*, v. 31, no. 3, pp. 96–102.

Unger-Shayesteh, K., Vorogushyn, S., Farinotti, D., Gafurov, A., Duethmann, D., Mandychev, A., and Merz, B., 2013, What do we know about past changes in the water cycle of Central Asian headwaters? A review: *Global and Planetary Change*, v. 110, pp. 4–25.

Wang, Q., Yi, S., Chang, L., and Sun, W., 2017, Large-scale seasonal changes in glacier thickness across high mountain asia: *Geophysical Research Letters*, v. 44, pp. 10427–10435.

Wang, S. and Zhang, T., 2013, Glacial lakes change and current status in the central Chinese Himalayas from 1990 to 2010: *Journal of Applied Remote Sensing*, v. 7, pp. 1–15.

WGMS, 2017, Global Glacier Change Bulletin No. 2 (2014–2015). In Zemp, M., Nussbaumer, S. U., Gärtner-Roer, I., Huber, J., Machguth, H., Paul, F., and Hoelzle, M. (eds), *Global Glacier Change Bulletin, Volume 2*. ICSU(WDS)/IUGG(IACS)/ UNEP/UNESCO/WMO, World Glacier Monitoring Service, Zurich, Switzerland, p. 244.

Würzer, S., Jonas, T., Wever, N., and Lehning, M., 2016, Influence of initial snowpack properties on runoff formation during rain-on-snow events: *Journal of Hydrometeorology*, v. 17, no. 6, pp. 1801–1815.

Xenarios, S., Gafurov, A., Schmidt-Vogt, D., Sehring, J., Manandhar, S., Hergarten, C., Shigaeva, J., and Foggin, M., 2019, Climate change and adaptation of mountain societies in Central Asia: uncertainties, knowledge gaps, and data constraints: *Regional Environmental Change*, v. 19, pp. 1339.

Yakovlev, A., 2005, Method of the Pyandzh river runoff simulation during flood season with use of numerical weather predicted data and satellite images: *Proceedings of NGIMI*, v. 5, no. 250, pp. 123–134.

Zemp, M., Haeberli, W., Hoelzle, M., and Paul, F., 2006, Alpine glaciers to disappear within decades?: *Geophysical Research Letters*. doi:10.1029/2006GL026319.

Zhou, H., Aizen, E. M., and Aizen, V. B., 2013, Deriving long term snow cover extent dataset from AVHRR and MODIS data: Central Asia case study: *Remote Sensing of the Environment*, v. 136, pp. 146–162.

# 9

# Transboundary water management

*Marton Krasznai*

## Key messages

- Narrowly defined national interests (e.g., irrigation versus hydropower) and lack of political will slowed down transboundary water cooperation in the quarter-century that followed the birth of independent states in Central Asia. Disputes over water inflict high economic costs on the countries and hinder environmental cooperation and climate change adaptation.

- Geopolitical and geoeconomic change puts increasing pressure on the countries of the Basin to end resource competition in order to open the way to closer cooperation and more effective promotion of their shared interest within broader Eurasian integration processes.
- Transboundary water cooperation must be placed within the broadest economic, political and strategic context. Participation of downstream countries in building large reservoirs in upstream countries would assure joint control and operation of strategically important regional infrastructure. Long-term regional strategies would more clearly and convincingly reveal shared interest than a focus on short-term issues, thus ensuring sustained support by governments and societies.
- Existing regional frameworks must either be reformed or replaced by new forms of cooperation in order to successfully translate political will into highly effective, integrated management of regional water resources.

## Introduction

There is hardly any other area of post-1991 regional cooperation in the Aral Sea Basin (ASB) that would offer a better opportunity to analyse the difficulties linked to the birth of the five countries in the region than water resources management. The political classes and societies of Kazakhstan, Kyrgyzstan, Tajikistan, Turkmenistan and Uzbekistan had to cope with an entirely new set of problems when water resources, earlier managed by the central authorities in Moscow, became the matter of inter-state relations and regional cooperation. Afghanistan, engulfed by an extended bloody conflict, remained a marginal participant in the management of shared water resources.

The economic slump that affected all post-Soviet states to various degrees after the dissolution of the Soviet Union and the difficult, often socially painful process of transition from centrally planned to market economy did not make it easier to find solutions to problems that inevitably emerged with the disappearance of a central authority and a central budget. Control over water has been an important pillar of state power for centuries (Menga 2017). The newly established states of Central Asia (CA) wanted to define their national interests and assume sovereignty over their resources. The early stage of nation-building was not an ideal moment for elaborating solutions to water problems that would have required mutual concessions or conceding part of their sovereignty to a regional organisation tasked with managing shared resources.

Water was a strategic issue for all five newly born states, as it played a crucial role in their social and economic development. Irrigated agriculture, covering more than 8.4 million hectares, uses 90 per cent of available surface water, contributes around 20 per cent to regional GDP and employs a high per centage of the total population (FAO 2012).

The Basin countries are connected by two large transboundary rivers, the Syr Darya and the Amu Darya. The Amu Darya riparians are Tajikistan and Afghanistan upstream and Uzbekistan and Turkmenistan downstream. The Syr Darya originates in Kyrgyzstan and crosses Uzbek and Tajik territory before flowing again into Uzbekistan and then into Kazakhstan. In two of the upstream countries, hydropower occupies a dominant place in the energy mix: in Kyrgyzstan almost 90 per cent and in Tajikistan over 90 per cent of electricity is produced by hydropower stations (HPPs) in the Amu and Syr Darya Rivers and their tributaries. Apart from these two major rivers, there are numerous smaller transboundary rivers in the ASB Basin, such as the Zerafshan, Chu and Talas Rivers (UNECE 2018). It is obvious that

Table 9.1 Water availability and use per country[1]

| Country | Internal renewable surface water per capita | Irrigation water as % of renewable resources | Population with access to improved water source | Agriculture contribution to GDP | Employment in agriculture | Storage capacity[2] (surface water) | Urbanization |
|---------|------|------|------|------|------|------|------|
| Units | (m3/cap/yr) | (%) | (%) | (%) | (%) | (million m3) | (%) |
| Kazakhstan | 3,722 | 13 | 93 | 5 | 42 | 12,000 | 53 |
| Kyrgyzstan | 8,385 | 30 | 90 | 15 | 40 | 20,260 | 36 |
| Tajikistan | 7,650 | 48 | 74 | 22 | 54 | 42,764 | 27 |
| Turkmenistan | 265 | 106 | 74 | 11 | 30 | 3,674 | 50 |
| Uzbekistan | 531 | 103 | 87 | 20 | 43 | 21,987 | 36 |

*Data*: (1) All data: 2016 World Development Indicators; FAO Aquastat (2013–2014 data); (2) Only reservoirs within the Amu and Syr Darya Basins, including ongoing construction in Rogun (TAJ), and comprising man-made structures and natural depressions

efficient and rational management of water resources in the ASB is possible only through close regional cooperation, including long-term strategic planning, which, until today, has remained beyond the reach of the governments.

## The evolution of regional frameworks for water resources management

As soon as CA states became independent and had to learn how to manage their economies under market conditions, competition for water between various sectors of the economy— irrigated agriculture, energy, communal and industrial use—became more acute. During Soviet times, constituent republics of the Union were expected to assure each other of their strong wish of brotherly (or at least friendly) cooperation[1] even if there were serious clashes of interests or economic or ethnic tensions among them (Lapidus 2018). This tradition continued in CA after 1991. However, below the surface of loud declarations of friendship, the growing number of problems accumulating in the water and energy sectors started to undermine bilateral and multilateral relations. Disputes over water became one of the most important stumbling blocks for regional cooperation in other areas.

The establishment and functioning of regional organisations for water resources management, like the International Fund for Saving the Aral Sea (IFAS) and its institutions, the Inter-State Commission for Water Coordination (ICWC), the Inter-State Commission on Sustainable Development, their secretariats and scientific and information centers, as well as the river basin organisations (BVO-s), reflect this political reality. Formally, an elaborated institutional structure for collaboration was set up.

Already in September 1991, the ministers of water resources of the newly independent CA states issued a statement in which they called for joint regional management of water based on the principles of equality and mutual benefit. In February 1992, the five countries signed the Agreement on Cooperation in Joint Management, Use and Protection of Interstate Sources of Water Resources (CAWA 2018). It kept the water allocation quotas among the riparian states, established by the Soviet Union in the 1980s, in place, and provided for the establishment of the Interstate Commission for Water Coordination (ICWC) for its implementation. The ICWC was the first regional institution set up after independence. Its members are the heads of the respective national water ministries or departments, who meet every quarter to agree on the exact water distribution. Its operative bodies are a secretariat in Khujand (Tajikistan), a scientific information centre in Tashkent (SIC ICWC) with branches in all member countries, and the two river basin organisations (BVOs), which had already been established by the Soviet government. The headquarters of the BVO Syr Darya is in Tashkent (Uzbekistan); the one of the BVO Amu Darya is in Urgench (Uzbekistan).

In 1993, the International Fund for Saving the Aral Sea (IFAS) was set up and all above-mentioned bodies were integrated into its structure. In addition, an Interstate Commission on Sustainable Development (ICSD) with a Scientific Information Centre (SIC ICSD) based in Ashgabat was established. The chairmanship of IFAS has rotated among the presidents of the five member states. The EC IFAS is accordingly located in the respective country. Thus, the Executive Committee has been located in Almaty (1993–1997), Tashkent (1997–1999), Ashgabat (1999–2002), Dushanbe (2003–2009; the planned move to Bishkek did not take place due to political turmoil in 2005), Almaty (2009–2012), Tashkent (2012–2016) and Ashgabat (2017–2019) (Diebold and Sehring 2012). Afghanistan, even though a major upstream riparian of the Amu Darya, is not involved in the cooperation arrangements—another legacy of the Soviet Union.

Most of the founding documents of the Fund are political declarations and statements expressing the intent to cooperate. The few legally binding agreements, ratified by parliaments, that constitute the legal basis of IFAS do not provide for real collective decision-making power and have not even established a modest operating budget for the Fund (UNECE 2010). The weak institutions and legal basis of IFAS were dictated by political reality: none of the member states was ready to accept collective decisions on the management of the strategic resources of water and energy, which would have constrained decision making at the national level. IFAS member states were unwilling to negotiate and agree upon any legally binding document that would have given decision-making power—at least on technical issues—to the Fund and its institutions. They insisted on decision-making by consensus by representatives of member states, even on simple procedural issues. The fact that the states were not able to agree on a permanent location of the Executive Committee (EC IFAS), accepting considerable financial and capacity-related losses with each relocation, indicates the low level of trust.

The effectiveness of decision-making by IFAS and its institutions was further weakened by the limited geographical scope of its mandate and by insufficient coordination among the Executive Committee and other institutions. The IFAS/UNECE study 'Strengthening the Institutional and Legal Frameworks of the International Fund for Saving the Aral Sea: Review and Proposals' (UNECE 2010) noted: 'Even though the 1999 Agreement included ICWC and ICSD in the structure of IFAS, the mechanisms for regional cooperation under IFAS require clarification. Having no representatives either from the energy or from the environmental sector, ICWC cannot coordinate the management of water resources effectively; in particular, the runoff pattern in the Amudarya and Syrdarya basins. The jurisdiction of its executive bodies, Amudarya Basin Water Organization and Syrdarya Basin Water Organization, covers only the middle and lower part of Amudarya and the middle part of Syrdarya. Although ICSD, whose scope of activities partially overlaps with that of ICWC, is formally a part of the IFAS structure, its activities are insufficiently coordinated with the regional organizations involved in water resources management'.

Nevertheless, the Fund has performed a useful role by providing a forum for consultations and some information exchange. In the case of the Inter-State Commission on Water Coordination (ICWC), the adoption of protocols on water releases increasingly served as loose guidelines rather than legally binding and enforceable decisions. In low-water years or periods of drought, protocols adopted by ICWC were often disregarded by upstream countries (Linn 2018). Concrete decisions and their implementation then became a result of informal (and usually bilateral) politics rather than of official agreements. The Committee was further weakened when in 2016 Kyrgyzstan suspended its participation in its work.

The entrenched positions of CA governments on cooperation on water resources management were dictated by their relatively narrow room for maneuver in the economic and social fields. The heavy dependence of most of the countries of the Basin on irrigated agriculture is well illustrated by data in the 2003 edition of the *Agricultural and Mineral Commodities Yearbook* (Lea 2003); as noted, 'In 1995 exports of cotton accounted for 67% of Uzbekistan's total export earnings . . . export of cotton fibre contributed some 27% of total export revenue in 2000'. Food security in all the countries of the region depended heavily on local agricultural production, as the disastrous consequences of the drought of 2007–2008 dramatically illustrated (Linn 2018). The Kyrgyz and Tajik economy depended almost entirely on hydropower for energy, as presented in Table 9.2.

The economic slump that followed independence—e.g., the Tajik GDP dropped by 50 per cent between 1992 and 1996 (Akiner and Barnes 1991)—did not allow for proper maintenance of irrigation canals and the accumulation of resources for investment in infrastructure that would have been necessary for the technical aspects of effective management of water resources at the regional level. A vicious circle developed: decaying infrastructure led to increasing losses of water, forcing governments to focus on short-term, national solutions rather than on joint, regional efforts that would have required lengthy negotiations and ensuring the cooperation of several partners.

Government policies on regional water resources management have been 'captured' by the sector that was considered the economically and socially most important for the respective country. In downstream countries this was irrigated agriculture, in upstream countries the energy sector. This made it very difficult to manage regional water resources in an integrated way, carefully and rationally balancing the interests of various sectors, like agriculture, energy, industry, communal use and taking into account environmental considerations too. The slow progress towards integrated management of water resources at the regional level was also hindered by the lack of regional organisations in other areas of cross-border cooperation—transport, trade, education or scientific research. It is more difficult to integrate broader economic considerations into water resources management when the region lacks the adequate framework for trade and economic cooperation. The ephemeral nature of the Central Asian Economic Union, which later became the Central Asian Cooperation Organisation (CACO 2018) demonstrated the difficulties of developing intra-regional trade and economic cooperation.

*Table 9.2* Fossil fuels and hydropower in the energy mix of CA countries and region (MW)

| Countries (regions) | Coal | Gas | Combined Cycle Gas Turbines (CCGT) | Fuel oil | Hydropower |
|---|---|---|---|---|---|
| Kyrgyzstan | 340 | 0 | 0 | 0 | 2,910 |
| North Kazakhstan | 13,618 | 846 | 0 | 0 | 1,734 |
| South Kazakhstan | 2,314 | 0 | 143 | 0 | 530 |
| Tajikistan | 0 | 198 | 0 | 198 | 4,706 |
| Turkmenistan | 0 | 4,536 | 0 | 0 | 9 |
| Uzbekistan | 480 | 9,659 | 0 | 480 | 1,420 |

Source: EBRD (2011)

In contrast, the Kazakh economy was driven primarily by rapidly increasing exports of oil, while 95 per cent of the Kazakh agriculture (by cultivated area) is rainfed—so dependence on the timely availability of irrigation water was important but to a somewhat lesser extent than in other downstream countries (FAO 2012). This had a visible influence on Kazakh policies. Kazakhstan has for decades been an active supporter of regional cooperation—as the proactive and successful Kazakh Chairmanship of IFAS between 2009 and 2012, the initiative to strengthen the legal and institutional basis of IFAS or repeated attempts to establish a water–energy consortium demonstrated. However, all these efforts ultimately failed due to the impossibility to reach a regional consensus.[2]

## Dealing with water: a quarter-century of accumulating problems

Disputes over shared water resources have for almost three decades remained a symptom of the challenges facing closer regional cooperation in CA. The costs have been staggering both in the water sector and in the economy as a whole. In reaction to a gradually collapsing Central Asian Power System and the failure of energy for water schemes, water was more and more often released by upstream countries in the winter period to generate badly needed electricity, thus reducing water availability in the vegetation period.

Due to the lack of resources and persistent objections by their downstream neighbours, upstream countries have been unable to build new hydropower stations to reduce the considerable unmet demand. As a consequence, they have suffered from repeated water, energy and food crises (Linn 2018). A vicious circle evolved: inadequate storage and generating capacity forced upstream countries to continue with winter releases that fuelled political tensions with downstream countries, strengthening their resistance to the idea of building new hydro-power stations—like Rogun (Tajikistan) or Kambarata I and II (Kyrgyzstan)—by their upstream neighbours. Regional energy trade has dropped by 90 per cent since the early 2000s, further worsening winter energy shortages in the upstream countries (World Bank 2014).

Both upstream and downstream countries suffered significant losses caused by the problems in regional water resources management. Insufficient water releases during low-water years or droughts negatively influenced yields, further reducing the profitability of irrigated agriculture. While statistics on crop yields in high-, medium- and low-water years in different CA countries are not always available or consistent, significant losses of productivity must have happened in low-water years according to relevant FAO studies (e.g., crop yield responses to water) (FAO 2012). Anecdotal evidence suggests up to 20-per cent reduction of yields of water-intensive crops (like cotton) in low-water years in downstream countries.

Upstream countries suffered serious economic losses due to unmet demand for electricity and power outages. According to the EBRD (2018), 'Electricity issues remained the top obstacle for Tajikistan . . . the share of firms experiencing power outages remained close to 60%'. Outages resulted in an almost 15-per cent loss of total annual revenue. In Kyrgyzstan 'the share of firms that experienced power outages increased to 72.9%, causing a 7% loss of total annual revenue' (EBRD 2018).

This negative trend has been reversed only recently. The efforts by President Mirziyoyev of Uzbekistan, who came into power after the death of former president Karimov—a firm opponent of regional cooperation—have already improved political relations with CA neighbours. Indicatively, in the energy field there was partial restoration of the Central Asia Power System (CAPS), which has considerably helped neighbouring countries to face seasonal deficits (AzerNews 2019).

The effective management of shared water resources after 1991 would have taken not only sustained political will but also significant investment in the maintenance and modernisation of national and regional water infrastructure and the introduction of modern agronomic techniques. CA countries inherited from the Soviet Union an agriculture with relatively low productivity. According to statistics by FAO and the World Bank, CA countries' water productivities (calculated as GDP in constant US$ equivalent divided by annual total water withdrawals) were among the lowest in the world after they became independent. In the case of Kazakhstan, it was 8.7, Turkmenistan 1.3, Uzbekistan 1.0, Kyrgyzstan 0.7 and Tajikistan 0.6 (World Bank 2018). For comparison, the water productivity is 42.1 in the Europe-Central Asia region, 28 in the Russian Federation, 24.4 in Turkey and 9 in Ukraine. Because of resource limitations and the slow spread of modern agronomic technologies, only modest improvements of water efficiency could be achieved.

Most of the CA countries undertook efforts to reform their systems of water resources management, introduce IWRM, improve water efficiency through the dissemination of modern agronomic and irrigation technologies and restore and modernise irrigation networks (FAO 2012). However, as in most cases, national programmes were underfunded; they have not been able to sufficiently reduce water withdrawals for agricultural use in order to mitigate competition among different sectors. The low profitability of irrigated agriculture hindered efforts to introduce market-based solutions to ensure the proper maintenance and modernisation of irrigation infrastructure. State budgets could not always provide sufficient resources to repair or modernise old irrigation canals and pumping stations. Modern technologies—like drip irrigation—are more expensive to install and operate than old-style furrow irrigation.

Previous studies have noted why efforts to improve regional cooperation could not have been really successful without national efforts to increase water efficiency, thus reducing

*Figure 9.1* An aerial view of the Amu Darya, a natural border between southern Uzbekistan and Afghanistan

Source: Andrew Potter/ Shutterstock.com

competition for water at the regional level: 'As population and industrialization increase, growing municipal and industrial water needs will compete with demands for irrigated agriculture. Increasing the efficiency of agricultural water use is essential for supporting rural livelihoods, producing sufficient food for the growing population and producing commodity crops that are important to the national economy, and continuing social and economic development' (USAID 2003).

Another reason for the slow improvement of water efficiency were efforts to achieve self-sufficiency and thus reduce mutual dependence. Even in cases when regional solutions were likely to produce better outcomes, national solutions were favoured. For example, water can be stored more efficiently in large reservoirs high up in the mountains than in shallow reservoirs in the desert with tremendous losses through evaporation and seepage. Still, instead of investing in infrastructure of regional use, more often than not national solutions were pursued, at times without careful feasibility studies to assess the long-term viability of these investments. 'Countries have unilaterally invested in additional infrastructure in order to increase self sufficiency in their water, agriculture and energy sectors' (Pohl *et al.* 2017). While governments made efforts to manage demand by improving water efficiency, they were swimming against the tide. Economic development and demographics contributed to a growing strain on regional water resources. The combined effects of climate change and the drying out of large parts of the Aral Sea—such as higher average temperatures, irregular weather patterns and earlier snowmelt—make it more difficult to manage shared water resources and ensure sufficient water flows in the growing season (Pohl *et al.* 2017).

The Scientific-Information Center of the Interstate Commission for Water Coordination in Central Asia (SIC ICWC) and the Kazakh Hydrometeorological Services, as well as several projects financed by international and bilateral donors, analysed the effects of climate change on water availability and prepared medium- and long-term forecasts and models. However, regional organisations and institutions—IFAS, ICWC and the Interstate Commission on Sustainable Development (ICSD)—lacked an effective mechanism to translate the results of scientific research and modeling into political decision-making. For example, low-level participation in ICSD or ICWC meetings made it difficult to push data and forecasts about climate change up to levels where decisions were made. As a result, transboundary water managers focused on resolving short-term problems, trying to cope, often unsuccessfully, with immediate challenges, like ensuring adequate water flows in low-water years or during droughts—without suggesting long-term solutions to a growing number of problems.

While there is still a lack of reliable data, considerable progress has been made by national as well as international research to provide a better data basis. However, there is reluctance by political decision-makers to actually use this for policy-making. Formation of a long-term vision for a water-secure CA and an agreed plan of action to develop regional solutions to looming challenges where national solutions are insufficient were never even discussed. Such plans would have required approval at the highest levels. The IFAS Summits of the Heads of States, which took place in April 2009 and August 2018, would have been the right place for this, but neither of them discussed the necessity to develop a long-term regional strategy on water resources management.

Climate action is going to be a defining feature of international cooperation in the coming decades. Downstream countries in the ASB, depending overwhelmingly on fossil fuels for their energy needs, would need to make increased efforts to shift to renewable energy sources. Not all CA countries were able to include meaningful reductions of $CO_2$ emissions in their nationally determined contributions put forward under the Paris Climate Agreement (NDC 2018).

It would require significant investment (including international assistance) to wean downstream countries from fossil fuels. An agreed, long-term regional water and energy strategy would offer an opportunity to increase the share of clean hydro-generated electricity in the energy mix of downstream countries. Large new reservoirs in upstream countries, if constructed after thorough environmental and social impact assessments and operationalised jointly, could help mitigate impacts of climate-induced flow variability and provide cheap, relatively clean electricity through a fully restored Central Asian Power System to downstream countries. The significant concessionary financing available for climate change-related investments would help ease financing problems for large-scale regional water infrastructure.

Several experts (UNDP 2018) argue that upstream countries should build numerous small HPPs, instead of trying to build huge reservoirs. Building small or run-of-the river hydropower stations has less severe environmental impacts and some other advantages: it does not require lengthy coordination with neighbouring countries, as small dams do not have any regulating capacity, and therefore do not influence flows. Building small dams would help provide electricity to far-away regions, like Gorno-Badakhshan in east Tajikistan, without investing too much into building new power lines. Building small HPPs would not put a huge financial burden on already indebted countries, as investment can be phased according to the actual financial situation of the country. Building large dams puts considerable financial burden on government, as the 2018 report of the IMF mission to Tajikistan warned: 'There are downside risks to the fiscal outlook owing to infrastructure projects . . . A growth friendly fiscal consolidation to reduce the risk of debt distress can be achieved by optimizing non-priority capital expenditures' (IMF 2018). At the same time, as experience of small HPPs built by international

*Figure 9.2* Nurek Reservoir in Tajikistan
Source: Ronan Shenhav, 2016

donors in upstream countries shows, these installations have serious limitations: the amount of electricity they produce is subject to seasonal change of river flows and some may freeze up in winter time, thus stopping the provision of electricity when it is most needed.

However, financial arguments are brought forward in favour of large dams. The World Bank-sponsored study 'Key Issues for Consideration on the Proposed Rogun Hydropower Project' has compared various Rogun dam design options with alternative means to meeting Tajikistan's energy needs. Its summary concludes 'The study first examines the combination of all possible sources – including import of electricity, new thermal generation from coal or imported gas and new hydro-generation – that would meet electricity demand in Tajikistan at the lowest cost (the "least-cost option plan") for 2020 through 2050 . . . Each of the nine Rogun scenarios included in the study meets demand at a lower cost than does any combination of the non-Rogun alternatives' (World Bank 2014).

For the two upstream countries, hydropower is not just about providing sufficient electricity to the economy and the population. Building large dams would create a significant number of jobs up front, and electricity exports to neighbouring countries and Pakistan would provide much needed income to the two least-developed economies of the region. Thus, they would contribute to economic and social stability of countries—an advantage small dams would not offer.

The question, therefore, is not easy to judge, since both solutions have their advantages. The key is the existence of trust and a well-functioning system of integrated water resources at the regional level. If neither trust nor reliable regional IWRM exists, the risks posed by large HPPs may indeed outweigh their advantages. If trust can be built and an effective regional system of IWRM developed, large HPPs may offer convincing advantages both to upstream and downstream countries.

The UN Sustainable Development Goals, to which all CA countries have subscribed, offer a good opportunity to develop a long-term vision for regional cooperation on water resources. Improved management of water resources would facilitate progress towards achieving the majority of these goals, including poverty reduction, food security, health, clean water and sanitation, affordable and clean energy, employment and economic growth, reduced inequalities, sustainable cities and climate action. The available financing mechanisms for SDG implementation would provide much-needed additional resources for investment in infrastructure and capacity-building.

## Emerging political will for cooperation in transboundary water management

After almost three decades of gradually accumulating problems in the area of transboundary water management, a window of opportunity to improve regional cooperation in the ASB is emerging as a result of a combination of regional and broader Eurasian geopolitical processes. The current warming-up of bilateral political ties among certain countries in the region is generating political will to resolve disputes and start improving regional water resources management. Geopolitical and geoeconomic change in the broader Eurasian space like China's Belt and Road Initiative and the Eurasian Economic Union offer not only risks but also new opportunities. Namely, increasing the profitability of irrigated agriculture and thus generating resources necessary to properly maintain and modernise water infrastructure, introduction of modern irrigation and agronomic technologies and thus the reduction of water use and improvement of water efficiency. The Kazakh press has already reported significant increases in high-value agricultural exports to China (*Astana Times* 2018). Accelerated economic development also reduces the relative weight of the export of water-intensive crops in the economies of the countries of the Basin, thus broadening the room for manoeuvre of economic decision-makers, e.g., the contribution of cotton to the total export revenue of Uzbekistan has dropped from 67 per cent in 1995 to below 10 per cent in 2017.

However, warming bilateral political ties and geoeconomic changes alone are not going to lead to improved regional water resources management. Governments need to put in place policies that maximise the positive effects of these changes. Equally important is that they agree on a broad, long-term strategy to jointly develop and manage water and energy resources in a rational and efficient way.

The development of transport corridors will not automatically make irrigated agriculture profitable or guarantee that increased income is used to maintain canals and modernise irrigation systems. Governments need to adopt a whole set of measures, including improvements to regulatory environments, targeted incentives like tax breaks and cheap credits, strengthening of the rule of law and and even targeted subsidies to facilitate the transition to less-water-intensive crops. Simultaneously, CA governments should continue their efforts to improve market access for their agricultural products. The negotiation and signing of Enhanced Partnership and Cooperation Agreements with the European Union are important steps in this direction.

Upstream countries that want to build large HPPs will need to further improve their economic policies in order to create a favourable investment climate, so they can attract the multi-billion-dollar investments necessary for these huge projects. The World Bank study 'Key Issues for Consideration. . .' emphasizes: 'International experience demonstrates the importance of transparency and financial soundness of the power sector in order to successfully undertake large scale projects such as Rogun. At present, Tajikistan's energy sector does not meet generally expected levels of institutional capacity and financial viability' (World Bank 2014). The transparency and financial soundness of the power sector in Kyrgyzstan is not better than in Tajikistan. Without the transparency and financial discipline dictated by partners within an international consortium, governments of upstream countries may find it difficult to meet financial obligations in the longer run.

## Conclusions

Increasing demand for water due to economic development and demographic growth and the effects of climate change and drying out of the Aral Sea will inevitably necessitate large investments in regional water infrastructure in the long run. As the authors of 'Rethinking water in Central Asia' remark: 'Flood control and water storage for instance are typically far more efficient in upstream, mountainous areas' (Pohl *et al.* 2017). They add: 'Given the significant inter-annual flow variability, only a multi-annual perspective can ensure appropriate management and hope to increase overall benefits for all riparians'. A multi-year storage/regulating capacity for the region would support stable economic development, food security and contribute to poverty reduction.

Developing a robust, modern regional water infrastructure—canals, pumping stations, reservoirs—would require huge investment. CA countries need to carefully examine available options for attracting investment into infrastructure of regional importance as below.

- Upstream countries may attempt to complete the building of large HPPs or start building new ones by borrowing significant amounts on international financial markets and try to pay it back by selling electricity abroad and at home. Tajikistan's issuance of 500-million-euro bonds in 2017 (*Financial Times* 2017) to finance the continuation of works on the Rogun HPS was a much-disputed step in this direction. Both Kyrgyzstan and Tajikistan are likely to face serious problems of debt sustainability even in the medium term if they chose this option. The IMF has already warned Tajikistan about the possibility of debt distress, should its external debt continue to rise (IMF 2018).

- Attracting investors from large neighbouring countries. The presidents of Kyrgyzstan and Tajikistan, when in Beijing on state visits, both spoke about attracting Chinese investment to build new hydropower stations. This solution is not without risks. Not all banks of the neighbouring countries meet international standards when it comes to responsible lending practices and transparency, therefore the risk of debt distress may arise if this solution is chosen.

- With the assistance of international financial institutions—like the World Bank, ADB or AIIB—form an international consortium, preferably with the participation of neighbouring countries. The World Bank paper 'Key Issues for Consideration. . .' underlines: 'Financing costs could be brought down significantly in the context of an international consortium approach. International experience shows prudent realization of such a large project usually involves shedding risk through equity participation of other countries, possibly including downstream riparians and commercial partners, and broad international support to help improve financing terms' (World Bank 2014). The participation of international financial institutions would help resolve the complex technical, legal and financial issues associated with the building and operation of large HPPs. The participation of downstream countries with more sustainable debt levels in an international consortium would significantly improve the feasibility of infrastructure projects from the financial point of view and provide reliable guarantees for the economically and environmentally responsible, coordinated operation of cascades of dams of regional importance. By having agreed to jointly build two hydropower stations on the Zeravshan River (Kazinform 2018), the governments of Tajikistan and Uzbekistan have demonstrated the feasibility and advantages of such cooperation.

Long-term solutions may prove elusive until regional water resources management becomes an integral part of sub-regional cooperation. Achieving efficient and rational management of water resources would require a long-term vision. The countries of the region are in the midst of profound geopolitical and geoeconomic changes. They need to strengthen their ability to jointly formulate and promote their shared interests. Their decision-makers should remember that the process of European integration, ultimately leading to the creation of the European Union, started with the establishment of the European Coal and Steel Community in 1951. The main objective of the Community was to prevent renewed competition among Western European nations for natural resources. Ending competition for the most precious resource in the ASB, water, would greatly facilitate closer cooperation among the countries of the region. A shared long-term vision for a water-secure region, to be implemented jointly with the help of efficient regional structures, could become the driver of sub-regional integration—like the European Coal and Steel community was in the middle of the last century, paving the way for the Rome Treaty and then the establishment of the European Union. The 'Nowruz meetings' of the five presidents could become the right framework to launch long-term strategic coordination to eliminate resource competition. Once trust among the countries of the region has been strengthened and they have agreed on a long-term strategic vision for water and energy cooperation, modernising existing regional organisations or replacing them with new coordination frameworks could be the next step.

# Notes

1 Statements by Soviet politicians of various levels often contained references to the brotherly friendship of Soviet citizens.
2 The issue of institutional and legal strengthening of IFAS till today remains on the agenda of various high-level meetings of the Fund.

# References

Akiner S and Barnes C (1991) *The Tajik Civil War: Causes and Dynamics.* www.c-r.org/downloads/Accord%2010_3The%20Tajik%20civil%20war_2001_ENG.pdf (accessed on 12 December 2018)

Astana Times (2018) Kazakh agriculture ripe for international investment, says official. *Astana Times.* https://astanatimes.com/2018/12/kazakh-agriculture-ripe-for-international-investment-says-official (accessed on 5 January 2019)

AzerNews (2019) Kazakh Energy Ministry: talks on restoring Central Asian Power System underway. *Azeri News.* www.azernews.az/region/135771.html (accessed on 2 December 2018)

CAWA (2018) *Interstate Council of the Republic of Kazakhstan, the Kyrgyz Republic, the Republic of Tajikistan, and the Republic of Uzbekistan. Council of Prime Ministers: Agreement between the Governments of the Republics of Kazakhstan, the Kyrgyz Republic, the Republic of Tajikistan, and the Republic of Uzbekistan on the Parallel Operation of the Energy Systems of Central Asia.* http://www.cawater-info.net (accessed on 5 December 2018)

Central Asian Cooperation Organisation (CACO) (2018) Info. www3.nd.edu/~ggoertz/rei/reidevon.dtBase2/Files.noindex/pdf/3/caco-info.pdf (accessed on 6 June 2019)

Diebold A and Sehring J (2012) *Water Unites. From the Glaciers to the Aral Sea.* Berlin: Trescher Verlag.

EBRD (2011) *The Low-Carbon Transition, Special Report on Climate Change.* www.ebrd.com/news/publications/special-reports/special-report-on-climate-change-the-low-carbon-transition.html (accessed on 5 March 2019)

EBRD (2018) *Business Environment and Enterprise Performance Survey (BEEPS).* https://ebrd-beeps.com/ (accessed on 12 December 2018)

FAO, Aquastat Survey (2012) *Irrigated Agriculture in Central Asia in Figures.* www.fao.org/3/i3289e/i3289e.pdf (accessed on 10 January 2019)

Financial Times (FT) (2017) Tajikistan enters bond market with $500 m. deal. *Financial Times.* www.ft.com/content/1dd9f200-93d5-11e7-bdfa-eda243196c2c (accessed on 5 January 2019)

International Monetary Fund (IMF) (2018) IMF staff concludes visit to Tajikistan. www.imf.org/en/News/Articles/2018/10/09/pr18386-imf-staff-concludes-visit-to-tajikistan (accessed on 5 January 2019)

Kazinform (2018) Uzbekistan & Tajikistan to build hydroelectric power stations. *Kazinform.* https://lenta.inform.kz/en/uzbekistan-tajikistan-to-build-hydroelectric-power-stations_a3359362 (accessed on 5 January 2019)

Lapidus GW (2018) *Ethnic Conflict in the Former Soviet Union.* Stanford, CA: CISAC. https://cisac.fsi.stanford.edu/research/ethnic_conflict_in_the_former_soviet_union (accessed on 5 December 2018)

Lea D (2003) *Agricultural and Mineral Commodities Yearbook.* London: Routledge

Linn JF (2018) The compound water-energy-food crisis risks in Central Asia. Brookings Institution. www.brookings.edu/.../the_compound_water_energy_food_crisis_risks_in_central_asia (accessed on 12 December 2018)

Menga F (2017) *Power and Water in Central Asia.* London: Routledge

NDC Registry (2018) Home page. www4.unfccc.int/sites/ndcstaging/Pages/Home.aspx (accessed on 4 December 2018)

Pohl B, Kramer A, Hull W et al. (2017) Rethinking water in Central Asia: the cost of inaction and benefits of water cooperation. www.adelphi.de/en/system/files/mediathek/bilder/Rethinking%20Water%20in%20Central%20Asia%20-%20adelphi%20carec%20ENG.pdf (accessed on 2 December 2018)

UENCE (2018) Transboundary cooperation in Chu and Talas river basin. www.unece.org/env/water/centralasia/chutalas.html (accessed on 10 January 2019)

UNECE (2010) Guide to public participation under the Protocol on Water and Health. www.unece.org/environment/water/2010 (accessed on 5 December 2018)

UNDP (2018) Home page. www.tj.undp.org (accessed on 4 December 2018)

USAID (2003) *Irrigation District Improvements in Uzbekistan.* United States Agency for International Development. www.usaid.gov/sites/default/files/acvfafullreptoct2003.pdf (accessed on 6 March 2019).

World Bank (2014) *Key Issue for Consideration on the Proposed Rogun Hydropower Project.* http://pubdocs. worldbank.org/en/339251488268367141/World-Bank-Note-Key-Issues-for-Consideration-on-Proposed-Rogun-Hydropower-Project-eng.pdf (accessed on 12 December 2018)

World Bank (2018) *Water Productivity in Europe and Central Asia.* www.worldbank.org/en/data/interactive/2018/02/22/eca-data-water-productivity (accessed on 2 December 2018)

# 10

# Local and national institutions and policies governing water resources management

*Manon Cassara, Jelle Beekma, Lucia de Strasser, Oyture Anarbekov,*
*Makhliyo Murzaeva, Sara Giska and Andrei Dörre*

## Key messages

- The countries of Central Asia (CA) are still transitioning from a centralized water management system inherited from Soviet times toward the establishment of national water institutions and policies. Common hurdles are met in all CA countries, such as that a multitude of entities are in charge of water resources management; implementation of water reforms is challenging in all CA countries; water resources management issues require continuous attention; local water management is a priority, but achievements are mixed as of today.

- While integrated water resources management (IWRM) is the main approach to undertake water reforms in each CA country, other objectives like the achievement of the Sustainable Development Goals (SDGs), the introduction of the green growth concept, the integration of climate change policies, and the adoption of the water–energy–food (WEF) nexus concept are also driving towards water reforms on a national level.
- Various initiatives have been taken for reforming the water sector in CA, like the gradual establishment of river basin management approaches in all CA countries, the WEF assessment of the Syr Darya River, the development of the green growth project in Kazakhstan for efficient water use, the creation of a national water information system in Kyrgyzstan, the engagement of development partners in each newly designed basin in Tajikistan, and the creation of a Multi-Partner Human Security Trust-Fund for the Aral Sea region (MPHSTF) in Uzbekistan.
- However, reforming the water sector goes well beyond adopting new policies and initiatives, updating the legislative framework, and building new institutions. The key challenges are the continuous and strong high-level political engagement of all CA countries to better coordinate water management at national and transboundary level, the active participation of stakeholders on basin level, and technical and financial support of the relevant authorities and organizations that are implementing the water reforms.

## Introduction

The countries of Central Asia (CA) have been in a continuous state of transition since their independence in 1992, shifting from the Soviet centralized economic system to identifying their own priorities. Early independence days were marked by a progressive degradation of essential infrastructure for water, sanitation, energy, central heating, and solid waste, due—among others reasons—to limited financial resources (World Bank 2011; Chatalova *et al.* 2017). This also impacted institutions in charge of managing and regulating natural resources and led to a gradual decline of their technical capacities.

The first part of this the chapter retraces this transition, which is characterized by the integration of different paradigms, including integrated water resources management (IWRM), the Sustainable Development Goals (SDGs), climate resilience policies, and the WEF nexus approach. The second part of the chapter provides key highlights for each country on the current institutional developments in the water sector. The study acknowledges the particularities of each country, taking into consideration the latest water policy reforms. The chapter looks also into the overall development of the water sector by focusing on the overarching institutions in charge of IWRM implementation on a national level. At local level, the study focuses on the inclusion of key stakeholders in river basin management and the operation of local associations (e.g., water users' associations, water consumers' associations) in charge of water supply in agriculture, which is the highest water-consuming sector in CA.

## Common water resource challenges in Central Asia

Although the institutional framework is different for each country, we can observe features and challenges that are common to all CA water management systems.

*A multitude of entities are in charge of water resources management.* The mandate and authority of water resources is spread among different ministries and agencies. Surface water resources management (quantitative and water infrastructure) is usually attributed to ministries or committees related to agriculture or energy. Hydrological monitoring (rivers, glaciers, precipitation, etc.) is under the mandate of hydro-meteorological executive bodies. Environmental functions are attributed to ministries, committees and agencies in charge of environmental protection. Groundwater management (quantity and quality) functions are usually the responsibility of executive bodies in charge of geological exploration and subsoil natural resources management. Drinking water supply and sanitary and epidemiological functions are assigned to the Ministry of Health. Prevention and response to emergencies and hazards in relation to water are under the responsibility of the ministries of emergencies or the Cabinet of Ministers as well as local authorities (UNECE 2018).

*Implementation of water reforms is challenging in all CA countries.* Legal frameworks for reforming the water sector have been established and updated regularly. However, the operational and implementing status does not always match the provisions of regulations; overlapping of functions, imperfect procedures (e.g., gaps in the national strategies, etc.), weak information systems and monitoring networks, lack of scientific and technical capacities of staff within the institutions, insufficient mobilization of the public budget, and poor coordination and exchange of information are among the challenges faced across all CA countries (UNECE 2011; World Bank 2018).

*Water resources management issues require continuous attention.* Following the collapse of the Soviet Union, socio-economic and political issues were given priority over environmental and health problems. The water and environment sectors, but also public health, have undergone

*Figure 10.1* Pumping water irrigation in north Tajikistan, Ferghana Valley

Source: Stefanos Xenarios

budget cuts that have reduced investments in infrastructure and maintenance, and degraded monitoring networks (UNECE 2011). Increased soil salinization and chemical pollution from misuse of fertilizers and pesticides has further degraded surface and groundwater quality, with repercussions on human health, ecosystems, and agricultural productivity (Bekturganov *et al.* 2016; UNECE 2011). Groundwater quality had been receiving less attention than surface water quantity until recent years (UNECE 2011). Many CA governments are increasing their interest in water quality and groundwater management due to the anticipated loss of freshwater volume in CA resulting from climate change and the increasing demand for human consumption.

*Local water management is a priority, but achievements are mixed as of today.* At local level, reforms in the agriculture sector and irrigation in particular were focused on privatization. In this context, users were requested to contribute to the operational and maintenance costs of irrigation and drainage networks. The users were organized in new ways (e.g., water users' associations (WUAs) in Tajikistan and in Kyrgyzstan, water consumers' associations (WCAs) in Uzbekistan), with various degrees of success. Despite slight differences between national legislations for WUAs/WCAs, these institutions are generally implemented from the top down, and defined as membership-based, non-governmental and non-commercial bodies maintaining agricultural water supply systems (Asryan *et al.* 2018).

In recent years, we can also observe a new policy direction on local and decentralized development within the CA countries (i.e., Strategic Development Plan 2018–2025 in Kazakhstan, 'on the announcement of 2018 the year of Regional Development' in Kyrgyzstan, Rural Development Program 2019–2021 in Tajikistan, and a Regional and Rural Program to be developed in Uzbekistan). While water (e.g., access to drinking water, agriculture, etc.) has a special position in the above programs, the implications at institutional level are not yet clear. Nevertheless, the regional policies accentuate the efforts to delegate duties and responsibilities on local level, including water resources management.

## Reforming the water sector in Central Asia

Institutions and policies established as part of the Soviet system have been used as a baseline for national and interstate water resources management since independence in 1991. The concept of IWRM has been introduced mainly by development partners for the management of transboundary water resources in CA. IWRM implementation in CA was perceived as a process based on the coordinated use of all available water sources (surface, ground, and return waters) within the hydrological boundaries of a river basin, involving all stakeholders in decision-making and integrating the interests of different sectors (environment, energy, industry, etc.). While IWRM has been used as the principal framework to reform the water sector in all CA countries, other platforms have been also developed in parallel to attain water sector reforms but also face the major challenges of climate change on the hydrological systems of the region.

The CA countries are committed to tackling climate change repercussions and to strengthening resilience measures. All CA countries have signed the Paris Agreement and are on track to develop national policies to combat climate change. The *Nationally Determined Contributions* (NDCs), which 'Embody efforts by each country to reduce national emissions and adapt to the impacts of climate change' according to the Paris Agreement on Climate Change (2015), are also prioritizing water actions as a tool for adaptation in the inter-sectoral NDCs. Focus is given on the institutions and infrastructure as the primary drivers to reach climate-related goals (GWP 2018). In parallel, the CA countries are pursuing the fulfilment of the Sustainable Development Goals (SDGs) and especially SDG 6, which focuses on water resources (UNWATER 2018).

The SDGs and climate resilience frameworks have been recently embedded in related polices in CA and are still in the process of integration at the national level (UNECE 2016). Of late, the green growth concept has received increasing attention across the CA region (e.g., the Concept for Transition to a Green Economy 2013 in Kazakhstan; the National Council for Sustainable Development 2012 in Kyrgyzstan, etc.). The integral assessment of water–energy–food systems (the WEF nexus) has also been recently introduced into CA. The WEF nexus is one of the latest trends for identifying cross-sectoral impacts and exploring feasible solutions toward efficient, equitable, and sustainable use of resources, applied at all levels of governance (UNSGAB 2014). In Box 10.1 an assessment by the UN Economic Commission for Europe (UNECE) on the Syr Darya River is presented.

---

### Box 10.1 Nexus assessment for the Syr Darya River Basin

UNECE carried out a WEF nexus assessment for the Syr Darya River, published in 2017, which is part of the UNECE Water Convention's program of work. The assessment fostered transboundary cooperation by identifying inter-sectoral synergies as well as determining technical and policy measures. The process engaged diverse expertise in an inter-sectoral dialogue aimed at identifying key issues and possible solutions. One of the main recommendations from the assessment was that the riparian countries take a progressive approach to transboundary cooperation. Even if driven by national security concerns or self-interest, some national policies can have beneficial effects on regional relations by reducing pressure on shared resources and ecosystems. From this perspective, national policies can become instrumental to enhance cooperation in the short term. In the medium term, more and better access to environmental and hydrological data is very much needed to carry out integrated analyses. In the long term, a mutual trust is needed to be built among CA countries for increasing cooperation especially on hydropower and agricultural matters. The WEF nexus approach is expected to bring new inputs by reinforcing and complementing efforts to implement IWRM in the region, namely by: the establishment of definitions and indicators for water, food, and energy security in the countries; the alignment of policies for water–energy–food security (which requires an understanding of the cross-sectoral impact of pricing schemes and subsidies); the investment in ecologically sensitive protection systems (water risk), in particular by advancing multi-purpose water infrastructure.

Source: UNECE (2017)

---

## The institutional landscape in the water sector in the Aral Sea Basin

### *Key highlights: Kazakhstan*

*Status of the water sector in Kazakhstan.* The first steps of Kazakhstan toward developing IWRM were translated into a 'National Plan for IWRM and Water Efficiency in Kazakhstan'. The plan was adopted in 2004 (supported by the Government of Norway, UNDP, DFID, and GWP), acknowledging IWRM as one of the main pre-requisites for sustainable development. The key objectives of the IWRM National Plan align with the 'Strategy of Kazakhstan 2030' on the 'Conservation and Rational Use of Water Resources for the Health and Well-Being of Citizens' (President of the Republic of Kazakhstan 2010). A new Water Strategy for 2020 has been also adopted, outlining key principles for (i) the rational use and protection of water resources in the basins of major rivers, (ii) provision of safe drinking water, (iii) sustainable management of specific water bodies (i.e., Balkhash, Aral Sea and Caspian Sea), (iv) the introduction of water-saving technologies and water monitoring (GWP 2014).

In parallel, Kazakhstan has been a pioneer among CA countries in launching in 2013 the transition to a 'green economy' as a new paradigm for sustainable development. The main objective of a green economy is to contribute to the long-term environmentally sustainable and inclusive economic development of Kazakhstan through the introduction of a modern environmental governance system, state-of-the-art water management policies and practices, enhanced environmental impact assessment procedures, and economic incentives for the sustainable use of water resources. Kazakhstan is undertaking major efforts to diversify its economy, which has until now heavily relied on the extraction of fossil fuels and minerals. With the objective to meet 50 per cent of its energy needs from alternative and renewable sources by 2050, Kazakhstan's Green Economy Plan appears to be very ambitious.

The Green Economy Plan aims at bringing closer environment, water, food, and energy aspects into a nexus implemented at the national level. The nexus approach has also enriched the National Policy Dialogues (NPD) for water policy in Kazakhstan, which are held once a year. The National Water Policy Dialogues (NPDs) have been adopted in Kazakhstan since 2012 as an interdepartmental platform for a common and coordinated water policy nationwide. The concept of NPDs was initiated by the European Union Water Initiative (EUWI) in 2002 as a transnational, multi-actor partnership to support water governance reforms. A joint project was funded by the EU, UNDP, and UNECE to 'Support Kazakhstan for the transition to a Green Economy model' for the period of 2015–2018 (EU 2018), as shown in Box 10.2.

---

**Box 10.2 Green economy in the Aral Sea region**

An EU/UNDP/UNECE joint project, 'Support Kazakhstan for the transition to a Green Economy model' (2015–2018) is being implemented in partnership with the Committee for Water Resources and contributes to the implementation of the green economy concept, with a focus on water resources and climate change, demonstrating practical solutions through pilot actions. The pilot action 'Demonstration of Oasis Irrigation in the Kyzylorda Region' aims at rehabilitating saline lands of the Aral Sea region for agricultural production. The project was implemented on more than 100 ha. Using alternative energy sources, about 5 ha of saline land have been already rehabilitated and drip as well as subsoil and sprinkling irrigation techniques were introduced. The project contributed to saving water and energy supply costs of up to 29 million KZT (~US$76,000).

Source: EU (2018)

---

*Overview of the institutional water set-up.* In the Republic of Kazakhstan, the Committee for Water Resources (CWR) is the overarching institution in charge of water resources use and protection measures. Operating under the Ministry of Agriculture, the key tasks of the

Committee are to (i) coordinate the implementation of the national water management policy and (ii) conduct state control over water management (Water Code Art. 37). The activities of the CWR are extended at local level through eight Basin Water Inspections (BWI), in charge of the eight established river basins in the country (Figure 10.2). The basin management approach was adopted in 2003 through the ratification of the new Water Code on water governance and management aspects. Articles 42 and 43 of the Water Code were devoted respectively to the establishment of River Basin Organizations (RBOs) and River Basin Councils (RBCs) by signifying the needs to restore and protect water bodies in the country.

The BWIs implement state management policy in the sphere of use and protection of water resources, ensuring environmentally sustainable conditions of water bodies and water resources within the basin. The responsibilities of the Committee for Water Resources were originally held by the Ministry of Water Resources and then devolved to a State Committee placed under various ministries. The transfers of duties to different ministries have weakened the Committee's role, which, progressively, due to budget cuts, lost much of its competent staff by downgrading the overall technical capacity of the institution (World Bank 2001). This also impacted the management of the BWI at local level, with a lack of technical and human capacity. In this regard, full implementation of its mandate remains difficult as of today.

The establishment of water users' associations (WUAs) has also come along with the Water Code and basin reforms in Kazakhstan. Many of them are operating with economic and technical difficulties. The questions of accountability and transparency have also been triggered for WUAs in Kazakhstan (Zinzani 2015).

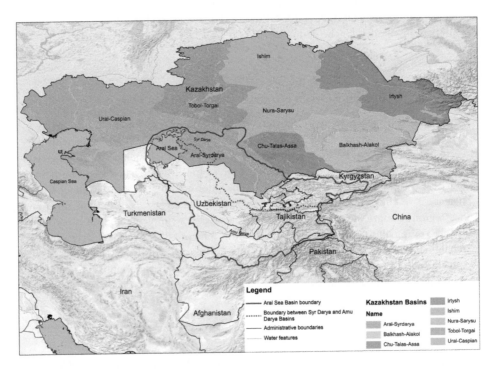

*Figure 10.2* River basins in Kazakhstan and renewable (internal and external) freshwater sources
Source: Constructed by Hamid Mehmood, UNU-INWEH, based on KRIG (2013)

## *Key highlights: Kyrgyzstan*

*Status of the Kyrgyz water resources sector.* Kyrgyzstan is endowed with quite a well-established legislative framework for water resources management. The Water Code was endorsed in 2005 based on IWRM principles, introducing basin management, decision-making transfer to the lowest appropriate levels through the establishment of WUAs, improving provisions for drinking water, dam safety, environment protection, and determining the economic value of water resources (GWP 2014). However, for a long time, Water Code principles have been poorly implemented across the country. Kyrgyzstan's National Water Council (NWC) was set up in 2012 for the implementation of the Water Code and the implementation of a roadmap in water policy. The roadmap outlined long-term (longer than ten years) and intermediate goals (for five years), activities, and actions. Within this roadmap, different projects were initiated in 2014, such as the National Water Resources Management Project, which is currently ongoing. This project is implemented by the World Bank and financed through a Trust Fund set up by the Swiss Development Cooperation (SDC). Since the beginning of the project, River Basin Councils were established, and River Basin Management Plans (RBMPs) were prepared for two basins. The project partners have also come out with ideas on further amendments in the Water Code for the inclusion of climate mitigation measures, environmental flows, and water quality standards (UNECE 2018).

In 2015, Kyrgyzstan also upgraded its regulations for groundwater, determining the use and protection of aquifers. At local level, the Decree 'On the Announcement of 2018 as the Year of Regional Development' was adopted as a primary document for the development of different policies, including water use. It is expected that the Decree will enhance the implementation of existent national programs like the 'Program for Providing Population with Clean Drinking Water' and the 'Irrigation Improvement Program' (Decree n-2377, 2018). Recently, Kyrgyzstan has cooperated with World Bank and the Swiss Development Cooperation to establish a National Water Information System for linking all the relevant agencies on water resources and land management within a digital information network.

---

### Box 10.3 The National Water Information System in Kyrgyzstan

The National Water Information System (WIS) is being developed as part of the National Water Resources Management Project (WB & SDC). The WIS will be a distributed database application, with text-based data and geo-referenced thematic map layers stored on different servers and linked together through a network system. The WIS will consist of a number of databases, including the National Water Cadastre, irrigation system passports, WUA information and water delivery data, as well as a water resources geospatial database for thematic map layers. A technical coordinator will set up to facilitate the development of this distributed database. A website will be created with different levels of secure access to different kinds of information. The main advantage of the portal application is that the agencies will be able to retain full control over data that are stored locally on their servers. The WIS will incorporate additional information on river discharges and discharge forecasts, weather, groundwater availability, and groundwater table elevations while documents, such as basin management plans and maps, will also be shared.

---

*Overview of the institutional water set-up.* Kyrgyzstan is the only country whose water resources are almost completely formed on its own territory. Kyrgyzstan has established two levels of management for water resources, separating policy and strategic issues from management and operation tasks. Established under the president, the National Water Council is the overarching institution in charge of coordinating the activities of the ministries, administrative agencies, and other state bodies responsible for water resources use and protection (Article 9, Water Code). It also prepares the National Water Strategy and supervises the activities of the State Water Administration. The Council comprises the heads of ministries as well as other relevant entities and is chaired by the Prime Minister of Kyrgyzstan.

The Department of Water Resources and Land Improvement (DWRLI) under the Ministry of Agriculture and Land Improvement is the State Water Administration and is responsible for the development, management, operation, and maintenance of water economy infrastructures and for the reliable and equitable delivery of water resources to different users. It is also responsible for implementing a single-state policy in the field of protection and management of water and acts as a secretariat for the National Water Council. While the implementation of Water Code is ongoing, River Basin Councils (RBCs) have been established and two RBMPs have been developed in accordance with Article 20 of the Water Code. The basin management plans are developed for a five-year cycle and then will be renewed by the National Water Council.

The DWRLI also has the mandate to provide water service delivery to all WUAs and farmers to meet the irrigation needs of their crops. Despite the fact that WUAs are not yet implemented throughout Kyrgyzstan and discrepancies exist among WUAs (i.e., capacities, resources, operating efficiency), DWRLI has set up clear policies on its operational status (contractual basis, appliance of the water fee to district administrations, transparency of WUA budgets, etc.). As of today, Kyrgyzstan has about 486 WUAs serving 70 per cent of the irrigated land, totalling about 166,000 members (South-South World 2017).

## Key highlights: Tajikistan

*Status of the water resources sector in Tajikistan.* In 2002, a new Water Code laid the foundations for an entirely new water management system, including modern approaches to governance. Indicatively, the involvement of the private sector and business principles for the provision of water-related services was adopted. An overall process of modernizing legislation and institutions has been launched since 2009 in Tajikistan. Reforms have been implemented in all sectors of governance, including water resources management. After the last amendment of Water Code in 2012, the basis for IWRM and river basin management was scheduled. Further, through the 2015 resolution 'On the Water Sector Reform of the Republic of Tajikistan for the Years 2016–2025', a nationwide water sector reform was initiated based on IWRM and basin management principles. Also, in 2017, the National Development Strategy 2030 (NDS) for Tajikistan was promulgated by also mentioning the need for a new Water Code in alignment with NDS principles. The Ministry of Energy and Water Resources (MEWR) is leading the overall water reform processes.

The water reforms that were mainly earmarked in the resolution of 2015 are being implemented in four out of the five basins that comprise the overall basin management in Tajikistan. The reforms in the fifth basin are scheduled for 2019 onwards. International cooperation is supporting the elaboration of basin plans and the formation of basin and sub-basin authorities and councils. MEWR is also in charge of coordinating and facilitating basin planning in the country. The first version of the basin plan in the Tajik part of the Syr Darya Basin has been finalized with the support of the Swiss Development Cooperation (SDC); three more basin plans are under development.

In recent years, the government of Tajikistan has also focused on water supply and sanitation issues, with the adoption of 15 programs, strategies, and plans of action to address drinking water sanitation and hygienic conditions across the country (World Bank 2018). These efforts are translated into the realization of public and donor-funded investments (e.g., rehabilitation of urban water systems and the installation of latrines, boreholes, and pumps in rural areas and small towns). Tajikistan is also engaged in meeting Sustainable Development Goal 6 (SDG 6), related to the water sector, by launching the plan International Decade for Action: Water for Sustainable Development (2018–2028). In regard to local development, a rural development program has been launched throughout the period 2019–2021 to improve livelihoods in the most vulnerable areas of the country. Box 10.4 outlines the water sector reform initiatives in Tajikistan in coordination with different development partners.

---

## Box 10.4 Water sector reform in Tajikistan and coordination with development partners

Tajikistan has a history of more than ten years of intensive and effective development coordination in reforms and is an example of well-coordinated development support in the water sector. The water sector reforms process is arranged by MEWR through a special coordination unit. This unit conducts regular meetings with various development partners where the strategy, progress, and milestones are discussed, monitored, guided, and adapted if necessary. The development partners have been requested to concentrate on different basins and focal areas of support. The development partners align their approaches to basin development and exchange experiences in a water sector group known as the Development Coordination Council (DCC). At present, the Swiss Development Cooperation (SDC) concentrates its support to the reforms in the Syr Darya Basin, the European Union (EU) in the Zarafshon Basin, the World Bank (WB) in the Kofernihon Basin, and the Asian Development Bank (ADB) in the Pyanj Basin. Support for the Vakhsh Basin has not been launched yet. It will most likely be supported by a combination of SDC, focusing on watershed management in the upper sub-basins of the Vakhsh, the EU, focusing on basin planning and institutional development, and the World Bank, focusing on infrastructure development and an irrigation information system. The European Bank of Reconstruction and Development (EBRD) and the World Bank also support domestic water supply in major cities, towns, and rural districts in close coordination with the local authorities, also coordinated by MEWR. The water reform experience in Tajikistan has demonstrated that political will and leadership are essential aspects for coordinating such complex processes.

---

*Overview of the institutional water set-up in Tajikistan.* The Ministry of Energy and Water Resources (MEWR) was established in accordance with the Decree of the President of the Republic of Tajikistan of 19 November 2013, replacing the Ministry of Land Reclamation and Water Resources. The MEWR is the key institution in charge of water and energy resources management, carrying out water and energy policy and the development of strategies and programs. In addition, MEWR is in charge of bringing water and energy resources under a single entity by demonstrating the strategic character of water resources for the energy sector—98 per cent of electricity in Tajikistan is generated through hydropower.

Provision of hydropower, irrigation water supply, and drinking water supply and sanitation are managed as services and are still fragmentarily supervised by different state organizations. Indicatively, the Geological Survey of Tajikistan is in charge of groundwater and the Committee for Environment is in charge of licensing water permits. Environmental issues are managed through independent local-level offices in most districts. These issues are addressed in the draft version of the new Water Code that will be finalized by 2020. The main points in the new Water Code will include the formation of state basin and sub-basin authorities under MEWR's supervision. Basin councils will be established by the basin and sub-basin authorities and will include the participation of the mainline departments, water users, and civil society organizations.

At local level, WUAs have been formed since the 1990s following the implementation of the first phase of Land Reforms in 1998–2000 through the support of different non-governmental organizations (NGOs) and projects. One of the main purposes was to operate, maintain, and use on-farm irrigation systems through an adequate and reliable water supply. The Regulation of the Government of Tajikistan #281 from 25 June 1996, on the 'Assertion of Regulation on Order of Fee Collection for Water Delivery Service to Water Users from State Water Management Systems' was decreed for this purpose.

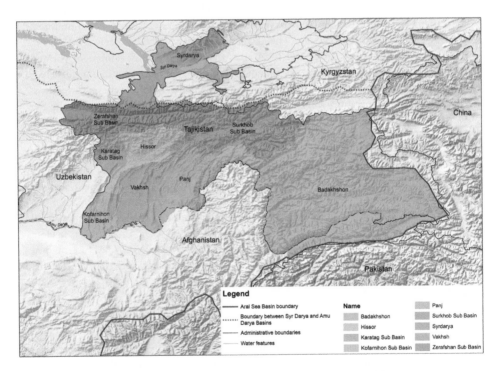

*Figure 10.3* River basin allocation plan in Tajikistan according to IWMR principles

Source: Constructed by Hamid Mehmood, UNU-INWEH, based on Government of Tajikistan Resolution No. 791 (2015)

WUAs were responsible for ensuring the optimal operation of the water sources within their jurisdiction for the benefit of their members. A WUA should exercise fair, effective, and timely distribution of water between farms, collect payments for the water supply, and settle disputes related to the distribution and use of water. In many WUAs the payment for water service delivery remains challenging. In small WUAs, there are still informal arrangements on the election of the board and management of the associations. There are also other organizations dealing with water management in rural areas, such as pasture management groups and catchment associations. However, the presence of numerous and different entities makes it challenging to follow a consistent, inclusive method on coordinating stakeholder participation on a basin level.

## Key highlights: Turkmenistan

*Status of the water resources sector.* Turkmenistan has been focused on reforming the water sector as part of a broader context of diversifying its economy. Water and food security have been on high priority in the government agenda for improving agro-industrial activities and introducing modern crop production techniques. The reform also includes the diversification of agricultural production as well as the introduction of financial instruments such as payments for water services and water pollution charges. Reforms were also conducted on an institutional level by merging the Ministry of Agriculture and the Ministry of Water Economy into a single institution in 2016: the Ministry of Agriculture and Water Resources. Development partners have also been active in the reform process, aiming at the introduction of IWRM principles.

National Policy Dialogues (NPDs) have been introduced in 2010 to support the revision of Water Code regulation. An inter-ministerial expert group was established, and a new Water Code was adopted in October 2016. A pilot project was launched in the Murgab River Basin in 2012 by a program on 'Water Management and Basin Organizations in Central Asia (WMBOCA)' implemented by the German Society for International Cooperation (GIZ) and financed by the European Union. The program aimed at supporting the Ministry of Water Economy (at that time) to introduce a basin-planning approach for the Murgab River Basin. This initiative was further continued as part of the Smart Waters Project,[1] which launched the River Basin Council (RBC) for the Murgab River. This RBC was the first management body in Turkmenistan, established on the basis of hydrographic principles, and aligning with the newly adopted Water Code.

*Overview of the water institutional set-up.* A system of administrative-territorial management of water resources is still prevalent in the Republic of Turkmenistan, except for in the case of Murgab River. The Ministry of Agriculture and Water Economy of Turkmenistan is the main executive body to implement changes in the national agricultural sector and enhance efficient use on water sources. It is also responsible for the construction and operation of irrigation and drainage systems, and the delivery of water to various users. Cooperatives and unions represent the land users (i.e., farmers, etc.), who are also involved in the local management of water resources. This is not, however, foreseen in the Water Code and the cooperatives currently decide arbitrarily on issues related to irrigation and drainage networks. The Water Code states that WUAs shall be established and provided with rights and responsibilities as an inherent part of the agricultural sector (e.g., irrigation system operating procedures, financial compensations, etc.).

## Key highlights: Uzbekistan

*Status of the water resources sector in Uzbekistan.* Since 2017–2018, the water sector has been experiencing a deep structural transformation in Uzbekistan. During this period, policy developments have triggered rapid institutional changes. The Presidential Decree from 17 April 2018 'On Measures for Improving the System of Public Administration Agriculture and Water Management' introduced an organizational restructuring within the original Ministry of Agriculture and Water Resources, which has been split in two separate ministries. The decree provides separate agriculture and water management policy directions to procure specific attention to both sectors as part of the overall modernization process undertaken by the government of Uzbekistan.

The Decree includes two resolutions pertaining to the organization of both ministries. Both include organizational schemes, guidance for the staffing of the ministries, and roadmap activities. The changes underlined in the Decree redefine the water sector vision for Uzbekistan, separating the operations for water resources management and agricultural productivity. No further legislative reforms have been planned at the moment within the sector. Nevertheless, new strategies and policies at national level are being developed in the extended water sector (agriculture, drinking water supply and sanitation, environmental flows) reflecting the need for integrating approaches. Food security remains a priority for Uzbekistan according to the functions outlined in the new resolution for the Ministry of Agriculture as (i) a focus on food security, (ii) development of the economic potential in the agriculture sector, and (iii) sectoral and territorial development programs. In 2015, Uzbekistan proposed the set-up of a Multi-Partner Human Security Trust-Fund for the Aral Sea region (MPHSTF) to alleviate the environmental impacts and improve local livelihoods, as presented in Box 10.5 and Figure 10.4.

---

### Box 10.5 Multi-Partner Human Security Trust-Fund for the Aral Sea region (MPHSTF)

In September 2015, at the UN Sustainable Development Summit in New York, the delegation of Uzbekistan submitted a proposal to establish a special trust-fund under the auspices of the UN for the Aral Sea region. With the funding support of the UNTFHS, the Multi-Partner Human Security Trust-Fund for the Aral Sea region was established. The Trust-Fund serves as a platform for mobilizing the international donor community's efforts and resources to improve the human security response framework for the Aral Sea region, focusing on environmental, economic, food, social, and health security for affected communities. The structural and operational scheme of the Fund is presented in Figure 10.4. On 22 November 2018, the governments of Japan, Nigeria, Norway, and Uzbekistan joined forces to organize a high-level event for mobilizing partners for the Trust-Fund. The government of Uzbekistan was committed to allocate about US$2 million, while members of the delegation of Norway confirmed the government's decision to allocate US$1.2 million. The Fund will prioritize the improvement of welfare in the most vulnerable and remote areas of the Aral Sea region through direct work with communities.

Source: UN (2018)

---

Attention is also currently given to water supply and sanitation; in 2017–2018, public funds totalling over UZ$ 1.2 trillion (~US$1.4 million) were allocated from the state budget to improve the delivery of utility services to the population. At local level, there

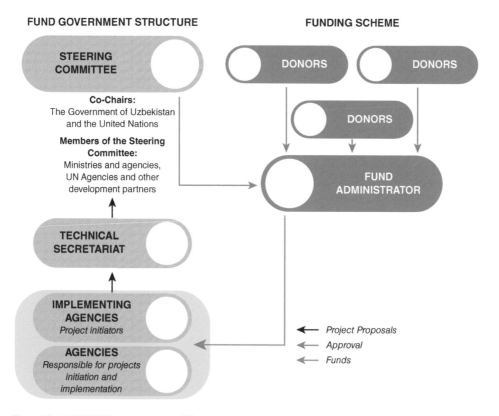

*Figure 10.4* MPHSTF governance and financing

Source: http://wecoop2.eu/sites/default/files/documents/VI%20EU-CA%20HLC_Jan%202019/1%20-%20Session%
20A%20Presentation%201%20Helena%20Fraser%20ENG.pdf

is a governmental effort to increase decentralization by also attracting foreign invest-ments in the water sector. Major foreign donors like the European Union and the Japan International Cooperation Agency are currently investing in rural development, water and environment security, as well as agricultural development.

*Overview of the institutional water set-up in Uzbekistan.* As of today, the Ministry of Water Resources of the Republic of Uzbekistan is one of the key institutions in charge of water resources management. In particular, and among other functions, it is responsible for (i) devel-oping and coordinating water resources management policy, (ii) the operation and maintenance of irrigation and melioration systems, reservoirs, pumping stations, and hydro-technical struc-tures, and (iii) transboundary water management and use (Government of Tajikistan 2018). In 2003, the Decree N. 320 of the Cabinet of Ministers of July 21 enforced the transition from administrative-territorial to basin principle management of irrigation systems. As a result, the Basin Irrigation System Authorities (BISA) were created. The BISAS are in charge—among others things—of providing forecasts on water use and water consumption, setting limits for water withdrawals from the main canals, controlling surface water resources, and main-taining the State Water Cadastre. The BISAs further include the Main Canal Management Organization (MCMO) and Irrigation System Authorities (ISAs).

The management of irrigation and drainage system on the local level is the responsibility of the Water Consumer Associations (WCAs), which are almost equivalent to WUAs in other CA countries. Similar to the problems faced by WUAs in CA countries, the WCAs in Uzbekistan depend on external funding provided by development partners, which makes their operating capacity and performance vulnerable (Zinzani 2015; World Bank 2016). In addition, Uzbekistan's agricultural and irrigation sector is characterized by strong governmental interventions to meet state quotas of cotton and wheat production. Although measures are being taken to diversify the agriculture sector by including higher-value crops and providing more flexibility for the farmers, top-down organization is still followed (Swinkels et al. 2016). The overall poor state of the irrigation system, which has suffered from substantial underfunding over the past 20 years, adds to the challenges and difficulties deriving from top-down organizational structures.

## Indigenous water management in Central Asia

The political discourse and the scientific debate on the water problem in CA are dominated by two focal points. The first one is the spatiotemporally unequal water availability, insufficient transboundary cooperation, and inadequate coordination of the interests of the states of the region, and multidimensional ecological challenges that are a consequence of the Soviet era and the current global climate change. The second is the addressing of the above complexities and their components, usually by means of external intervention measures and top-down politics. With regard to the actors, the focus is on multilateral bodies, national governments, authorities and administrations, international development organizations, and implementation institutions at the local level. From an institutional point of view, the focus is primarily on international agreements, national development plans, legislation and other formal regulations, and financing measures by external donors, as well as program packages and individual projects.

What has been overlooked in the course of this approach are indigenous forms of local and regional water management, which are well-adapted to the respective social and ecological processes and contexts because they are based on long-term experience and have been developed as non-formalized bottom-up approaches by communities, largely or entirely without external influences. Historically, self-organized water management approaches have been around for a long time and have been documented for different regions of the Aral Sea Basin in varying depth and detail (Le Strange 1905; O'Hara 2000). Some examples of these include: the Oasis of Merv in what is today Turkmenistan, which is fed by the Murghab River (Bartol'd 1965 [1914]); the middle Serafshan Valley in today's Uzbekistan (Radloff 1871; Rassudova 1987); the Ferghana Basin (von Middendorf 1882; Rassudova 1987) shared today by Tajikistan, Kyrgyzstan, and Uzbekistan; and small river oases in the Badakhshan Province of Afghanistan (Kussmaul 1965).

Autonomous approaches to resource governance and management were largely suppressed in the course of the Soviet modernization of society and the centralization of the national economy. After the dissolution of the USSR in 1991, and against the background of the post-socialist states' limited ability to act, resource governance and management approaches initiated by local communities can again be observed, especially in peripheral rural areas. Particularly impressive examples are the small-scale autonomous water management and irrigation arrangements in arid mountain regions of Central Asia, such as the Western Pamirs in Tajikistan.

As different as the local water management arrangements documented so far may appear at first glance, due to different natural conditions, social organization, and infrastructural endowment (Hill 2013; Bossenbroek & Zwarteveen 2014; Füssel & Schaddenhorst 2016; Shabdolov 2016; Dörre & Goibnazarov 2018; Dörre 2018), they share some fundamental characteristics. Developed from historically grown knowledge, these approaches are adapted to the corresponding socio-ecological contexts, starting with infrastructural installations adapted to the respective physical-geographical conditions, through patterns of water distribution among the users, reflecting the uneven spatiotemporal availability of the liquid resource, to the sharing of the material and immaterial costs arising from the operation of such arrangements, and the joint division of management responsibilities within the community.

Functional hierarchical systems of management responsibilities are typical for such arrangements. Corresponding to the flow of water, they start with the role of the water master, widely referred to as *mirāb*, who is responsible for the entire local arrangement encompassing both material (e.g., infrastructure and its maintenance) and immaterial (e.g., rules and regulations, and compliance with agreements) components. At the next level are the people responsible for individual (e.g., secondary and tertiary) canals, widely referred to as *mirjui* locally. These are followed by those responsible for the destinations of the water intended for consumption (e.g., irrigation, drinking, and other household purposes). A prominent example is the WUA 'Ob Umed' ('water is hope') of Porshnev Municipality (Shugnan District of the Gorno-Badakhshan Autonomous Province of Tajikistan), which was initiated in a grassroots effort led by local activists after experiencing previously unknown conflicts over water distribution in the course of the dissolution of the state farm and land privatization in the 1990s. Historical experiences and memories of the older inhabitants were used for the development of the inclusive and participatory irrigation water management approach of the WUA (MSDSP 2009), which was awarded the Equator Prize of the United Nations Development Programme (UNDP) in 2014 for its efforts 'to meet climate and development challenges through the conservation and sustainable use of nature' (UNDP 2016: 2).

Such indigenous arrangements are usually flexible enough to quickly adapt to changing circumstances. Due to the central importance of water for sheer existence and the broad participation of local communities, these arrangements are usually based on common decision-making, a basic consensus principle, as well as cooperation and mutual support amongst the water users. In this way, indigenous water management approaches not only contribute to securing the livelihoods of the members of rural communities, but also enable a widely accepted division of the considerable material burden and immaterial costs arising from the implementation of such arrangements. Being a means for social organization, they serve to balance interests and strengthen social cohesion within the communities in the socio-economically difficult period of national independence.

## Conclusions

The policy and institutional landscape for the water sector in CA is nowadays influenced by rapid transformations of the socio-economic and political context in the region. The meetings between CA countries are becoming more frequent. Indicatively, the heads of the five countries met twice in 2018, discussing multilateral and bilateral policies on water, environment, trade, energy, and other fields. At the same time, most of the CA countries have gone through tough economic times in recent years, which has entailed higher utilization and control of natural resources in their territory, including freshwater. The region also nowadays experiences an increase in global connectivity mainly through the Chinese 'Belt Road Initiative', which has increased the possibilities for cooperation and development, but which might also put additional pressure on natural resources. Afghanistan, as the most upstream riparian country on the Amu Darya River, is also increasing its participation in the discussions and activities of water management in the region.

Despite the desire of the CA countries to operationalize the concept of IWRM and even more the WEF nexus approach, the technical capacity of the involved institutions remains limited to the implementation of the relevant legal and institutional frameworks. Strengthening the water management agencies through training programs, technical assistance, staffing, and technical equipment are necessary for attaining IWRM objectives and fulfilling the ongoing reforms in each country. Decentralization of water governance on a basin level is still in its initial stage. Higher engagement of local communities and more powerful basin organizations are essential components for better basin and sub-basin management. Rural communities often dispose of versatile capabilities to meet the challenge of water abundance or water scarcity within challenging social and environmental contexts, which should always be examined before considering political interventions and implementing development measures. Political actors and development practitioners often lack detailed knowledge about already existing indigenous water management approaches, which enjoy legitimacy amongst the local population and are finely attuned to the socio-environmental conditions of the respective locality. This gives these approaches considerable potential for decentralized and participatory handling of the varied challenges of our time, which deserves greater consideration.

Overall, the CA countries have been proactive in defining the legislative and institutional framework for water resources management. However, the reform of the water sector is a long-term goal that demands the cooperation of local communities, several national authorities, and interregional agencies due to the transboundary nature of most freshwater systems.

# Note

1 The USAID-funded Smart Waters project was launched in October 2015 until September 2020, with a total budget of US$9.5 million, to enhance regional cooperation on shared water resources in CA.

# References

Asryan A, Baialieva G, Cassara M, Doerre A, Féaux de la Croix J, Menga F, Samakov A, and Weiss A (2018) Actors, approaches and cooperation related to water management and natural hazards under climate change in central Asia and the Caucasus. In: *Managing Disaster Risks and Water under Climate Change in Central Asia and the Caucasus*, September 2018, Khorog, Tajikistan.

Bekturganov Z, Tussupova K, Berndtsson R, Sharapatova N, Aryngazin K, and Zhanasova M (2016) Water-related health problems in Central Asia—a review. *Water* 8, 219.

Bartol'd V V (1965 [1914]) Kistorii orosheniya Turkestana. In: Belenickii AM (ed.), *V. V. Bartol'd. Sochineniya. Vol. III. Raboty po istoricheskoi geografii.* Nauka, Moscow, Russia, 95–233.

Bossenbroek L and Zwarteveen M (2014) Irrigation management in the Pamirs in Tajikistan: a man's domain? *Mountain Research and Development* 34(3), 266–275.

Chatalova L, Djanibekov N, Gagalyuk T, and Valentinov V (2017) The paradox of water management projects in Central Asia: an institutionalist perspective. *Water* 9, 300.

Dörre A and Goibnazarov Ch (2018): Small-scale irrigation self-governance in a mountain region of Tajikistan. *Mountain Research and Development* 38(2), 104–113.

Dörre A (2018) Local knowledge-based water management and irrigation in the Western Pamirs. *International Journal of Environmental Impacts* 1(3), 254–266.

European Union (EU) (2018) *Supporting Kazakhstan's transition to a Green Economy Model.* Retrieved 10 March 2019, https://eeas.europa.eu/delegations/tajikistan/47350/supporting-kazakhstans-transition-green-economy-model_en.

Füssel G and Schaddenhorst H (2016) *Understanding Community-Managed Hill Irrigation Systems in the Tajik Pamirs.* Lecture at the Workshop Persistence and Change of Institutions in Natural Resources Management in Central Asian Context, 26–29 January 2016. Humboldt University, Berlin, Germany.

Global Water Partnership (GWP) (2014) *Integrated Water Resources Management in Central Asia: The Challenges of Managing Large Transboundary Rivers.* Retrieved 10 March 2019, www.gwp.org/globalassets/global/toolbox/publications/technical-focus-papers/05-integrated-water-resources-management-in-central-asia.pdf.

Global Water Partnership (GWP) (2018) *Preparing to Adapt: The Untold Story of Water in Climate Change Adaptation Processes.* Retrieved 10 March 2019, www.gwp.org/globalassets/global/events/cop24/gwp-ndc-report.pdf.

Government of Tajikistan (GoT) (2015) Resolution No. 791, On the Water Sector Reform of the Republic of Tajikistan for the Years 2016–2025. 30 December 2015, Dushanbe, Tajikistan.

Hill J (2013) The role of authority in the collective management of hill irrigation systems in the Alai (Kyrgyzstan) and Pamir (Tajikistan). *Mountain Research and Development* 33(3), 294–304.

Kazakhstani Research Institute of Geography (KRIG) (2013) *National Atlas of Republic of Kazakhstan.* Environment and Natural Resources. Almaty, Kazakhstan.

Kussmaul F (1965) Badaxšan und seine Tağiken. Vorläufiger Bericht über Reisen und Arbeiten der Stuttgarter Badaxsan-Expedition 1962/63. *Tribus. Veröffentlichungen des Linden-Museums* 14, 11–99.

Le Strange G (1905) *The Lands of the Eastern Caliphate: Mesopotamia, Persia and Central Asia from the Moslem Conquest to the Time of Timur.* Cambridge University Press, Cambridge, UK.

MSDSP Mountain Societies Development Support Programme (2009) *Tavsiyanoma oidi istifodabarii samaranoki zamin va ob dar dzhamoati Porshinevi nokhiyai Shughnon.* MSDSP Dushanbe, Tajikistan.

O'Hara SL (2000): Lessons from the past: water management in Central Asia. *Water Policy* 2(4/5), 365–384.

President of the Republic of Kazakhstan (2010) *The Strategy for Development of the Republic of Kazakhstan.* Retrieved 10 May 2019, www.akorda.kz/en/official_documents/strategies_and_programs

Radloff FW (1871) Das mittlere Serafschanthal. *Zeitschrift der Gesellschaft für Erdkunde zu Berlin* VI, 401–439, 497–526.

Rassudova RY (1987) Semeinye Gruppy: Odna iz Form Organizacii Truda v Oroshaemykh Raionakh Srednei Azii (XIX – pervaya polovina XX v.). *Strany i Narody Vostoka. Geografiya, Etnografiya, Istoriya* XXV, 68–88.

Shabdolov A (2016) *Incomplete Reforms and Institutional Bricolage in Community-Based Governance of Mountain Irrigation Systems in Tajikistan: A Case Study in the Pamirs.* Michael Succow Stiftung zum Schutz der Natur, Greifswald, Germany.

South-South World (2017) *Water Users Associations (WUAs) in Kyrgyz Republic.* Retrieved 10 March 2019, www.southsouthworld.org/blog/46-solution/203/water-users-associations-wuas-in-kyrgyz-republic.

Swinkels, RA, Romanova, E and Kochkin, E (2016) *Exploratory assessment of factors that influence quality of local irrigation water governance in Uzbekistan* (English). Central Asia Energy-Water Development Program. Washington, D.C. : World Bank Group. http://documents.worldbank.org/curated/en/590421472098503155/Exploratory-assessment-of-factors-that-influence-quality-of-local-irrigation-water-governance-in-Uzbekistan

UN (2018) *The Multi-Partner Human Security Trust Fund for the Aral Sea Region: Advancing Regional and International Cooperation Towards Comprehensive Strategies in Support of Sustainable Development.* Background Note. Retrieved 10 March 2019, www.un.org/humansecurity/wp-content/uploads/2018/10/Final-FINAL-Background-Note-for-the-HLE-MPHSTF.pdf.

UNDP United Nations Development Programme (2016) *WUA "Ob Umed", Tajikistan.* Equator Initiative Case Study Series. UNDP, New York, USA.

UNECE (2011) *Development of Regional Cooperation to Ensure Water Quality in Central Asia – Diagnostic Report and Cooperation Development Plan.* Retrieved 10 March 2019, http://wedocs.unep.org/handle/20.500.11822/7533.

UNSGAB (2014) *Nexus Concept, The Nexus Approach Vs. IWRM – Gaining Conceptual Clarity.* Retrieved 10 March 2019, https://www.water-energy-food.org/news/nexus-concept-the-nexus-approach-vs-iwrm-gaining-conceptual-clarity/.

UNECE (2016) *Country-Level Needs for SDG Implementation in Europe and Central Asia.* Retrieved 10 March 2019, www.unece.org/fileadmin/DAM/sustainable-development/SDG-Needs-Assessment_RCs-and-UNCTs.pdf.

UNECE (2017) *Reconciling Resource Uses in Transboundary Basins: Assessment of the Water-Food-Energy-Ecosystems Nexus in the Syr Darya River Basin.* Retrieved 16 April 2019, www.unece.org/index.php?id=45042.

UNECE (2018) *Surface Water Quality Monitoring Needs Assessment.* Retrieved 10 March 2019, www. unece.org/fileadmin/DAM/env/Projects_in_Central_Asia/SURFACE_WATERS_QUALITY_ MONITORING_SYSTEMS_IN_CENTRAL_ASIA_NEEDS_ASSESSMENT.pdf.

UNWATER (2018) *SDG6 Synthesis Report 2018 on Water and Sanitation.* Retrieved 10 March 2019, www. unwater.org/publication_categories/sdg-6-synthesis-report-2018-on-water-and-sanitation/.

World Bank (2001) *Project Appraisal Document, Syr Darya Control and Northern Aral Sea- Phase1.* Retrieved 10 March 2019, http://documents.worldbank.org/curated/en/617961468752125207/pdf/multi0page.pdf.

World Bank (2011) *Central Asia – Hydrometeorology Modernization Project.* Retrieved 10 March 2019, http://documents.worldbank.org/curated/en/579861468242390986/pdf/593840PAD0P12010 only1910BOX0358354B.pdf.

World Bank (2018) *Glass Half Full: Poverty Diagnostic of Water Supply, Sanitation, and Hygiene Conditions in Tajikistan.* Retrieved 5 March 2019, https://openknowledge.worldbank.org/bitstream/handle/ 10986/27830/W17023.pdf?sequence=2&isAllowed=y.

von Middendorf AF (1882) *Ocherki Ferganskoi doliny.* Tipografiya Imperatorskoi akademii Nauk, St Petersburg, Russia.

Zinzani A (2015) *The Reconfiguration of Water Policies in Central Asia: A Reflexion on the IWRM Implementation in Uzbekistan and Kazakhstan.* Retrieved 5 March 2019, https://www.genevawaterhub.org/sites/default/ files/atoms/files/2015.07.07.pb_iwrm_centralasia_zinzani_eng.pdf.

# 11

# International actors and initiatives for sustainable water management

*Jenniver Sehring, Dinara R. Ziganshina, Marton Krasznai
and Thijs Stoffelen*

## Key messages

- International actors over the past 30 years have supported sustainable water management in Central Asia at various levels and in different fields—ranging from technical projects and infrastructure investments to institutional reforms, capacity development, scientific cooperation and political dialogue.
- Main actors include multilateral organisations like the World Bank, the Asian Development Bank, the Organization for Security and Co-operation in Europe and several UN entities as well as the European Union, Germany, Switzerland, Finland, the USA, Japan and other countries.

- The impact of these activities to date is modest—partially due to the nature of donor policies and projects, their lack of expertise in the region, application of blueprint models and insufficient coordination; and partly because the socio-political context in Central Asia (e.g., the lack of political will, hierarchical decision-making culture, capacity gaps) limits the effectiveness of international activities.
- International engagement can have more impact by (1) building upon and further developing local knowledge, (2) establishing a long-term programmatic perspective and (3) taking into account the technical-political interdependencies of water management in Central Asia.

## Introduction

During the past 30 years, international actors have had a keen interest and major role in promoting sustainable water management in Central Asia (CA). After the independence of the Central Asian republics, the Aral Sea catastrophe got worldwide attention, with experts warning about water wars in the region, and CA countries faced tremendous water investment needs at times of social, economic and political transformation and crisis. All these aspects together provided a fertile ground for both demand from CA for support to address water challenges in the region, and willingness of international actors to provide such support. This chapter gives an overview of the main areas international actors have focussed on, ranging from technical rehabilitation and investment projects to institutional reforms from local to regional level. It also discusses coordination and competition among donors and assesses the impacts of their engagement.[1]

The involvement of international actors in water management in CA started in the very last days of the Soviet Union, when the Aral Sea catastrophe was officially acknowledged, and when the Central Committee of the Communist Party adopted a resolution with measures for its restoration in 1988.[2] Based on this, the Soviet Union started in 1990 to cooperate with the UN Environmental Programme (UNEP) for conducting an international expert mission, and for the development of the *Diagnostic Study for the Development of an Action Plan for the Conservation of the Aral Sea* (UNEP 1992). It outlined measures to reduce cotton cultivation and restructure the economy in order to save water. When the study was presented in September 1992, however, the Soviet Union had already ceased to exist and no efforts to implement these measures were taken (Micklin 1998).

With the establishment of the independent CA states, their governments got cut off from the funding that Moscow used to provide. Facing immediate water management challenges, the countries had to source funding internationally (Weinthal 2002) and the involvement of international actors with the newly independent states began.

## The Aral Sea Basin Programme (ASBP)

Already in 1992, the CA governments signed the 'Agreement on Cooperation in Joint Management of Use and Protection of Water Resources from Interstate Sources' as the basis for the establishment of the Interstate Commission for Water Coordination (ICWC). Subsequently, the CA governments sought assistance from the World Bank. Anticipating a looming threat of water-related regional conflicts due to the transboundary nature of the challenges, the World Bank made the establishment of a regional institutional framework a condition for financial assistance to address the Aral Sea crisis. UNEP's initial study was followed by a World Bank mission in September 1992, a first donor meeting in Washington in April 1993 and a joint

World Bank/UNDP/UNEP mission in May 1993. The Diagnostic Study and the World Bank mission's report and recommendations were the basis for the development of the Aral Sea Basin Programme (ASBP). It was approved by the five CA countries in January 1994 and presented at a donor conference in June of the same year (Micklin 1998; World Bank, UNDP, UNEP 1994). It has been continuing until today and is now in its fourth phase. The CA states established the International Fund for the Aral Sea (IFAS) in 1993. Accounts on the extent to which the latter was a result of World Bank pressure or of national decisions and commitment of the states vary (Kirmani and Le Moigne 1997; Weinthal 2002; authors' interviews with key actors). In any case, support from the World Bank was significant for the initial phase.

In 1995, UNDP organised the Nukus Conference, where the seminal Nukus Declaration was signed by all CA presidents, the World Bank and the UN. The Declaration includes a commitment to regional cooperation and emphasises the need for an international convention on sustainable development of the Aral Sea Basin, requesting the UN to assist in the development of such a convention. It can thus be said that international partners backed up the countries' initiatives in signing new agreements and establishing regional institutions (Micklin 1998, Nukus Declaration 1995).

The ASBP was originally planned for three phases covering 15–20 years. It developed into a major coordinating framework among the CA states and international actors beyond the three original partners UNEP, UNDP and the World Bank, receiving support from ADB, the EU, UNESCO, the USA, Canada, the Netherlands, Switzerland and others (see Table 11.1). Donors are closely involved in defining the priorities and funding the implementation. Nevertheless, coordination and implementation of the ASBP has been all but smooth. Information on projects actually funded and implemented in the various phases is contradictory. Table 11.1 is an attempt to give a general overview, taking into account partly contradictory information.

While the World Bank was the leading international actor in the first years of independence, its engagement decreased after 1998, partly driven by the wish to strengthen regional responsibility. It remained an important donor, but left implementation and coordination to other agencies (Micklin 1998; Stoffelen 2017). During the late 1990s and the 2000s, many more actors got involved in water management at different levels.

## Main areas of international engagement

Many international actors engage to pursue the sustainable use of the region's water resources. In doing so, each of them has their own particularities and individual approach that, put together, targets a wide array of areas. These areas can roughly be grouped into four themes: (1) technical assistance and infrastructure investment, (2) political-institutional support, (3) improving data availability and (4) capacity development.

### *Technical assistance and infrastructure investment*

A major narrative in the discourse on water challenges in CA is about deteriorated or ill-designed water infrastructures being responsible for inefficient water usage. A water sector dominated by engineers has been pointing out that decaying infrastructure and damaged canals, gates and pumps are in desperate need of repairs or replacement. Donors have responded with support for rehabilitation projects for water infrastructure (irrigation systems, water supply, dam safety, etc.) and the modernisation of on-farm technologies in an attempt to increase water use efficiency.

Table 11.1 Phases of the Aral Sea Basin Programme (ASBP)

| Phase | Timeframe | Priorities | Funds (planned) | Donors |
|---|---|---|---|---|
| **ASBP-1 approved in 1994** | 1994–1996 preparatory phase | • Stabilising the environment in the Aral Sea Basin<br>• Restoring the disaster zone around the Sea<br>• Improving management of transboundary waters in the Basin<br>• Developing the capacity of the regional organisations to plan and implement the Programme | US$32 million for regional projects | USAID (US$7 million), the Netherlands (US$6 million), EU TACIS (US$7 million), World Bank (US$5.5 million), UNDP (US$2 million), other donors such as Canada, Finland, Switzerland, the UK, Italy, Denmark, Sweden, Japanese PHRD funds and the Kuwait Fund (US$6 million) |
| | 1998–2003 Water and Environmental Management Project, WEMP | | US$21.2 million | GEF (US$12.2 million), five Central Asian countries (US$4.1 million), the Netherlands (US$2.3 million), EU TACIS (US$1.4 million), SIDA (US$0.3 million), unidentified (US$0.9 million) |
| **ASBP-2 approved in August 2003** | 2003–2010 | • Water management<br>• Socio-economic development<br>• Ecology<br>• Environmental monitoring | Between US$2 million and US$1 billion (different sources), of which US$35 million by IFAS member countries | IFAS members, UNDP, World Bank, Asian Development Bank, USAID, Switzerland, Japan, Finland, Norway and others |
| **ASBP-3 approved in 2009** | 2011–2015 | • Integrated use of water resources<br>• Environmental protection<br>• Socio-economic development<br>• Improving institutional and legal instruments | US$15.4 billion, including US$15 billion for 501 national and US$386 million for 90 regional projects | National budgets (invested US$13.3 million for 374 projects as of 2016), World Bank, GIZ, ADB, SDC, EU, USAID (invested and/or committed to US$96.79 million as of 2016) |
| **ASBP-4 mandated in 2018** | Under development | • Integrated use of water resources<br>• Environmental protection<br>• Socio-economic development<br>• Improving the institutional and legal instruments | T.b.d. | T.b.d. |

Sources: Table constructed by authors based on Dukhovny *et al.* 2015, World Bank (1998, 2006), EC IFAS (2008, 2016)

*Figure 11.1* Germany's 'Berlin Process' includes political-institutional as well as technical projects, like this rehabilitation of an irrigation canal in Uzbekistan

Source: Shavkat Rakhmatullaev/ GIZ

While many international actors implement infrastructure projects as part of their assistance programmes (often focused on rehabilitation or maintenance), there are also commercial investment projects, in particular linked to hydropower facilities. Some examples are as follows.

- The Asian Development Bank (ADB) has implemented numerous projects for the rehabilitation of irrigation and drainage infrastructure and improvement of potable water supply systems with the aim of improving living standards and to support agricultural development (ADB 2018a, 2018b, 2018c).
- China, Russia and Iran have invested in the construction of new dams and hydropower plants, such as Sangtuda-1 and 2 in Tajikistan, Kambarata-1 in Kyrgyzstan and Moinak in Kazakhstan, as well as of many smaller ones (Abdolvand *et al.* 2015; Peyrouse 2007).

## Political-institutional support

While still engaged in technical assistance, awareness has risen among international actors that tackling water challenges effectively—in CA as well as generally—requires not only better infrastructure, but better management and governance. Therefore, beginning with agricultural restructuring in the 1990s, support of institutional reforms has been a major focus of donor policies. These range from local-level interventions (establishment of water users' associations) to the development and implementation of integrated water resources management (IWRM) and basin planning at local and national level, support of national water sector reforms and national legal frameworks up to platforms for transboundary dialogue and development of regional legal and institutional frameworks for joint water management. Some examples are as follows.

- From 2001 to 2012, the Swiss Agency for Development and Cooperation (SDC) with a long-running programme supported the introduction of IWRM and the establishment of water users' associations (WUAs) in the Ferghana Valley, implemented by the Scientific Information Centre of ICWC, in cooperation with the International Water Management Institute (IWMI) in the Kyrgyz, Tajik and Uzbek parts of the valley (see also below) (Swiss Agency for Development and Cooperation 2012).
- In April 2009, the CA Heads of States met for an IFAS summit in Almaty and, in their joint statement, expressed their readiness to improve its organisational structure. In the framework of Germany's Berlin Process, UNECE and GIZ supported efforts to reform and strengthen the institutional structure and legal framework of IFAS. Considerable progress was achieved in developing the respective documents and strategies, but in the end no consensus among all five IFAS member countries was reached (Ziganshina 2014).

### *Improving data availability and exchange*

One major obstacle for effective water management in the region is the lack of reliable data at transboundary level, aggravated by sensitivities in data sharing (Sehring and Diebold 2012). Along the rivers, many gauging stations are inoperable, inaccessible or have been removed completely. Besides that, there is no regionally accepted system to exchange hydrological data, and data sharing among the countries is highly contested. Sharing data and planning are often considered first steps in building up trust and initiating cooperation among riparian states, therefore they are also part of projects targeted at transboundary water management and diplomacy.

Some examples include the following.

- Since 2003, the Central Asia Regional Water Information Base (CAREWIB), which was financed by SDC and implemented by SIC ICWC, in cooperation with UNECE and GRID Arendal, has maintained an internet portal with information and databases on water and the environment in Central Asia, including up-to-date flow data (www. cawater-info.net). The CAWater-Info Portal embraces a knowledge base and regional information system (Figure 11.2). The main purpose of the information system is to build up a single system for accounting for the land and water resources in the Aral Sea Basin with the possibility of assessing the effectiveness of their use and to make flow forecasts. However, lack of data transparency, resistance of riparian countries to data sharing and lack of trust in the reliability of the information have limited the value of the portal. Since 2012, it operates without donor support and more efforts are needed to establish and improve communicative, informative and analytical tools at the local, basin, national and transboundary levels.
- One main component of the German Berlin Process is the regional research network Central Asian Water (CAWa), coordinated by the German Research Centre for Geosciences (GFZ) (CAWa 2018). Its objective is to contribute to a sound scientific and

reliable regional database for the development of sustainable water management strategies by, among other things, installing hydro-meteorological stations and conducting analysis of the impacts of climate change. Under this project, several tools for water-related monitoring and modelling were developed to support transboundary water management (e.g., WUEMoCA, MODSNOW).

## Capacity development

Capacity development in the water sector entails a variety of different elements. These elements together contribute to the overall objective of improving the skills, knowledge and tools to enhance the water management capacity of the regional countries, an important aspect to make technical and political changes sustainable. With this aim, international actors have initiated and supported many capacity development activities. Most of them are part of projects in the areas mentioned above, but there are also a few distinct programmes worth mentioning.

- In the early 2000s, the Canadian International Development Agency (CIDA) helped to revitalise the Central Asian water sector professional system under the umbrella of ICWC, which was taken up in 2009–2012 by the Netherlands (through then UNESCO-IHE). These two projects were instrumental in establishing a Regional Training Centre and

*Figure 11.2* The CAWater-Info Portal

Source: CAWa, www.cawater-info.net

maintaining regular contacts among water professionals and developing capacity in the areas of water resource management in a systematic and holistic way (CaWater-info.net, 2018).

- Since 2011, the German-Kazakh University (DKU) in Almaty runs an interdisciplinary regional master's programme on Integrated Water Management. Being part of Germany's Berlin Process, it also receives funding (student scholarships) from other donors, such as USAID.

## Priorities and policies of international actors

We have already mentioned several concrete activities of some international actors. Providing a comprehensive overview on all international actors engaged in fostering sustainable water management in Central Asia is beyond the scope of this chapter. We therefore present in more detail those actors and projects that are most relevant in terms of political leverage and financial contributions.

### *The World Bank*

One of the first international players in addressing the Aral Sea disaster and its region-wide roots and implications was the World Bank. Its impact and leverage over the Central Asian countries has been substantial from the onset of its involvement, as has already been described above. The World Bank was a driving force in mobilising support for the Aral Sea Basin Programme (ASBP). However, its own Implementation Completion Report (ICR) assessed the WEMP project (under ASBP-1) as 'unsatisfactory'. Identified weaknesses were internal ones such as poor financial management, implementation and leadership, but also insufficient participation and representation by the concerned countries and communities, lack of attention to regional and local peculiarities, and weak design of regional institutions. The main goal to reduce water use and increase conservation was not achieved (World Bank 2006).

Following its engagement with the ASBP, more recent commitments have been channelled through the World Bank's Central Asian Energy and Water Development Program (CAEWDP), a Multi-Donor Trust Fund (MDTF). The CAEWDP comprises knowledge development and technical assistance to build regional energy and water security by enhancing cooperation among the five CA countries. It is also supporting hydrometeorology services in Kyrgyzstan and Tajikistan, as well as their regional coordination. Since its inception in 2010, the programme has received funds from various donors, including the European Commission (EC), the United Kingdom's Department for International Development (DFID), the United States Agency for International Development (USAID), the Government of Switzerland's State Secretariat for Economic Affairs (SECO) and the World Bank Group (CA Water Future 2017). CAEWDP explicitly aims to support national priorities. It helps the individual countries to extract their resources while at the same time honouring regional requirements and sentiments. In doing so, the World Bank has scaled back its emphasis on regional cooperation. CAEWDP now aims for coordination across national borders, instead of full-scale cooperation, and for addressing the individual needs and interests of the respective countries first, to ensure that conditions conducive to regional integration will be cultivated.

A specific activity within CAEWDP was the independent assessment by international experts of the technical feasibility and the social and environmental impacts of the planned Rogun dam in Tajikistan. The World Bank decided in 2010 to finance and coordinate this in light of the raising political tensions stemming from this project (see Chapter 5). Being a highly controversial undertaking, it took four instead of two years until the results were released in 2014 after a consultative regional process (World Bank 2014). The studies found that, subject to design changes and mitigation measures, it was possible to safely build and operate a dam at the Rogun site. There are differing assessments of the value and impact of these studies; some authors (Kleingeld 2016; Stoffelen 2017) argue that the World Bank engagement helped to soften the tensions between Uzbekistan and Tajikistan. Others see the conflict primarily rooted in the personal relations among the two presidents and the World Bank being overwhelmed with the political repercussions. Consequently, the recent changes in bilateral relations and Uzbekistan's approach to Rogun are attributed chiefly to the change in Uzbekistan's leadership and not to the World Bank involvement.

## The Asian Development Bank

The ADB started its activities in the region in 1994, mostly targeting individual countries. The ADB's loans and technical assistance projects in the water sector included promotion of national water sector reforms, strengthening water management organisations, rehabilitation of key irrigation infrastructures, urban and rural water supply and sanitation as well as climate resilience/DRR (ADB 2018a, 2018b, 2013c). The ADB was also among the first donors to support specific gender-related activities on water, by supporting the establishment of the Gender and Water Network, hosted by SIC ICWC, in 2004–2010 (SIC ICWC n.d.).

During the Regional Consultative Meeting of Senior Officials of Central Asian Republics in August 2001 in Manila, the governments of the Central Asian republics formally requested the ADB's assistance for regional water management issues. The ADB responded by supporting high-level policy dialogues, forums and consultations on regional water management issues on the Amu Darya, Syr Darya and Chu-Talas Rivers (ADB 2003). Most prominently, in 2005–2008 the ADB supported the CA countries' negotiations on transboundary water issues under the umbrella of the ICWC with the aim of revising the 1998 Syr Darya Agreement (RETA 2007). Although a new Syr Darya agreement has not been signed, the process of bringing together all interested sectors from all riparian countries has illustrated that solutions can be found in joint exercise. Political approval, however, is still to follow. The ADB's most recent commitment to help strengthen the capacity of the Amu Darya River Basin countries for water balance and water productivity measurements did not materialise due to lack of timely response by EC IFAS.

## The United Nations Regional Centre for Preventive Diplomacy in Central Asia

The mission of the United Nations Regional Centre for Preventive Diplomacy in Central Asia (UNRCCA), established in 2007, is to enhance dialogue and to build confidence among the five CA countries as a response to the threats and challenges that these respective countries presently face (Draganov 2017). One of their three focus areas is the management of common natural resources. In view of the fact that uneven distribution and use of water resources fuel tension and present a basis of conflict, the UNRCCA's objective is to turn water resources into a basis for cooperation (UNRCCA 2015). The UNRCCA engages in political dialogue as an impartial third party and is mandated to liaise between the respective governments, to provide monitoring and analysis, to maintain contact with regional organisations and to facilitate information exchange and coordination of actions. Having access to the highest level of decision-making, the UNRCCA has facilitated high-level political dialogue on water and the development of two basin conventions for transboundary water management in the Amu Darya and Syr Darya, which, however, did not find support from all riparian countries.

## *The UN Economic Commission for Europe (UNECE)*

The UNECE's work on transboundary water management in CA is mainly built on the promotion and implementation of the UNECE environmental conventions. The UNECE Convention on the Protection and Use of Transboundary Watercourses and International Lakes (Water Convention), its Protocol on Water and Health, as well as other UNECE environmental conventions are an important foundation that guides national and transboundary developments in this field. The UNECE works both with parties to the Water Convention (Kazakhstan, Uzbekistan and Turkmenistan) to help them implement its provisions, and also with those parties that have not ratified the Convention (Afghanistan, Kyrgyzstan and Tajikistan) to help them apply its provisions and attract them to the Water Convention's family.

Over the years, the UNECE has been implementing projects to enhance regional dialogue and to strengthen the capacity of regional institutions for water resources management; to improve regional information exchange; and to support Chu and Talas River commission between Kazakhstan and Kyrgyzstan. It has also worked on issues related to water quality, dam safety, the water–energy nexus and strengthening cooperation on hydrology and environment between Afghanistan and Tajikistan in the upper Amu Darya (UNECE n.d.).

The UNECE also works together with the United Nations Economic and Social Commission for Asia and the Pacific (UNESCAP) to support the United Nations Special Programme for the Economies of Central Asia (SPECA), launched in 1998, and facilitated the process of developing a regional water and energy strategy within the framework of the SPECA.

## *The Organization for Security and Co-operation in Europe (OSCE)*

Being a regional security organization, whose mandate also includes tackling environmental aspects of security, the OSCE has been involved in promoting water cooperation and good water governance in Central Asia both at regional level as well as nationally through its field operations in all five countries. The OSCE sees itself as a platform for dialogue and regularly brings its Participating States, including all five Central Asian states, together for political dialogue and exchange of experiences and best practices. In addition, concrete project activities are implemented to foster transboundary water cooperation (e.g., between Kazakhstan and Kyrgyzstan on the Chu-Talas Basin or between Tajikistan and Afghanistan), to prevent local-level water conflicts (e.g., support of WUAs and basin councils) and to develop capacities. Afghanistan is sometimes also involved as a Partner for Cooperation of the organisation.

## *The European Union*

The combined assistance of the individual EU Member States and the European Commission makes the European Union one of the major donors in the region. Regional cooperation and the rational use of natural resources as a key factor for the development and political stability of the region is one of the priorities of EU engagement. Activities started in the mid-1990s under the larger Technical Assistance to the Commonwealth of Independent

*Figure 11.3* An OSCE training workshop on gender mainstreaming and conflict resolution in water
governance

Source: OSCE/ Hedda Femundsenden

States (TACIS) programme. Based on its 2007 Strategy for a New Partnership with CA, the
European Union has established an EU–Central Asia Platform on Environment and Water
with regular high-level conferences as well as working group meetings for senior officials.
At a national level, under the EU Water Initiative, National Policy Dialogues (NPDs),
jointly facilitated by the OECD and UNECE, have been implemented in Kazakhstan (since
2013), Kyrgyzstan (since 2008), Tajikistan (since 2009) and Turkmenistan (since 2010).
NPDs are platforms where key national stakeholders meet regularly to discuss and advance
policy reforms on IWRM, in some countries also including transboundary water management. However, the programme is also facing funding challenges, which affect its impact.

## Switzerland

Swiss cooperation in CA dates back to the early 1990s, focussing mainly on Kyrgyzstan,
Tajikistan and Uzbekistan. Throughout the years Switzerland has built its reputation as a con-
structive donor, with lasting commitments to introduce positive changes on the ground. The
involvement of Switzerland in water-related activities in these countries evolved around the
idea of integrated water resources management (IWRM) and was tailored to local conditions
and circumstances. IWRM in the Ferghana Valley project (2001–2012) has been implemented
by national teams from Kyrgyzstan, Tajikistan and Uzbekistan in partnership with the Scientific
Information Centre of the Interstate Commission for Water Coordination (SIC ICWC), the
International Water Management Institute (IWMI) and the Swiss Development Cooperation
(SDC). As a result, water withdrawals in the pilot area have decreased, agricultural productivity
has increased and water services have improved—consequently improving farmers' incomes
and food security for millions of rural people (Dukhovny *et al.* 2008). WUAs as well as canal

management organisations have been established along hydrographical borders. SDC also funded the establishment and maintenance of the Central Asian Regional Water Information Base and Portal (see above) and many other water management projects.

The Swiss Cooperation Strategy in Central Asia 2012–2015 (Swiss Agency for Development and Cooperation 2012) introduced a re-orientation of its Regional Water Resource Management programme that sought to complement achievements at the grassroots level by an enhanced policy dialogue, encompassing institutional changes and agricultural policy reform. At this stage, Swiss support at the regional level was mostly delivered through the CAEWDP (see above). In 2017, Switzerland launched the 'Blue Peace' Initiative for CA, which is based on the principles of informed dialogue, trust building and taking informed decisions. It is a regional effort that combines platforms for political dialogue at the governmental level, implementation of technical projects for data and knowledge management and training and networking among the young generation of water professionals.

## *Germany*

In 2008, the German Federal Foreign Office launched its water initiative for CA as an integral component of the 2007 EU Strategy for a New Partnership with Central Asia. The Berlin Process, as it became known, is an offer by the German government to the countries of CA to support them in water management and to make water a subject of intensified transboundary cooperation. The primary goal is to start a process of political rapprochement in CA that leads to closer cooperation on the use of water. It consists of three components and is meanwhile running in its third phase, which will end in 2019.

- The Transboundary Water Management in Central Asia Programme, which the Deutsche Gesellschaft für Internationale Zusammenarbeit (GIZ) is carrying out in close collaboration with national and international partners such as the United Nations Economic Commission for Europe (UNECE). Under that programme, measures have been implemented since 2009 at regional, bilateral and national levels aiming at institutional reform and political dialogue. The programme has also attracted co-funding of the European Commission.
- The regional research network Central Asian Water (CAWa), coordinated by the German Research Centre for Geosciences (GFZ) in close cooperation with the Central Asian Institute of Applied Geosciences (CAIAG) in Bishkek. Its objective is to contribute to a sound scientific and reliable regional database for the development of sustainable water management strategies by, among other things, installing hydro-meteorological stations, analysis of the impacts of climate change and application of remote sensing for monitoring of water use efficiency and overall water management (CAWa 2018).
- The regional master programme in Integrated Water Management at the German-Kazakh University in Almaty contributing to capacity development and education of the next generation of water experts in the region (DKU 2019).

The process also involved several high-level conferences that brought together political decision-makers from all five countries. Its combination of (small) technical investment projects at national level and capacity development with longer-term regional political and institutional processes promised to leverage political will be providing short- and long-term benefits (Sehring and Diebold 2012).

*Figure 11.4* Researchers of the CAWa network install high-altitude meteorological observation stations
Source: Abror Gafurov/ GFZ Potsdam

## Other countries and actors

*Finland* also has a long track record in water sector support to the EECCA countries under its Wider Europe Initiative (WEI), implemented since 2009. In CA, the work is focussed on Kyrgyzstan and Tajikistan and aims to enhance water security with a rights-based approach and includes components ranging from IWRM and transboundary cooperation to water quality and health, and to climate change adaptation (SYKE 2018).

In 2004, *Japan* started the 'Central Asia plus Japan' dialogue process, which addresses water and energy as well as water and environment issues among its priorities for intra-regional cooperation. Turkmenistan participates only as an observer in these regular political meetings. The Japanese Development Agency, JICA, has been funding and implementing projects on infrastructure maintenance, water supply and irrigation system maintenance, but also in support for institutional reforms, like the establishment of water users' associations, with Uzbekistan being the leading recipient in the region (Dadabaev 2016, 2018; Ministry of Foreign Affairs of Japan 2006).

*USAID* runs many water projects in the region, initiated already in the early 1990s by then Vice-President Al Gore, who had visited the region in 1990. An Aral Sea programme with a budget of over US$20 million was set up and run for a few years, but now continues at a much smaller scale (Micklin 1998). Currently, the biggest regional project is the Smart Waters Project, implemented by the Regional Environmental Centre for Central Asia (CAREC) in all five Central Asian countries and Afghanistan. It includes capacity building and academic exchange, networking among water organisations and water professionals and the promotion of IWRM and basin planning.

Russia, China and Iran also play a role in Central Asian water politics, mainly through investments in water infrastructure, in particular in dams. *China* has emerged as the dominant actor in the region's energy sector, including hydropower. China has financed several dams, and its engagement is expected to increase further with its Belt and Road Initiative. *Iran* and

*Russia* have also invested in hydropower infrastructure. Russia's investment plans in controversial projects like the Rogun dam in Tajikistan and the Kambarata-1 dam in Kyrgyzstan were cancelled due to political and financial challenges. A few more UN agencies and programmes are also supporting sustainable water management in Central Asia, like *UNESCO* and *UNDP.*

## Assessment of the impact of international engagement

International organisations, bilateral agencies, development banks and other development partners have provided significant assistance to Central Asian governments to achieve sustainable management of their water resources at the national and regional levels. The international engagement has to a certain extent contributed to technical improvements in water efficiency, improved data and progress in institutional and governance arrangements. However, nearly three decades of joint work have produced relatively modest results. The situation remains critical in many important areas of water resources management. Why has assistance from international actors not produced better results? The reasons can be traced back both to the CA countries themselves as well as to their international partners.

### *Competition and coordination among international actors*

There are many coordination efforts, collaborations and uses of synergies, e.g., through co-funding of activities and regular exchange through various platforms at national, regional and international levels. An example of effective donor coordination is the Kazakh IFAS Chairmanship in 2009–2012. In preparing the third phase of the ASBP, EC IFAS established close coordination with the donor community in developing the programme through regular meetings with donors and international organisations. International actors on their side established a Donors' Consultative Group for their own coordination, and several donors also assisted EC IFAS by providing experts (EC IFAS 2010). Some key donors also issued a joint statement in support of the Third Aral Sea Basin Program (ASBP-3), where they emphasised the need to raise donor coordination to a new level and called for a more prominent role of EC IFAS in this process through information exchange, research, monitoring and evaluation, according to its mandate.[3] Similar calls have been received from national and regional institutions. Nevertheless, the impact has not been sustainable due to the regular rotation of EC IFAS.

Designation of a 'lead' donor representing development partners in one particular sector is a common feature in many countries and regions. At the beginning of the 1990s the World Bank was playing this role in regional water cooperation in CA rather successfully (Dukhovny *et al.* 2016). However, already from the very beginning, there were discrepancies between UNEP and the World Bank in the preparation of the ASBP-1 regarding its thematic priorities (Sehring 2002). In 2010, the World Bank in partnership with the European Commission, Switzerland, the United Kingdom, and the USA initiated the already-mentioned Central Asia Energy-Water Development Program (CAEWDP). So far, however, the CAEWDP has not managed to align other development partners in driving forward a coordinated regional water and energy agenda.

Could the CA countries themselves take over this coordinating role? Central Asian countries, after gaining independence, were not fully prepared to manage donors or large projects, e.g., by setting up a regional organisation (IFAS and its institutions). The Aral Sea Basin Programme

*Figure 11.5* Coordination meeting between EC IFAS and international organisations dedicated to elaborate solutions concerning issues of the Aral Sea Basin (organised by GIZ TWMCA Programme), 9–10 May 2018, Ashgabat, Conference hall hotel 'Archabil'

Source: Caroline Milow

(ASBP) was intended to be the main instrument of development coordination in the region and also for linking donor activities with the national activities across different sectors to deliver development results. However, the management of the First Aral Sea Basin Program (ASBP-1) failed to do that (ICG 2002; World Bank 2006). Relatively large projects (e.g., by GEF) have produced few tangible improvements. As a result, ASBP-2 received much less support from international partners. The Kazakh Chairmanship of IFAS (2009–2012) made significant efforts to improve donor coordination (see above). In 2017, the SDC and the World Bank established monthly regional Water Partners coordination meetings, but both efforts remained by and large donor-driven processes with limited success. A good example for a country being in the driver's seat for coordination is the Water Sector Reform in Tajikistan, in which different donors take the lead for the implementation in the different river basins, while overall coordination is in the hands of the government.

## Blueprint models or long-term institution building?

International engagement has contributed to strengthening regional arrangements that were established by riparian countries like the International Fund for Saving the Aral Sea (IFAS), but also to reforms at national level, like the development of new water laws, the introduction of IWRM and the establishment of WUAs. Assessing these impacts based on the formal policy outputs, they might seem successful. Looking at actual implementation, ownership and informal processes, the picture becomes less rosy. Donor-driven organisations and policies in some cases lacked ownership. Consequently, organisations have only a weak mandate (like the IFAS, see Sehring 2002), or policies have not been effectively implemented, like new Water Codes, IWRM plans or WUAs (Sehring 2009; Veldwisch and Mollinga 2013; Wegerich 2008).

Successful projects in the region, like the Swiss IWRM project in the Ferghana Valley (see above), followed a comprehensive, context-sensitive approach. Experience has also shown that such projects cannot be simply replicated and upscaled in their technical elements without taking the full political and socio-economic context into account, but that all efforts need to address the interconnected issues of managing water and mobilisation of key stakeholders (Dukhovny *et al.* 2014).

At regional level, development partners play a crucial role in strengthening and supporting the efforts of existing regional water institutions such as the IFAS and the Interstate Commission for Water Coordination (ICWC). However, due to lack of effectiveness and transparency of regional institutions, donors have significantly decreased their support in recent years, or shifted it from one institution to another. Many partners moved away from implementing regional projects focusing on main rivers to local and bilateral projects on small rivers. On the one hand, this acknowledges the deficiencies and lack of truly regional bodies; on the other hand it might risk further weakening regional structures.

Still, at times when intercountry relations with regard to water had seriously declined, institutions like the IFAS and its various platforms helped to prevent conflicts. International actors are also the driving force behind many conferences and meetings that serve as platforms for political dialogue. Even though these conferences do mostly not produce binding agreements, they contribute to regional confidence building, the reduction of mistrust and the promotion of international principles and best practices.

### *International expertise versus local knowledge*

Related to this is an often-voiced critique of the excessive reliance of international consultants and ignorance of local knowledge and capacities. Many CA water professionals remember countless presentations by Dutch water experts on the famous Dutch flood management experience, which is, however, only of limited value in the Central Asian context. In the 1990s in particular, many international consultants were not familiar with the region and donors tended to ignore the importance of knowledge of the post-Soviet societies and cultures in order to plan meaningful interventions (Dukhovny *et al.* 2016; Sehring 2002), which was also one of the identified problems in the first phase of the ASBP (World Bank 2006). This also means that a considerable part of the project funding goes to international consultants and not to the countries themselves, which has significantly diminished the capacity of technical and research institutions that are not able to compete with international firms, while often similar reports are produced by the same consultants for different projects.

Also with regard to regional water cooperation, international actors have not always managed to take into account the obvious differences between Western and Central Asian political cultures and institutions. One example is UNRCCA assistance to implement the decision of the 2009 IFAS summit on the institutional and legal strengthening of the Fund. UNRCCA hired foreign consultants, who were all excellent experts with impeccable professional credentials, but who had little understanding of CA legal systems, traditions and political culture. They developed proposals, including on third-party arbitration, which were mostly based on North American legal practice. Unsurprisingly, CA experts opposed the presentation of such a proposal to high-level decision-makers. Subsequently, UNRCCA involved a new consultant with in-depth knowledge of the legal and political culture of CA. There are manifold examples like the above where donors prefer the engagement of consultants from their own country or region. They can bring in different knowledge and new ideas. But the eventual lack of understanding of historic, economic, political and cultural peculiarities of the receiving country or region might reduce the effectiveness of project implementation when it comes to improving regulatory frameworks, governance or regional cooperation.

## Is it really about water? Clashing and short-term interests

It also has to be noted that engagement in water was never only about water; right after the dissolution of the Soviet Union, Western actors entered the field with the aim to support the political and economic transformation and institution building. The new CA states, on their side, were looking for new partners and funding sources as an alternative to Moscow. The Aral Sea crisis provided a good opportunity to build trust in Western actors and attract money. However, only a small portion of the funds of the ASBP are actually spent for addressing environmental issues (6.8 per cent of ASBP-1); much more is spent on building institutions— showing the interest in regional stability rather than in environmental recovery (Sehring 2004; IFAS and UNDP 2008). Later, the war in Afghanistan, fears of regional destabilisation and geopolitical competition between the West, China, Iran and Russia were factors that guided the engagement of Western actors. In particular with regard to transboundary water cooperation, some international actors state clearly that their main reason for getting engaged is in promoting stability in the region, not in sustainable water management.

On the side of the Central Asian countries different interests have been at stake. After gaining independence in 1991, they engaged in the difficult process of nation building. One (perhaps inevitable) consequence of the efforts to build new nations was the narrow definition of national interest. Efforts by international partners to help Central Asian countries develop a cooperative, efficient and rational regional management of water resources were often stymied by national policies that focused only on the interest of one sector (in the case of downstream countries, irrigated agriculture, and in the case of upstream countries, hydropower) that was considered the most important for the national economy. The joint development and management of infrastructure of regional importance (like large reservoirs in upstream countries), promoted by international partners, would have required a broader definition of national interest, including the recognition that in certain cases regional solutions can be more efficient and produce better results.

A narrow focus on national interest was aggravated by the lack of a long-term vision shared by governments and their international partners. Major long-term trends in CA—increasing pressure on water resources due to demographic and economic growth, combined effects of climate change and desiccation of Aral Sea, including higher average temperatures, earlier spring runoffs and greater frequency of low-water years and droughts—all call for regional solutions. Shallow reservoirs in downstream countries with tremendous losses through evaporation and seepage will hardly be capable of mitigating the effects of climate change for long. However, international actors and their CA partners have never managed to forge a shared long-term vision, despite efforts in this respect (see, e.g., UNESCO 2000). A shared vision would have helped identify areas where regional cooperation offers better long-term outcomes than can be achieved only through national solutions. The recognition of the advantages of cooperation will help bridge political (and personal) differences.

## Political dynamics and timing of interventions

The initiatives and programmes proposed by international actors have sometimes suffered from problems of timing. Between 2009 and 2012 the GIZ and UNECE programme on the institutional and legal strengthening of IFAS had accomplished tremendous work by elaborating a legal basis for a modern and efficient organisation. Yet, at that time, due to the intransigence of President Karimov, the Uzbek government had blocked any real progress towards reforming IFAS. Today, President Mirziyoyev of Uzbekistan is undertaking a series of steps to improve political relations with the country's neighbours and one of the preconditions of Kyrgyzstan's return to active participation in IFAS is real progress on its institutional and legal strengthening. Yet international partners have not been in a hurry to come up with resources and expertise to support the successful conclusion of a programme that started in 2009. The political support stated by the CA heads of state at the IFAS Summit in August 2018 might help revitalise the program, but its impact and translation into actual policies still has to materialise.

Perhaps the most politicised and securitised issue of water and energy cooperation in CA is the Rogun hydropower station. Both Tajikistan and Uzbekistan have published numerous statements, articles and analyses on the project—in most cases with contradictory conclusions. In 2014 the World Bank published a Final Report on the analyses of technical, economic, social and environmental feasibility of the project. The study was intended to help narrow the gap between the positions of the Tajik and Uzbek sides, but overlooked the deeply personal-political nature of the conflict. The political climate between the two presidents at that time was less than favourable for serious, result-oriented negotiations between the two parties. Today, when, thanks to regular meetings of presidents, Tajik–Uzbek relations are rapidly improving, and even the topic of collaboration on the Rogun HPP is on the negotiation table, the World Bank (as well as other donors) displays a more cautious attitude to the project, supposedly due to problems it encountered when financing large dams in other regions.

Aligning mid- or long-term planning processes with short-term political changes is a challenging task. Nevertheless, even if projects might not have achieved their full impact due to an unfavourable political environment, they still—through capacity development, for example—provided a basis that served to be useful once the political climate changed.

## Conclusions

International actors have supported sustainable water management in CA at various levels and in many different fields—ranging from technical projects and infrastructure investments to capacity development, scientific cooperation and political dialogue on transboundary issues. They have done this with the aim of mitigating one of the world's worst environmental disasters (Aral Sea), but also with own interests in mind—sometimes competing, but also collaborating and trying to create synergies. The impact of their activities is modest, however. This is only partly related to the nature of donor policies and projects. Beyond that, the political context in CA also limits the feasibility and effectiveness of international activities. Reluctance to cooperate, lack of political will and hierarchical decision-making cultures in the CA countries are obstacles to comprehensive, inclusive and collaborative water management policies on which international actors have only limited impact. However, the scope of the environmental challenges in the region, and their potential political repercussions, make international support a continued necessity.

How can international engagement become more impactful? When looking at examples of donor interventions that are considered successful by both international actors as well as local partners, a few crucial factors can be identified: (1) building upon and further developing local knowledge, (2) a long-term programmatic perspective, and (3) taking into account the technical-political interdependencies of water management in CA. At the regional level, the current political dynamics in the region offer a new window of opportunity to foster water cooperation. When considering the lessons learnt in the past, international actors can provide meaningful support to the Central Asian societies on their way to a more sustainable future.

## Notes

1  The authors thank Bo Libert for his comments on an earlier version of this chapter.
2  Resolution of the Central Committee of the Communist Party and the Council of Ministers of the USSR No. 1110 of 19 September 1988 'On Measures for Fundamental Improvement of the Ecological and Sanitary Conditions in the Aral Sea Region, and for Increased Effectiveness in the Utilisation and Strengthening Conservation of Water and Land Resources in this Basin'. Available in Russian at: www.cawater-info.net/bk/water_law/pdf/ussr-1110-1988.pdf.
3  'Statement by the Donors and Implementing Agencies on the Occasion of the Presentation of the Third Aral Sea Basin Program (ASBP-3)', Almaty, Kazakhstan, 9 December 2010. www.unece.org/fileadmin/DAM/env/water/cadialogue/docs/Donors_9-15Dec/Donors_statement_Eng.pdf.

## References

Abdolvand, B., Mez, L., Winter, K., Mirsaeedi-Gloßner, S., Schütt, B., Tilman Rost, K., Bar, J. (2015) 'The Dimension of Water in Central Asia: Security Concerns and the Long Road of Capacity Building'. *Environmental Earth Sciences* 73:897–912.

ADB (2018a) 'Asian Development Bank and the Kyrgyz Republic: Fact Sheet'. www.adb.org/publications/kyrgyz-republic-fact-sheet (accessed 5 May 2019)

ADB (2018b) 'Asian Development Bank and Tajikistan: Fact Sheet'. www.adb.org/publications/tajikistan-fact-sheet (accessed 5 May 2019)

ADB (2018c) 'Asian Development Bank and Uzbekistan: Fact Sheet'. www.adb.org/publications/uzbekistan-fact-sheet (accessed 5 May 2019)

ADB (2003) 'Technical Assistance for Improved Management of Shared Water Resources in Central Asia'. www.adb.org/sites/default/files/project-document/70176/tar-oth-36516.pdf (accessed 5 May 2019)

CA Water Future (2017) 'Central Asia Energy-Water Development Program – Central Asia Water Future'. https://www.worldbank.org/en/region/eca/brief/cawep (accessed 19 July 2017).

CAWa (2018) 'Regional Research Network "Water in Central Asia" (CAWa)'. www.cawa-project.net/ (accessed 6 May 2019)

CaWater-Info.net (2018) 'Training'. www.cawater-info.net/training/ (accessed 6 May 2019)

Dadabaev, T. (2018) *Japan in Central Asia: Strategies, Initiatives, and Neighboring Powers.* Palgrave Macmillan, New York, USA.

Dadabaev, T. (2016) 'Water Resource Management in Central Asia, A Japanese Attempt to Promote Water Resource Efficiency'. *Journal of Comparative Asian Development.* DOI: 10.1080/15339114.2015.1115745.

Diebold, A., Sehring, J. (2012) *From the Glaciers to the Aral Sea: Water Unites.* Trescher Verlag, Berlin, Germany.

DKU (2019) 'Regional studies: IWRM'. https://dku.kz/en/content/programm-view/?id=51 (accessed 6 May 2019).

Draganov, P. (2017) 'Message from the SRSG | UNRCCA'. https://unrcca.unmissions.org/message-special-representative-un-secretary-general-and-head-unrcca (accessed 7 August 2017).

Dukhovny, V., Sokolov, V., Manthrithilake, H. (eds) (2008) 'Integrated Water Resources Management: Putting Good Theory into Real Practice Central Asian Experience'. SIC ICWC and GWP-CACENA, Tashkent, Uzbekistan.

Dukhovny, V., Sokolov, V., Ziganshina, D.R. (2014) 'Integrated Water Resources Management in Central Asia: The Challenges of Managing Large Transboundary Rivers'. Global Water Partnership Technical Focus Paper. www.gwp.org/globalassets/global/toolbox/publications/technical-focus-papers/05-integrated-water-resources-management-in-central-asia.pdf (accessed 10 May 2019).

Dukhovny, V., Sokolov, V., Ziganshina, D.R. (2015) 'The Role of Donors in Addressing Water Problems in Central Asia'. *Irrigation and Drainage* 65(Supplement S1):79–85. (Special Issue: Selected Papers of the ICID Gwangju Congress by Asian Authors.)

EC IFAS (2008) 'Report on the Activities of the Interstate Fund for Saving the Aral Sea in 2002–2008 (under chairmanship of Tajikistan)'. Available in Russian.

EC IFAS (2010) 'ASB Program'. http://ec-ifas.waterunites-ca.org/pbam/index.html (accessed 10 May 2019).

EC IFAS (2016) 'Report on the Activities of the Interstate Fund for Saving the Aral Sea in 2013–2016 (under chairmanship of Uzbekistan)'. Available in Russian at: www.cawater-info.net/library/rus/ifas/report_ifas.pdf (accessed 10 May 2019).

IFAS, UNDP (2008) *Review of Donor Assistance in the Aral Sea Region (1995–2005)*. Tashkent, Uzbekistan.

Kirmani, S., Le Moigne, G. (1997) 'Fostering Riparian Cooperation in International River Basins: The World Bank at its Best in Development Diplomacy'. World Bank Technical Paper no. 335. World Bank, Washington, DC, USA.

Kleingeld, E. (2016) 'The Rogun Dam in Tajik-Uzbek Official Discourse'. MSc Thesis, Leiden University, the Netherlands.

Micklin, P. (1998) 'International and Regional Responses to the Aral Crisis: An Overview of Efforts and Accomplishments'. *Post-Soviet Geography and Economics* 7:399–416.

Ministry of Foreign Affairs of Japan (2006) 'Central Asia Plus Japan Dialogue – Action Plan'. www.mofa.go.jp/region/europe/dialogue/action0606.html (accessed 10 May 2019).

Peyrouse, S. (2007) 'The Hydroelectric Sector in Central Asia and the Growing Role of China'. *China and Eurasia Forum Quarterly* 5(2):131–148.

RETA (2007) 'Improvement of Shared Water Resources Management in Central Asia'. www.cawater-info.net/reta/index_e.htm (accessed 10 May 2019).

Sehring, J. (2009) *The Politics of Water Institutional Reform in Neopatrimonial States: A Comparative Analysis of Kyrgyzstan and Tajikistan*. VS Verlag, Wiesbaden, Germany.

Sehring, J. (2004) 'The Role of International Organisations in Fostering Regional Cooperation in Water Management'. In: Seidelmann, R., Giese, E. (eds), *Cooperation and Conflict Management in Central Asia*. Peter Lang Verlag, Frankfurt, Germany, 237–249.

Sehring, J. (2002) 'Kooperation bei Wasserkonflikten. Die Bemuhungen um nachhaltiges Wassermanagement in Zentralasien'. Institut fur Politikwissenschaft, Abteilung Politische Auslandsstudien und Entwicklungspolitik (Dokumente und Materialien Nr. 30), Mainz, Germany.

SIC ICWC (n.d.) 'СЕТЬ «GWANET — Гендер и вода в Центральной Азии»'. www.gender.cawater-info.net/ (accessed 10 May 2019).

Stoffelen, T. (2017) 'Water Diplomacy, Neutral Brokerage or Institutional Entrepreneurship? A Central Asia Case Study on External Actors in the Rogun Dam Controversy'. MSc Thesis, Wageningen University, Wageningen, the Netherlands.

Swiss Agency for Development and Cooperation (2012) 'Swiss Cooperation Strategy in Central Asia (2012–2015)'. https://www.eda.admin.ch/dam/countries/countries-content/tajikistan/en/swiss-cooperation-strategy-central-asia-2012-2015_en.pdf (accessed 10 May 2019).

SYKE (2018) 'Programme for Finland's Water Sector Support to Kyrgyzstan and Tajikistan (FinWaterWEI II)'. www.syke.fi/en-US/FinWaterWEI_II (accessed 10 May 2019).

UNECE (n.d.) 'Projects in Central Asia'. www.unece.org/env/water/centralasia.html (accessed 10 May 2019).

UNEP (1992) *Diagnostic Study for the Development of an Action Plan for the Aral Sea*. UNEP, Nairobi, Kenya. https://searchworks.stanford.edu/view/4473262 (accessed 10 May 2019).

UNESCO (2000) *Water Related Vision for the Aral Sea Basin*. UNESCO, Paris, France.

UNRCCA (2015) 'United Nations Regional Centre for Preventive Diplomacy for Central Asia (UNRCCA) Programme of Action for 2015–2017'. https://unrcca.unmissions.org/ (accessed 10 May 2019).

UNRCCA (2017) 'United Nations Regional Centre for Preventive Diplomacy for Central Asia'. https://unrcca.unmissions.org/ (accessed 7 August 2017).

Veldwisch, G.J.A., Mollinga P. (2013) 'Lost in Transition? The Introduction of Water Users' Associations in Uzbekistan'. *Water International* 38(6):758–773.

Weinthal, E. (2002) *State Making and Environmental Cooperation, Linking Domestic and International Politics in Central Asia*. The MIT Press, Boston, MA, USA.

Wegerich, K. (2008) 'Blueprints for Water Users' Associations' Accountability versus Local Realities: Evidences from South Kazakhstan'. *Water International* 33(1):43–54.

World Bank (2014) 'Final Reports Related to the Proposed Rogun HPP'. www.worldbank.org/en/country/tajikistan/brief/final-reports-related-to-the-proposed-rogun-hpp (accessed 10 May 2019).

World Bank (2006) *An Independent Evaluation of the World Bank's Support of Regional Programs. Case Study of the Aral Sea Water and Environmental Management Project*. The World Bank, Washington, DC, USA.

World Bank (1998) 'Aral Sea Basin Program (Kazakhstan, Kyrgyz Republic, Tajikistan, Turkmenistan and Uzbekistan): Water and Environmental Management Project'. Project Document. documents.worldbank.org/curated/en/282981468768265004/pdf/multi-page.pdf (accessed 10 May 2019).

Ziganshina, D.R. (2014) *Promoting Transboundary Water Security in the Aral Sea Basin through International Law*. Brill Nijhoff, Leiden, the Netherlands; Boston, MA, USA.

# 12

# The future of water resources

*Saghit Ibatullin and Dinara R. Ziganshina*

**Key messages**

- Water is a key driver for food, energy, environmental security and social stability in the Aral Sea Basin (ASB). With the prospective economic and population growth of the basin countries, reliance on water resources will increase, urging the cooperation in jointly exploiting benefits and reducing costs.

- Piecemeal responses cannot address existing and future needs and challenges. Holistic and participatory approaches are needed to enable *identification* of workable and acceptable solutions through research and on-the-ground work and their *implementation* through engineering, institutional and other measures.
- The future directions on water resources management in the ASB must be built upon innovative thinking and best practices in the region and around the globe. The key areas for the future directions include (1) developing joint vision and strategic planning, (2) improving legal frameworks and institutions, (3) strengthening data, information and capacity, (4) promoting evidence based decision-making and water diplomacy, (5) harvesting the possibilities offered by infrastructure, technology and innovation, (6) enabling multi-sectoral and participatory governance arrangements at multiple scale, (7) paying more prominent attention to water quality and environmental degradation and (9) recognising multiple facets and values of water.

## Introduction

Water is a key driver for food, energy, environmental security and social stability in the Aral Sea Basin (ASB). Most of Tajikistan, Turkmenistan and Uzbekistan, over half of Kyrgyzstan, close to 40 per cent of Afghanistan and 10 per cent of Kazakhstan are located in the basin (Dukhovny and de Schutter 2011). Over 8 million hectares of irrigated agriculture contribute 20 per cent of GDP and employ 40 per cent of the total population. Forty-five large hydropower stations in the region generate 37 GWh/year of hydro-electricity that helps drive the economy. The fast-growing population and increasing demand for natural resources including water systems will be apparent in the following decades.

Climate change, population growth and continued environmental degradation are expected to exacerbate the situation. At the same time, there are also counterarguments stating that the rivers of the ASB can provide adequate water for irrigation, sustainable provision of ecological services of the delta lake systems, drinking water of adequate quality and all other uses if large-scale water protection and water saving measures are taken. To find solutions for ensuring the equitable and reasonable use of water and energy resources, a range of different measures are needed for environmental stability and food security for people in the Basin.

This chapter seeks to outline the main challenges and opportunities that countries of the ASB will face in their endeavours to live in a more secure and prosperous water future. In doing so, the chapter will draw information from the work conducted by an Expert Group (EG) commissioned by the World Bank in 2015–2016 on the future management of water resources in Central Asia (CA) with a time horizon of 2030. The findings of this EG are still relevant and up-to-date.

## Needs and challenges

### *Water for people*

Central Asia (CA), including Afghanistan, is home to more than 100 million people. Most are young and living along the main rivers or oases. About 60 million people—or 55 per cent of the total population—live in the Aral Sea Basin. The rural population accounts for 52 per cent in CA and 72 per cent in Afghanistan. The population has been growing at a rate of 1.3–2.7 per cent per year, with Afghanistan exhibiting the highest growth rates. Over the past 40 years, the population of CA has increased by 3.5 times. According to United Nations Department of Economic and Social Affairs (UNDESA 2017) projections, the population of the five CA countries will grow from 72 million today to 82 million by 2050, with 55.2 per cent living in urban areas. The population of Afghanistan is expected to reach 62 million by 2050. The rural population growth rate is expected to decline from the current 1.1 per cent to 0.46 per cent in 2025 and to 0.77 per cent in 2050. The most rapid population growth is expected to occur in Afghanistan, Uzbekistan, Tajikistan and southern Kyrgyzstan, while increases in Kazakhstan should be relatively small.

Population growth has already led to intensification of economic activities and increase in human-induced impact on water resources as well as reduction in water supply per capita. Over the past 40 years, the average water supply in CA has declined from 8,400 m³/person/ year in 1960 to 2,500 m³/person/year (Ibatullin 2009) (Figure 12.1). With the current rate of population growth in CA, by 2030 this reduction will be reaching a critical value of less than 1,700 m³/year. Drinking water consumption per capita in the five CA countries has decreased from 220 m³/person in 1990 to 106 m³/person. It is still necessary to annually provide an additional 500–700 million m³ of water to sustain the population of CA at even very low levels of consumption (Ibatullin *et al.* 2009). The domestic sector accounts for 5 per cent of total water use in the ASB. And yet, in CA, an estimated 8 million people remain without access to an improved source of water.

Over the past 20 years, the proportion of urban population increased by 10.2 per cent in Uzbekistan, by 5 per cent in Kazakhstan and by 2 per cent in Turkmenistan. At the same time, this indicator in Kyrgyzstan has decreased by 2 per cent and in Tajikistan by 9 per cent. According to a study conducted in 2013 by the Center for Economic Research based in Uzbekistan, for most of the CA countries, urbanisation challenges derive mainly from high

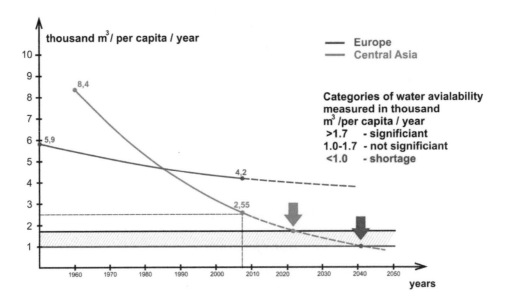

*Figure 12.1* Changes in water availability in the countries of CA

Source: Ibatullin (2009)

population growth concentrated in a few cities (Center for Economic Research 2013). These pressures are intensified by shrinking employment opportunities in rural areas due to limited land and water resources. Migrants from rural areas mostly settle in large cities, where urban housing, infrastructure and employment opportunities are not meeting their expectations. The urban infrastructure is straining under the demographic burden without the necessary fiscal support and investment. For example, at the current rate of population growth and urbanisation, most cities in Uzbekistan need to expand their water infrastructure by 2025 by 1.2 billion cubic meters of water supply per year, the replacement of 3,700 km of water pipes and 1,700 km of sewage pipes and increases of 200 per cent in urban waste disposal (Center for Economic Research 2013).

Another demographic threat that will need more attention is the population density, which is significantly higher in the basin (38.2 per km$^2$) than the average of CA (18 per km$^2$). Population density of Afghanistan has increased from 14 persons/km$^2$ in 1961 to 54.4 persons/km$^2$ in 2017 with an average annual growth rate of 2.53 per cent (World Bank Data 2018). The Ferghana Valley—shared by Uzbekistan, Kyrgyzstan and Tajikistan and located in the Syr Darya Basin— has the highest rural population density in CA, ranging from 300 to 660 persons/km$^2$. These trends illustrate that the effect of demographic factors on water security in this region will be even greater in the future. Population growth, along with rising incomes, changing dietary patterns, urbanisation and industrial development, will increase water demand but also carry us to a future that is characterised by growing competition for water for domestic, industrial, agricultural and environmental purposes.

## *Water for agriculture and energy*

Agriculture remains the biggest water user in ASB and plays a critical role in the economies of all Basin countries, with a share of GDP in Afghanistan of 21 per cent, in Tajikistan of 21.1 per cent, in Uzbekistan of 17.3 per cent and in Kyrgyzstan of 12.9 per cent (but with an amount of 22.3 per cent in the areas within the ASB) (World Bank Data 2018; CaWater-info. net 2018a). Agriculture also accounts for a significant share of employment in the region, with the highest per centage in Afghanistan (62 per cent) and Tajikistan (52 per cent). The FAO (2017) estimates that about 60 per cent more food will be needed by 2050 to meet the food requirements of a growing global population. It is also projected that irrigated food production will increase by more than 50 per cent by 2050. However, the amount of water withdrawn by agriculture can increase by only 10 per cent, provided that irrigation practices are improved, and yields increase (FAO 2017). Given that water from the Amu Darya and Syr Darya is nearly reaching its maximum potential, an increase in water demand for food can be met only through increased water and agriculture productivity (see Table 12.1).

The countries of the Aral Sea Basin are endowed with significant energy resources, which, however, are distributed unevenly. While the downstream countries have large reserves of oil, gas and coal, hydropower generation potential is high in the upstream countries. The main challenge in the allocation of water resources of the ASB is the contradiction between the irrigation regime in downstream countries (Uzbekistan, Kazakhstan, Turkmenistan) versus the energy-production regime in upstream countries (Tajikistan, Kyrgyzstan) (Dukhovny and de Schutter 2011).

Prior to 1991, in the context of a single-state (USSR) and planned economy, the inter-republican system of water distribution and exchange of electricity generated by upstream countries—Tajikistan and Kyrgyzstan—had functioned rather effectively (Dukhovny 2017).

Table 12.1 Water availability and use in the Amu Darya and Syr Darya Rivers

| | Afghanistan | Kazakhstan | Kyrgyzstan | Tajikistan | Turkmenistan | Uzbekistan | Total |
|---|---|---|---|---|---|---|---|
| *Water resources formation, km3/year* | | | | | | | |
| **Amu Darya** | 22.9 | | 1.60 | 49.9 | 0.12 | 4.82 | 79.22 |
| **Syr Darya** | | 2.78 | 27.6 | 1.00 | | 6.17 | 37.67 |
| **Total in ASB** | 22.9 | 2.78 | 29.2 | 50.9 | 0.121 | 10.9 | 117 |
| *Water withdrawals from surface water sources, km3/year (five-year average)* | | | | | | | |
| **Amu Darya** | 3 | | | 6.79 | 27.0 | 30.4 | 67.19 |
| **Syr Darya** | | 7.48 | 3.12 | 4.25 | | 21.8 | 36.65 |
| **Total in ASB** | 3 | 7.48 | 3.12 | 11.04 | 27.0 | 52.2 | 103.84 |
| *Irrigated lands, thousand ha (2017)* | | | | | | | |
| **Amu Darya** | 488 | | | 472. | 1553 | 2412 | 4925 |
| **Syr Darya** | | 727 | 432 | 287 | | 1879 | 3325 |
| **Total in ASB** | | 727 | 432 | 759 | 1553 | 4291 | 8250 |
| *Population, millions (2017)* | | | | | | | |
| **Amu Darya** | 5.48 | | | 2.56 | 5.02 | 15.86 | 28.92 |
| **Syr Darya** | | 3.65 | 3.24 | 6.18 | | 16.8 | 29.87 |
| **Total in ASB** | 5.48 | 3.65 | 3.24 | 8.74 | 5.02 | 32.66 | 58.79 |

Source: Ibatullin (2009); CaWater-Info.net (2018). Data on Afghanistan: expert estimation (water resources and withdrawals), Central Statistics Organization (2016–2017)

The reservoir hydropower systems built in the Soviet period mainly operated in irrigation mode, accumulating water in the autumn and winter period and emptying the reservoirs in summer for irrigation in downstream countries of Kazakhstan, Uzbekistan and Turkmenistan. On the other hand, the deficiency in electricity was replenished with energy supplies (coal, gas, oil) from the hydrocarbon-rich downstream countries. With the collapse of the Soviet Union and its centralised management of water and energy resources, this balanced scheme fell apart. To address the consequences of this disruption, countries made several attempts to create a mutually acceptable mechanism that would govern the use of transboundary water and energy resources. They produced significant research and data, involved leading CA and international experts and spent a large amount of funds; however, an effective mechanism is still to be developed.

In the future, we can foresee increased competition for water between countries in the region and aggravation of interstate relations on water-related matters, if coordinated and cooperative mechanisms are not in place. The countries of CA will need to find mutually acceptable mechanisms to enable a balanced use of water and energy resources of transboundary rivers that optimises the benefits for the region and maximises impacts on economic growth, employment and food and energy security. Coordinated actions across sectors, countries and even regions (such as CA and South Asia, for instance) should be combined by transformative change in agriculture and food systems as well as efficient water and energy policies at the national level. If current trends continue, environmental boundaries may well be surpassed.

## Water for ecosystems

Water-related ecosystems in the ASB are under increasing pressure from unsustainable use of resources and other threats including conversion of lands, pollution and expansion of infrastructure. The most extensive example of the environmental degradation in the basin is the shrinking of the Aral Sea, which has led to negative impacts on livelihood of the population, public health, large-scale desertification, salinisation, reduction of biodiversity and change of local climate.

The mountain ecosystems of the Pamir, Altai and Tien Shan in the eastern part of the ASB supply the main water sources of the entire basin. They are experiencing degradation processes such as deforestation and erosion, pollution, contamination with wastes and reduction of grazing lands. Since the middle of the last century, the forest area in CA has been decreased by four to five times already (ASBP-3 2013). Anthropogenic impact has been especially hard on Saksaul and floodplain forests (tugais). In the floodplain of the Amu Darya alone, the forest area over the past decade has shrunk from 150 ha to 22–23 ha. This process still continues to date. Degradation of floodplain forests comes with disruption to the hydrological regime of rivers.

The destruction of ecosystems has led to significant reduction of biodiversity. The number of species and plants that have disappeared or are endangered by extinction grows constantly. In some cases, these processes are irreversible. Future water and land management policies have to be more sustainable and environmentally friendly.

## Water quality and environmental degradation

Water quality and other environmental problems associated with water use still receive inadequate attention, although there are also encouraging examples. For example, Kazakhstan successfully implemented the Syr Darya Control and Northern Aral Sea project (World Bank 2012).

The construction of the Kok-Aral dam has prevented further desiccation and preserved part of the Northern Aral Sea at the volume level of 27–28 km³. In 2002, bifurcation of the delta of the Syr Darya was stopped. The first phase of the project brought some good results for the Kazakhstani part of the sea, with the volume of the Northern Aral Sea maintained at a level of 27.5 km³ within the last eight years and water salinity reduced by five times. As a result, more than 11 fish species were re-introduced and fish catches increased 10 to 12 times by attaining 7000 tons per year. Local population returned to the places of traditional residences. The second phase of the project is planned to increase the volume of the Northern Aral Sea by 1.5 times and bring the level of the lake up to the shores of Aralsk, the former port and fishing town. This will contribute to employment growth, economic revival and development of relevant infrastructure, benefiting not only the fishing industry, but also tourism purposes (Karlihanov and Ibatullin 2016).

The situation of the Large Aral Sea is less encouraging: its water level is low (23 m) and salinity very high (>200 g/l) (CaWater-Info.net 2018b). About 4.5 million ha of the seabed has become exposed and has turned into a vast desert—the Aralkum, which has changed the local climate and intensified desertification in the region. Uzbekistan has recently established afforestation projects along the former seabed to help stabilise soil. Interestingly, the vegetation cover has increased by half of the planted area due to natural regeneration.

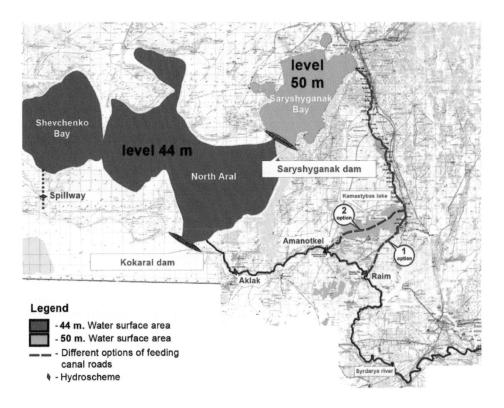

*Figure 12.2* Scheme for increasing the size of the Northern Aral Sea

Source: After Karlihanov and Ibatullin (2016)

Climate change is expected to have severe consequences for the Aral Sea Basin, as discussed in Chapter 9 of this book. The future of the Basin will heavily depend on how effectively CA countries will adapt to those consequences. More investments are need for research and measures that will enhance adaptation at the regional, national, watershed and farmers levels (Stulina and Solodkiy 2015).

At the recent IFAS summit in Turkmenistan, the president of Uzbekistan came up with several initiatives related to the Aral Sea (UzA 2018). He proposed transforming the Aral Sea zone (*Priaralie*) into a zone of ecological innovation, intensifying afforestation in the dry bed of the Sea and establishing protected areas in the *Priaralie* region. These innovative initiatives deserve widespread support.

## Valuing water

In 2017, a High-Level Panel on Water established by the UN Secretary-General and the World Bank Group President released its recommendations, whereby water valuation as a means for sustainable use and efficient allocation was mentioned (SDG Knowledge Platform 2017). Valuing water recognises and considers all the benefits provided by water that encompass economic, social and ecological dimensions. This concept takes a more multifaceted and multidimensional approach than the 1992 Dublin Principle on the economic value of water. In the ASB there are plenty of opportunities to reconcile and take into account the different values of water, both at national and regional scales. Economic incentives for better water management have to be developed at national levels. At the regional level, assessments of all benefits and costs of cooperation and non-cooperation are also essential.

The president of Kazakhstan, in his speech at the 2018 IFAS summit, reiterated the need to establish a water and energy consortium as a financial mechanism to balance water and energy interchange in the Syr Darya Basin (President of Kazakhstan 2018). This idea was also suggested in 2004, when CA countries approved the concept for the Consortium in IFAS; however, due to disagreements it remained unaccomplished. Due to lack of such financial mechanisms, the Syr Darya Agreement in 1998 turned out to be inoperative. An attempt to set up an International Water Energy Consortium for the Syr Darya and Amu Darya River Basins was made within the framework of the Eurasian Economic Union in 2007–2008. A draft blueprint for ensuring the efficient use of water and energy resources in CA was developed. Again, however, because of the lack of agreement between CA countries, this agreement was also not accomplished.

## Peace and security

The growing demand for water, combined with ecosystems degradation, water pollution and climate change, contribute to increased water insecurity across the globe that might undermine peace and security. In fragile contexts, water insecurity can heighten social disruption, intensifying conflicts and migration. The countries of the ASB have to continue working together to prevent conflicts over water. This will need political dialogues as well as technical-level interventions. The recent renewal of the political elite in CA seems to provide new prospects for the region. This can be demonstrated by the recently appointed president of Uzbekistan and his focus on supporting solutions in sensitive problems such as border delimitations and joint water management infrastructure (such as the Kasansai and Farkhod reservoirs).

Re-approachment with Afghanistan is an essential element of the water and peace pathway. The situation in Afghanistan raises concerns about threats to security for all the countries of CA.

By and large, Afghanistan has stood aside from regional water interactions in the ASB while several attempts have been made to build some connections. During Kazakhstan's chairmanship of IFAS in 2008–2013, it was proposed that Afghanistan be invited as an observer to the IFAS, which did not materialise due to lack of support by the CA countries, and probably also Afghanistan. Since 2012, Afghanistan and Tajikistan have cooperated actively on a bilateral basis on hydrology and environment-related issues in the upper Amu Darya, with support from UNECE, Russia and Finland (UNECE 2018).

Today the term 'water diplomacy' is often used in reference to water security and peace. Although some uses of the term are mere attempts to reframe old ideas under a new label, the real cause lies in the need for new policy dimensions and modes of peaceful interaction. Cooperation cannot entirely rely on technical and scientific approaches; policy- and stakeholder-driven approaches are essential for closer cooperation of CA countries in the area of water resources management (Ignatieff 2012).

It is of paramount importance for water diplomacy to enable continuous interactions at the professional level in order to produce knowledge and understanding, and to build durable relationships and a habit of cooperation. For instance, discussing the role of diplomacy for water security, Zeitoun (2012) focuses on knowledge, objectivity and the quest for sustainable water security as the essential elements that might assist policy-makers and diplomats in addressing the need 'to de-securitise and de-politicise environmental issues, without a-politicising them'. Throughout history, water has reshaped landscapes and boundaries as well as influencing national and foreign policies. Today water is pushing towards a reshaping in the mindset of CA decision-makers by acknowledging that water, peace and security are common goods, values and aspirations in the region.

## Legal frameworks

The CA countries have developed a fairly stable, although not perfect, legal framework of interstate cooperation in the management and use of transboundary water resources. It includes both the mandatory tools and numerous semi-formal arrangements and legal instruments, which are called 'soft law' instruments. The main agreements (Agreement on Cooperation in the Field of Joint Management of the Use and Conservation of Water Resources of Interstate Sources 1992; Agreement on Joint Actions for Addressing the Problems of the Aral Sea and Its Coastal Area, Improving the Environment, and Ensuring the Social and Economic Development of the Aral Sea Region 1993; Agreement on the Use of Water and Energy Resources of the Syr Darya Basin 1998) and declarations (Nukus Declaration 1995; Almaty Declaration 1997; Ashgabat Declaration 1999, Dushanbe Declaration 2002) have played a positive role and have laid the foundation for conflict-free regulation of water resources in the region. The relevant treaties and declarations are presented in the Annex to this chapter. In terms of geographic coverage, the existing system of international legal regulation of transboundary water cooperation is a two-level system, where, along with regional agreements of a more general nature, there are a number of bilateral agreements on practical issues relating to specific watercourses or areas of interaction.

However, these agreements could not solve all problems; given their disconnectedness, there was a failure to incorporate contemporary principles of international water law and—most importantly—failure to ensure that a compliance monitoring and detecting system is in place and working well. Transboundary water cooperation until today has revealed political, economic and legal weaknesses between CA countries.

The vision for the future development of the ASB should aim at ensuring sustainable water security in compliance with international law. New agreements might be necessary to develop a new consensus on the use of water resources in the Syr Darya as a replacement of the Agreement in 1998; a basin agreement on the Amu Darya; an 'institutional' agreement, which systematises and streamlines the work of regional institutions; an agreement to set up an international water and energy consortium as a financial and insurance mechanism; and an agreement to create a single regional information network on transboundary watercourses.

Such a framework would also contribute to the creation of a coherent multilevel legal framework for regional cooperation, which would include the following.

- At regional level, the participation of all ASB stakeholders, which lays down the basic principles of cooperation, general commitments and common institutional arrangements.
- At basin level, the participation of states sharing transboundary watercourses, which clarifies the special rules, procedures and institutional mechanisms.
- At bilateral level, where necessary, establishment of the specific mechanisms of cooperation on individual watercourses of interest to two states.

Agreements should be complemented and supported by detailed methodologies, standards and regulations to ensure the coordinated use of water resources in the region, in particular with respect to cost estimates for water management services; evaluation of ecosystem services; evaluation of damage due to non-compliance with mutual obligations; benefit sharing among the participants in joint water management projects; procedures and mechanisms

*Figure 12.3* The Kichi Naryn River in the Tien Shan Mountains in Kyrgyzstan, a source of the Syr Darya River

Source: Jakub Czajkowski/ Shutterstock.com

for making mutual payments for water management services; economic evaluation of water used by different economic sectors; and development of market conditions for capital assets, water management and maintenance services, design, engineering, construction, repair and other types of services. Also, it would be necessary to develop compliance monitoring mechanisms (Ziganshina 2014).

Interstate agreements have to trickle down to national level. Combination with sound national legislations and implementation practices, international norms and advanced water management practices can bring the desired change. Laws on the safety of hydraulic facilities, drinking water supplies and land reclamation, among others, need to be adopted in the countries that lack such regulations. However, the most significant drawbacks of the legal framework are the difficulties in ensuring its implementation.

For example, the Water Code of Kyrgyzstan adopted in 2005 has not yet been put into effect because of the lack of human, administrative and technical resources. Other CA countries are facing similar problems. It is necessary to enhance legal conditions to promote the comprehensive and efficient use of land and water and the protection of ecosystems. Laws on water users' associations (WUAs) need to be adopted or updated where they exist—in order to ensure more efficient water management at lower levels. To ensure the sustainability of WUAs, economic conditions should be created to make agricultural production profitable. To this end, farmers' rights to freely market their produce should be recognised by law, and the pricing of agricultural products should move towards market-driven approaches.

## Institutions

To jointly deal with water- and environment-related challenges, the CA countries established the International Fund for Saving the Aral Sea (IFAS) in 1993. The main goal of the Fund is to finance practical joint activities, programmes and projects aimed at environmental rehabilitation of the ASB, as well as at improvement of the socio-economic situation in the region. In accordance with the Agreement On the Status of IFAS and its Organizations (1999), IFAS comprises a Board, Audit Commission and Executive Committee (EC IFAS) with national branches and a Regional Center of Hydrology, an Interstate Commission for Water Coordination (ICWC) with Secretariat, a Scientific Information Center (SIC) and Basin Water Organizations (BWO) for the Amu Darya and Syr Darya, as well as an Interstate Commission for Sustainable Development (ICSD) with a Secretariat and Scientific Information Center (SIC).

IFAS is the only regional organisation in which all five CA countries cooperate. It has a broad mandate, from developing programmes to implementing joint activities in the Basin. Over the years, IFAS and its organisation have become a platform for negotiations between countries and the development of bilateral and multilateral instruments. In December 2008, IFAS was granted UN observer status.

IFAS and its subsidiaries have become an irreplaceable platform for negotiations between CA countries, the preparation and adoption of bilateral and multilateral agreements for efficient management and the use and preservation of transboundary water resources. The following results of joint work of regional institutions have been largely achieved: the preparation of three Aral Sea Basin programmes (ASBPs) and the Regional Environmental Action Plan in CA (ICSD 2018); the improvement of information and data transfer and exchange systems; the creation of a system of training of employees of water organisations and water users; the development of main provisions of the strategy to solve the problems of the Aral

Sea; the improvement of the work of hydro-meteorological services; joint regional projects designed to introduce the IWRM principles; testing market approaches in the process of reforming the irrigated cropping management system (e.g., the WUFMAS project designed to study water use and management of agricultural production); the improvement of ecological conditions in river estuaries; and the maintenance of dialogue between countries in order to strengthen the legal basis for regional cooperation. IFAS and its organisations have played and continue to play an important role in transboundary water governance, but changing political and economic situations also require changes in institutional set-up to better respond to current and future needs.

In 2009, the governments of CA countries called for the further improvement of the organisational structure and the contractual and legal framework of IFAS in order to raise the effectiveness of its operation (UNRCCA 2009). To achieve this goal, EC IFAS in Kazakhstan coordinated the work of national teams and international experts to develop proposals on improving IFAS structure and regulatory framework. The proposals include three possible options as to how to improve its functioning (UNECE, GTZ and EC IFAS 2010). The first option proposed the strengthening of the existing structure of the regional cooperation mechanism. The second option proposed the creation of a new regional organisation on the basis of the existing structures of cooperation. The third option proposed the establishment of international river-basin commissions for the Amu Darya and Syr Darya. However, neither option found support of all countries and approval of the IFAS Board.

In May 2016, Kyrgyzstan decided to 'freeze' its participation in the activities of IFAS and its bodies because 'the reforms of the IFAS repeatedly proposed by Kyrgyzstan were not carried out' and, in particular, IFAS 'does not take into account hydropower aspects of water use and the requirements of individual countries of Central Asia' (MFA KG 2016). At the meeting of the heads of five Central Asian countries in Turkmenbashi on 24 August 2018, the president of Kyrgyzstan again put forward a set of measures to reform the IFAS structure (President of Kyrgyzstan 2018). There is a hope that riparian countries will work together to strengthen the institutional framework to better address current challenges.

The institutional framework for water resources management at the national level is also under reform. In Kazakhstan, the Ministry of Agriculture (and its Water Resources Committee) is currently performing strategic, regulatory, realisation, control and supervision functions in the sphere of water resources use and conservation.

In addition, basin water departments and basin councils with advisory roles have been established. In Kyrgyzstan, institutional reform took place on water-related functions to different bodies with the National Water Council responsible for oversight and coordination for all agencies involved in water resources management. However, the Council convened only in 2013. A new body, the State Water Administration, has not yet been established as an independent administration, and since 2012 its responsibilities have been held by the Department of Water Economy and Melioration of the Ministry of Agriculture, Melioration and Food Production.

In Tajikistan, the water resources management role of the Ministry of Land Reclamation and Water Resources was merged with the Ministry of Energy and Industry in 2013 to form the Ministry of Energy and Water Resources. Today, policy and regulatory functions are carried out by two bodies: the National Water and Energy Council, which consists of heads and experts of various ministries and state agencies, and the Ministry of Energy and Water Resources. The Ministry of Water Management was established in Uzbekistan in 2017. On January 2019, the State Committee on Water Management was formed in Turkmenistan out

*Figure 12.4* The Orto Tokoy Reservoir in Kyrgyzstan in August 2018
Source: Lukas Bischoff/ Shutterstock.com

of the Ministry of Agriculture and Water Management. National institutional reforms and inter-sectoral coordination processes at the national level must promote water, energy and environment cooperation at the regional level.

## Data, information, capacity

Water-related data availability and accuracy remain a big challenge. In the last 20 years a total of 237 flow measuring stations (42 per cent) were closed in CA countries (Ibatullin 2016). The largest number of stations (almost half) were closed during the last 15 years in the countries that host most of the water sources, Kyrgyzstan and Tajikistan. The network of weather stations, glacier observations and chemical composition and quality-of-water observations has shrunken. The existing network needs rehabilitation and re-equipment.

An interstate agreement on the exchange of hydrological information in CA has not been in place for the last 20 years. These problems are connected with a range of factors, including: insufficient human resources and funding for collecting and managing information at national and regional levels (monitoring networks, specialists); unresolved political and institutional problems in the area of information exchange at the regional level; challenges of achieving regular information exchange within the country, not even mentioning the regional exchange; and distrust in information sources and in organisations engaged in the provision and consolidation of information. More consistent efforts are needed to ensure that data are collected, verified, processed and exchanged on a regular basis.

Nonetheless, the regional organisations have established a framework for data and information exchange as part of their mandates. Data on trends in water quantity and quality in the transboundary Amu Darya and Syr Darya Rivers are collected by the two BWOs and compiled by SIC ICWC on a ten-day, monthly and annual basis. A decision support system for water and environmental issues is being developed in cooperation with national water management organisations, as well as regional and international partners. This system includes a knowledge portal, a regional information system and a set of analytical models. In 2011, an EC IFAS working group proposed the development of a joint information space on water resources, the environment, power and hydro-meteorology, with information to be divided by themes and sectors.

The strengths of water cooperation in the ASB include technical and professional communication between the countries, especially regarding the operational management of water resources. Despite the often divisive discourse at the political level, CA water specialists maintain contact and exchange information by implementing joint projects and activities.

## Joint vision and strategic planning

Piecemeal responses cannot address existing and future needs and challenges. Holistic and participatory approaches are needed to enable *identification* of workable and acceptable solutions through research and on-the-ground work and *implementation* through engineering, technological, governance and other measures. The future directions have to be informed by and build upon innovative thinking and best practices around the globe. This section will outline some key areas for these future directions.

The greatest challenge to be faced in the ASB is the development of an integrated water resources management (IWRM) strategy and the task of harmonising national water policies. CA countries need to find long-term solutions to existing and emerging problems and develop a realistic long-term strategy for sustainable development of the Basin, taking into proper account all interests and the environment. Achieving an optimum balance between economic, social and environmental factors through IWRM at Basin level and at the regional level should become the core of this strategy.

The strategy should be reasonable, transparent and clear. It should include criteria for evaluating water resources management; consider sustainable development scenarios for the region, as well as implementation and monitoring procedures. 'Key Provisions of the Regional Water Strategy' that were approved in 1997 and contain a detailed list of necessary organisational, technical, financial, legal and administrative measures can be a good starting point (Key Provisions of the Regional Water Strategy 1997). This document highlighted the need to:

- clarify and strengthen a set-up for the regional organisations, including their mandate, status, financing and scope of activities;
- ensure common and integrated management of surface water, groundwater and return water and the need to pay much more attention to water quality issues;
- elaborate joint sound and practical rules and procedures for managing water quality and quantity along all transboundary rivers and their tributaries;
- ensure that the riparian countries are oriented towards rational water use, water storage and water saving at all levels of the water hierarchy;

- ensure that representation and financial contribution for regional organisations and the activities of these organisations are more equitably distributed and efficient; and ensure that water for hydropower requirements is aligned with water requirements for other sectors, notably irrigated agriculture and ecosystems maintenance

Aral Sea Basin Programmes (ASBP) also can be seen as a good step in Basin-wide planning and efforts to coordinate development projects (CaWater-Info.net 2018c). Three regional comprehensive programmes have been developed and implemented already, including ASBP-1 for 1995–2001, ASBP-2 for 2002–2010 and ASBP-3 for 2011–2015. On 23 August 2018, the IFAS board approved a conceptual basis for development of the ASBP-4, which will keep the core four directions of the ASBP-3: (i) integrated water resources use, (ii) the environmental component, (iii) the social and economic component, (iv) improvement of institutional and legal mechanisms.

## Evidence-based decision-making and water diplomacy

The process of ensuring sustainable water security should be based on a clear understanding of the existing situation and development prospects and comparisons of alternative options. It is important to assess the expected risks, the persons and institutions which will bear them and how to mitigate or avoid risks without ensuring water security of one to the detriment of others' security.

It is necessary to develop a continuous dialogue in the form of water diplomacy between CA countries and Afghanistan based on scientific and technical knowledge and laws and taking into account national and regional security. This was also highlighted by the president of Turkmenistan in his address at the 2018 IFAS summit. Also, under the aegis of IFAS and its regional organisations ICWC and ICSD, it is necessary to renew the practice of creating task forces with representatives of the line ministries and authorities, as well as independent experts in water management, energy and environmental protection to discuss and agree draft agreements and other documents.

In the context of ongoing and planned large-reservoir construction in the upstream areas of the Amu Darya and Syr Darya Rivers, common agreements should be conducted on mutually beneficial conditions for all CA countries. This issue has been excessively politicised, which hampers the development of conditions that are favourable for the creation of an up-to-date mechanism to govern the joint use of water and energy resources in CA. Decision-makers must maximise the utilisation of the positive experience of other countries and regions to coordinate the management of water resources in the ASB.

## Infrastructure, technology and innovation

Issues related to the use of existing and the construction of new water infrastructure are at the core of future developments in the Basin. Large-scale measures and investments are needed to reduce water consumption and improve the efficiency of water use at irrigated farms. Hydraulic infrastructure (including water and power facilities), irrigation canals, gauging stations, roads, flood protection facilities, fences and other structures that were built 40–50 years ago are now ageing and need reconstruction or, in some cases, replacement. The efficiency of many canals barely reaches 50–60 per cent since they lack anti-seepage lining; many mechanical components do not function; canals get quickly covered with aquatic vegetation; drainage systems and wells are not operational; many irrigated areas have become prone to salinisation (Ibatullin 2016).

During the 2018 IFAS summit, the presidents of Tajikistan and Uzbekistan emphasised the need for improved water use and conservation techniques (EC IFAS 2018). The president of Uzbekistan proposed the development of a regional programme for rational water use.

Future tasks should include new technological approaches in all sectors, including irrigation, aimed at reducing freshwater intake from sources and eliminating return water discharge into surface and underground sources. Innovative methods of rainwater harvesting, desalinisation, improving water use efficiency, wastewater treatment, recirculation and recycling technologies shall be utilised. These measures should be enhanced with a broad set of water and environment protection measures (including air, lithosphere, flora and fauna protection). The above interventions need modernisation of existing hydraulic structures; reservoir and canal lining; construction of pump stations and water supply systems; reduction of water use in the industry by switching to water saving and dry technologies; improvement of the accuracy of water measurements and transmission of online data to all users; equipment with modern water measurement devices (level gages, flow meters, SCADA) for main canals; and other issues.

Environmental innovation is another area that seems to be emerging in the region. The president of Uzbekistan has recently called for transforming the area adjacent to the Aral Sea (the so-called *Priaralie*) into a zone of environmental innovations, which clearly signals the shift of focus from discussion of problems to a search for workable solutions.

## *Multisectoral and participatory governance arrangements*

Over the years, riparian countries of the Amu Darya and Syr Darya Basins have been struggling to find mutually acceptable mechanisms for water and energy resources use

*Figure 12.5* Reflection in a puddle of water at the Khast Imam Mosque in Tashkent
Source: Mehmet/ Shutterstock.com

(Dukhovny *et al.* 2018). However, these discussions did not include all stakeholders and sectors. Also, regional institutions in the ASB have been criticised for the failure to include main stakeholders such as the environment and the energy sector into their decision-making frameworks. Intentions to establish a Basin-wide advisory body, bringing together key stakeholders for each RBO, have been incorporated into a Draft Agreement on Strengthening the Organizational Structure for Transboundary Water Management in the Basin (Version 1/2007). This Draft Agreement has not been signed yet, nor have enabling activities for establishing such a body been put in place.

Stakeholder engagement has gained traction as a key principle of good governance and a quasi-prerequisite to bottom-up, place-based and inclusive policy-making (OECD 2015). The complex web of water, energy and environmental problems in the Basin cannot be addressed without strong and resilient transboundary water governance principles and institutions (potential key change agents). Improved multisectoral dialogue coordinated across borders could help finding solutions and avoid negative impacts of unilateral decisions on other sectors, countries and the environment. More active participation of all key stakeholders and the public in planning and management of water resources at all levels shall be strongly encouraged.

## Multilevel governance

The complexity of water-related issues makes the case for managing water at different levels. Water management reforms and improvements in the Aral Sea Basin have to be dealt with at multiple levels simultaneously. Different scales and levels will have their own tasks and functions (Ibatullin 2010). *At interstate level*, it is necessary to agree on Basin-wide developments and arrangements, gradually aligning riparian countries' positions and possibly developing regional conventions on sustainable development in the ASB. It is also important to work together on multilateral environmental agreements such as the UNECE Convention on the Protection and Use of Transboundary Watercourses and International Lakes (1992) and the UN Convention on the Law of the Non-Navigational Uses of International Watercourses (1997).

*At interstate basin management level*, the main task is to introduce IWRM principles for the Amu Darya and Syr Darya Rivers. It is also necessary to provide transboundary water resource monitoring systems and a unified information database with a network of hydro-meteorological services and gauging stations, ensuring observations from glaciers to the mouths of rivers. The smaller rivers and tributaries also need their own arrangements to be aligned with the upper level and meet specific demand at their scale level.

*At the national level*, national water agencies and basin councils have to be strengthened to respond effectively to changing needs and constraints. It is also critically important to improve the technical condition of water systems, control the problem of soil salinity and reclaim salinised agricultural land, and develop a smart system for water tariffs.

## Roadmap to sustainable water management in the Ara Sea Basin

It is safe to conclude that the future prospect of the Aral Sea Basin will depend on a set of comprehensive measures that encompasses social, economic and environmental domains. Based on the strengths and weaknesses of water cooperation in the ASB and discussions of possible solutions, a number of proposals have been made in the preceding section as to how to achieve sustainable water security in the future. Those proposals are fully consistent with the statements agreed by the heads of states in 2009 (Joint Statement) and in 2018 (Communiqué)

(ES IFAS 2009 and 2018). A roadmap is proposed towards an improved water management in ASB, which shall be based on the following principles.

- Respect for, and compliance with, international law and global-level obligations (e.g., the Sustainable Development Goals (SDG), Paris Agreement, global conventions, etc.).
- Building upon existing agreements, decisions of the heads of Central Asian states, approved programmes and strategies and regional institutions.
- The adoption of mutually agreed implementation at regional and national levels.
- The adoption of a comprehensive systematic approach and being result-oriented.

Its main objectives shall be rigorous, with high-impact evaluation as follows.

- The setting of agreed benchmarks (indicators) in water resources use and management in the ASB until 2030.
- The creation and development of efficient mechanisms of regional cooperation.
- Improvement to the population's welfare and economic and environmental situation by securing a sustainable water supply.

The outcomes of the roadmap could be synopsised on the following.

- The development of common indicators for all countries to be used in the future to compare progress in efficient water use, protection of access and ensuring access to water resource.
- The reduction of total water consumption to satisfy a growing demand and a dwindling supply of resources, and to unlock adequate resources for ecological purposes.
- Proper risk management and preparedness to overcome extreme threats such as floods and droughts.
- Satisfaction of the basic needs of humans and ecosystems.

It is proposed to synchronise the roadmap activities with the implementation of the SDGs, which will serve as guidelines at the global, regional and national levels. In this context, regional programmes should be developed in which implementation of national measures will be monitored at a regional level using agreed regional indicators. The period from 2020 to 2030 can become the decade of Sustainable Water Security and Partnership in the ASB, if all existing opportunities and potentials are used in a timely manner. The roadmap may help in achieving sustainable water security in CA in the best interests of humans, ecosystems and the socio-economic development of the countries and the region.

In order to monitor and evaluate progress in the implementation of the roadmap, the development of indicators is proposed. Until now, the Basin countries have lacked common indicators to compare their individual efforts in achieving sustainable water supply and quality, as well as institutional efforts and improvements in the national water law in line with international best standards. In addition to establishing indicators, it is important to determine targets for each of the indicators as part of the roadmap, which will help to follow action (or inaction) on the part of social and political institutions in achieving the goals set. The examples of such indicators could be expenditure (share of GDP) on the protection and restoration of water ecosystems; expenditure on improvement measures; subsidies for water saving technologies; the rate of increase in the use of drainage water; participation in the work of regional organisations; participation in agreements related to water resources; reform of water management policies etc.

## Conclusions

The importance of transboundary water resources in the ASB has provided compelling incentives for regional cooperation on water, in particular within the framework of IFAS. However, despite certain progress in improving water relationships in the transboundary basins of the Amu Darya and Syr Darya Rivers, there are still important challenges and unresolved tasks in amplifying the impact of improved water resources management on income, employment, food security and economic growth. To find solutions for ensuring the equitable and reasonable use of water and energy resources, environmental stability and food security are needed in the Aral Sea Basin.

An introduction of IWRM in the ASB should be elevated, essentially taking into account the multilevel approach, building capacity of all stakeholders, and raising public awareness in water and environmental issues. It will be necessary to elaborate a regional water conservation programme that contains precise indicators for each basin as well as to make wider use of water saving technologies in agriculture, industry and the public utilities sector. In the wider context, a regional water strategy that takes into account water–energy–food–environment linkages might be useful for coordinating CA towards a more water-secure and peaceful future. It is also necessary to establish mutually acceptable economic arrangements of joint management of water and energy resources in the enclosed river basins within the ASB and to set up a water and energy consortium as a financial and insurance mechanism of cooperation.

Improved and mutually acceptable mechanisms of cooperation should be cemented in agreements that in turn are better monitored in full compliance with all CA countries. To reach this end, continuous dialogue based on scientific and technical knowledge as well as detailed procedures of interaction will be necessary. A comprehensive programme for research and development should be elaborated and implemented. Finally, data and information exchange and joint-scenario development should be further facilitated to bring together meteorological, hydrological, ecological, demographic and economic datasets recognised, trusted and used by all countries of the region.

## Reference

CaWater-Info.net (2018a) Database. www.cawater-info.net/bd/index_e.htm (accessed 3 December 2018).

CaWater-Info.net (2018b) Database of the Aral Sea. www.cawater-info.net/aral/data/index_e.htm (accessed 3 December 2018).

CaWater-Info.net (2018c) Aral Sea Basin Programs. www.cawater-info.net/library/agreem_e.htm (accessed 3 December 2018).

Center for Economic Research (2013) *Urbanization in Central Asia: Challenges, Issues and Prospects.* Analytical Report 2013/03. Tashkent, Uzbekistan.

Central Statistics Organization (2016–2017) Islamic Republic of Afghanistan. http://cso.gov.af/en/page/1500/4722/2016-17 (accessed 13 June 2019).

Dukhovny V and de Schutter J (2011) *Water in Central Asia: Past, Present, Future.* Taylor & Francis Group, London, UK.

Dukhovny V. (ed.) (2017) *The Aral Sea and the Priaralie.* Baktria Press, Tashkent, Uzbekistan.

Dukhovny V et al. (eds) (2018) *The Future of the Amudarya Basin under Climate Change.* SIC ICWC, Tashkent, Uzbekistan.

EC IFAS/Executive Committee of Interstate Fund for saving the Aral Sea (2009) www.ec-ifas.water-unites-ca.org (accessed 3 December 2018).

EC IFAS (2018) www.ecifas.gov.tm (accessed 3 December 2018).

FAO/Food and Agriculture Organization of the United Nations (2017) *The Future of Food and Agriculture – Trends and Challenges.* FAO, Rome, Italy.

Ibatullin S (2009) *Aral Sea Basin: Time to Ordeals and Actions.* International Conference 'Aral 2009', St Petersburg, Russia.

Ibatullin S *et al.* (eds) (2009). *Impact of Climate Change to Water Resources in Central Asia.* Eurasian Development Bank Sector Report No 6, Almaty, Kazakhstan.

Ibatullin S (2009) *Aral Sea Basin: Time to Ordeals and Actions.* International Conference 'Aral 2009', St Petersburg, Russia.

Ibatullin S (2010) *Methodological Bases and Principles in Using the Water Resources of Transboundary Rivers for Sustainable Development.* n.p., Taraz, Kazakhstan.

Ibatullin S *et al.* (2016) *Managing Water Resources in the Aral Sea Basin: Current Situation, Vision and Roadmap.* World Bank Expert Group. http://documents.worldbank.org/curated/en/272511487775228782/pdf/112964-WP-PUBLIC-ADD-SERIES-P120142-CAEWDPARWeb.pdf (accessed 10 June 2019).

Ignatieff M (2012) The return of sovereignty. *New Republic*, 25 January. www.newrepublic.com/article/books-and-arts/magazine/100040/sovereign-equality-moral-disagreement-government-roth (accessed 3 December 2018).

ICSD/Interstate Commission for Sustainable Development (2018) www.mkurca.org (accessed 3 December 2018).

Karlihanov T and Ibatullin S (2016) *Aral: Past, Present, Future.* Foliant, Astana, Kazakhstan, pp 226–245.

Key Provisions of the Regional Water Strategy (1997) www.cawater-info.net/library/strategies.htm (accessed 3 December 2018).

MFA KG/Ministry of Foreign Affairs of the Kyrgyz Republic (2016) First Deputy Foreign Minister of Kyrgyzstan Dinara Kemelova took part at the third High Level Dialogue between the European Union and Central Asia and Afghanistan in Brussels, 19 May. www.mfa.gov.kg/news/view/idnews/2291 (accessed 3 December 2018).

OECD/Organization for Economic Co-operation and Development (2015) *Stakeholder Engagement for Inclusive Water Governance.* OECD Studies on Water, OECD Publishing, Paris, France.

President of Kazakhstan (2018) *Participation in the Meeting of the Heads of the Founding States, International Fund for Saving the Aral Sea.* 24 August. www.akorda.kz/en/events/international_community/foreign_visits/participation-in-the-meeting-of-the-heads-of-the-founding-states-international-fund-for-saving-the-aral-sea (accessed 3 December 2018).

President of Kyrgyzstan (2018) President Sooronbay Zheenbekov: Kyrgyzstan stands for comprehensive reform of the IFAS taking into account the interests of all Central Asian states (in Russian). www.president.kg/ru/sobytiya/12245_prezident_sooronbay_gheenbekov_kirgizstan_vistupaet_zakompleksnoe_reformirovanie_mfsa_suchetom_interesov_vseh_gosudarstv_centralnoy_azii (accessed 4 March 2019).

Stulina G and Solodkiy G (2015) The effect of climate change on land and water use. *Agriculture Sciences*, vol 6, pp 834–847.

SDG Knowledge Platform (2017) *High Level Panel on Water.* https://sustainabledevelopment.un.org/HLPWater (accessed 3 December 2018).

Third Aral Sea Basin Program (ASBP-3) for the period 2011–2015 (2011) Almaty, Kazakhstan.

UNECE, GTZ and EC IFAS (2010) *Strengthening the Institutional and Legal Frameworks of the International Fund for Saving the Aral Sea: Review and Proposals.* Discussion paper. www.unece.org/fileadmin/DAM/env/water/cadialogue/docs/Draft_Paper_united_FINAL_ENG.pdf (accessed 6 December 2018).

UNECE (2018) *Strengthening Cooperation on Hydrology and Environment between Afghanistan and Tajikistan in the Upper Amu Darya River Basin.* www.unece.org/environmental-policy/conventions/water/areas-of-work-of-the-convention/envwatercentralasia/strengthening-cooperation-on-hydrology-and-environment-between-afghanistan-and-tajikistan-in-the-upper-amu-darya-river-basin.html (accessed 13 February 2019).

UNDESA/United Nations, Department of Economic and Social Affairs, Population Division (2017) *World Population Prospects: The 2017 Revision, Key Findings and Advance Tables.* ESA/P/WP/248. https://population.un.org/wpp/ (accessed 10 June 2019).

UNRCCA (2009) UNRCCA takes part in the summit of the heads of states of the IFAS. https://unrcca.unmissions.org/unrcca-takes-part-summit-heads-states-ifas (accessed 10 June 2019).

UzA (2018) Speech by President of the Republic of Uzbekistan Shavkat Mirziyoyev at a meeting of the heads of the founding states of the International Fund for Saving the Aral Sea held in Turkmenbashi (in Russian), August 24. www.uza.uz/ru/politics/vystuplenie-prezidenta-respubliki-uzbekistan-shavkata-mirziye (accessed 3 December 2018).

World Bank (2012) *Kazakhstan: Syr Darya Control & Northern Aral Sea Phase I Project*. http://projects.worldbank. org/P046045/syr-darya-control-northern-aral-sea-phase-project?lang=en (accessed 3 December 2018).

World Bank Data (2018) https://data.worldbank.org (accessed 3 December 2018).

Zeitoun M (2012). Diplomacy for water security. In Tanzler D and Carius A (eds), *Climate Diplomacy in Perspective: From Early Warning to Early Action*. Berliner Wisserschafts-Verlag, Berlin, Germany, pp. 35–45.

Ziganshina DR (2014) *Promoting Transboundary Water Security in the Aral Sea Basin through International Law*. Martinus Nijhoff Publishers, Leiden, the Netherlands.

## Annex: Treaties and declarations

UNECE/United Nations Economic Commission for Europe Convention on the Protection and Use of Transboundary Watercourses and International Lakes, Helsinki (adopted 17 March 1992, in force 6 October 1996) 31 ILM 1312.

United Nations Convention on the Law of the Non-Navigational Uses of International Watercourses, New York (adopted 21 May 1997, in force 17 August 2014) 36 ILM 700.

Agreement between the Republic of Kazakhstan, the Kyrgyz Republic, the Republic of Tajikistan, Turkmenistan, and the Republic of Uzbekistan on Cooperation in the Field of Joint Management of the Use and Conservation of Water Resources of Interstate Sources, Almaty (signed 18 February 1992).

Agreement between the Republic of Kazakhstan, the Kyrgyz Republic, the Republic of Tajikistan, Turkmenistan, and the Republic of Uzbekistan on Joint Actions for Addressing the Problems of the Aral Sea and Its Coastal Area, Improving the Environment, and Ensuring the Social and Economic Development of the Aral Sea Region, Kzyl-Orda (signed 26 March 1993).

Agreement between the Governments of the Republic of Kazakhstan, the Kyrgyz Republic, and the Republic of Uzbekistan on the Use of Water and Energy Resources of the Syrdarya Basin, Bishkek (signed 17 March 1998, Republic of Tajikistan joined in 1999).

Agreement between the Republic of Kazakhstan, the Kyrgyz Republic, the Republic of Tajikistan, Turkmenistan, and the Republic of Uzbekistan on the Status of IFAS and Its Organizations Ashgabad (signed 9 April 1999).

Nukus Declaration of the Central Asian States and International Organisations on the Problems of Sustainable Development in the Aral Sea Basin (adopted 5 September 1995).

Almaty Declaration of the Heads of States of the Republic of Kazakhstan, the Kyrgyz Republic, the Republic of Tajikistan, Turkmenistan and the Republic of Uzbekistan (adopted 28 September 1997).

Ashgabad Declaration of the Heads of States of the Republic of Kazakhstan, the Kyrgyz Republic, the Republic of Tajikistan, Turkmenistan and the Republic of Uzbekistan (adopted 9 April 1999).

Dushanbe Declaration of the Heads of States of the Republic of Kazakhstan, the Kyrgyz Republic, the Republic of Tajikistan, Turkmenistan and the Republic of Uzbekistan, (adopted 6 October 2002).

# 13

# Implementing water-related Sustainable Development Goals

*Aziza Baubekova and Anastasia Kvasha*

## Key messages

- Access to safe drinking water and sanitation, the most basic of needs, is still not universally provided within the Aral Sea Basin. For instance, the share of population using basic drinking water services ranges from 63 per cent in Afghanistan to almost 95 per cent in Turkmenistan, with an even greater disproportion with regard to access to handwashing facilities. At the same time, limited water availability and poor water quality have a critical effect on local people's livelihood and health.

- Overall, the region can be characterized by uneven achievement of Sustainable Development Goal 6 'Clean Water and Sanitation'. This is demonstrated not only by each individual country's progress but much more significantly by the contrast between urban and rural population within each state. This is most noticeable in the case of Afghanistan but is a common feature of every other country in the Basin.
- Rigid centralization of political institutions inherited from the Soviet period, as well as depreciated infrastructure and old, inefficient technologies that are still typical for the region are causing countries to lag behind in SDG 6 progress. This is particularly noticeable in the case of Target 6.4, on water-use and scarcity (global average water-use efficiency equals 15 USD/m3, while for the Basin it is around 1.5 USD/m3). Furthermore, while countries in the region are willing to spend their own funds on profitable projects, in other sectors, including water and sanitation, they usually prefer to rely on international donors. Water-efficient practices and cost-effective technologies must be promoted to improve the availability of water resources in the region.
- Many water-related issues of the Aral Sea Basin must be addressed jointly by all involved states, otherwise none can be fully resolved. With support from international organizations, regional as well as national strategies in water sector are already developed, though not yet fully implemented. The motivation of governments could be enhanced by addressing existing problems through the water–energy–food nexus concept and by incorporating trading and investment elements.
- Data availability and reliability play a crucial role in ensuring that SDG 6 and corresponding targets will be met. While the lack of reliable information on some indicators (6.3.1 and 6.3.2 particularly) is recognized as a problem on a global scale, the absence of data in other cases remains a pressing issue that must be addressed, particularly for the reconstruction and upgrading of national monitoring systems.

## Introduction

The 17 Sustainable Development Goals (SDGs) of Agenda 2030 adopted in 2015 by 193 UN member states utilize the integrated approach toward environmental protection, social inclusion, development, and economic growth; all goals are interdisciplinary and interconnected (UN 2015). The Aral Sea Basin countries, namely Afghanistan, Kazakhstan, Kyrgyzstan, Turkmenistan, Tajikistan, and Uzbekistan, are committed to contribute to the Agenda 2030 and benefit from the implementation of the SDGs. In 2016, countries of the UN Special Programme for the Economies of Central Asia (SPECA) (Azerbaijan, Afghanistan, Kazakhstan, Kyrgyzstan, Tajikistan, Turkmenistan, and Uzbekistan) adopted the Ganja Declaration on 'Enhanced Implementation of Sustainable Development Goals (SDGs) through Cooperation'. This declaration promotes regional integration for the SDGs' implementation including transport links, management of water and energy resources, and ICT connectivity through economic cooperation, strengthening policy coherence, and collaboration.

Each country in the Aral Sea Basin is at a different stage of economic development—Afghanistan is currently the least developed one—and is focused on its own development priorities and trajectories. Taking into account the consequences of climate change under the predominantly arid conditions of Central Asia's landlocked countries as well as expected overuse of natural resources such as water, the region can be characterized as most vulnerable to a water crisis. Implementing SDG 6 is, therefore, among the priority goals for all riparian states (Alzhanova 2017). While it might be too challenging for the region as a whole to actually attain some of the formulated targets of *Goal 6 on ensuring availability and sustainable management of water and sanitation for all* (SDG 6), it is nevertheless necessary to establish a positive trend now. This chapter attempts to analyze the chances of the Basin countries of achieving each of SDG 6's targets (eight in total – six "Outcome" and two "Means of Implementation" (MoI) targets). The chapter examines each target's current status in the Aral Sea Basin region. Subsequent sections are dedicated to analyzing the interconnectedness with other water-related targets and indicators.

## Entering the SDG process

Central Asia can be characterized as a region most vulnerable to a water crisis. This is further aggravated by the potentially severe impact of climate change and the fact that one of the most well-known environmental disasters has happened in this region with consequences that have not been resolved. Starting from the 1960s, the Aral Sea was experiencing rapid desiccation and related salinization, and at the current stage the restoration of the lake's historical conditions is basically out of the question (Micklin *et al.* 2014). Numerous studies and projects were carried out on the Aral Sea disaster, years before the introduction of the SDGs. Other environmental, economic, and social problems in the Central Asia region have been receiving more attention recently, such as the lifestyle and health concerns of the local population (Crighton *et al.* 2003; Törnqvist *et al.* 2011; Aladin *et al.* 2016).

The economy and development of Afghanistan have been negatively affected by conflict and an unstable political situation for several decades. Currently, with the help of international organizations, the Government of the Islamic Republic of Afghanistan (GoIRA) has made considerable progress in its social, economic, and political transition and has started the process of SDGs implementation (ME of AF 2017). Afghanistan will require substantial support from development partners and international communities to improve data collection and monitoring as well as overall SDGs implementation.

In Kazakhstan, most of the SDGs' objectives are already covered by national strategic documents. The President's 'Kazakhstan – 2050' strategy was adopted to address, among others, sustainable development and ecological issues, as well as the country's new policy on use of water resources to meet rising needs of agriculture. Several instruments are promoting international dialogue, such as EXPO 2017, the Astana Economic Forum, and the United Nations–Kazakhstan 'Partnership Framework for Development, Kazakhstan, 2016–2020'. In addition, the government is currently preparing a Voluntary National Review (VNR), which will be presented in July 2019.

The Sustainable Development Strategy of the Kyrgyz Republic for 2018–2040 recognizes the special importance of issues related to ecology, environment, and climate change, which will have a significant impact on the livelihood and health of the country's population. Water resources are recognized as the most vulnerable asset of the republic (NCSD of KG 2018). In response to these challenges, the government plans to modify the current policies to ensure sustainable development, preservation of ecosystems, efficient water use, and modern water-saving technologies in agriculture. However, there is no comprehensive national water strategy at the moment, and the water sector is still fragmented.

Tajikistan remains one of the poorest states among the Basin countries and its economy is particularly susceptible to the effects of the global economic crisis. The president of Tajikistan has emphasized the importance of water issues and especially of the dispute on water reservoirs with Uzbekistan. Consequently, the agenda of SDG 6 is Tajikistan's priority, followed by the development of the agricultural sector, improving quality of social services, reducing unemployment, and improving access to universal education to eliminate poverty.

Turkmenistan is a regional leader in the adaptation of the Sustainable Development Goals at the national level, in particular, with respect to optimizing the use of water resources. In September 2018, President Gurbanguly Berdimuhamedov published a book, *On the Way to Achievement of the Sustainable Development Goals in Turkmenistan*, to present the enormous amount of work carried out in the country.

Uzbekistan works closely with the UN agencies to fulfill the SDGs. The country aims to improve the access of the urban and rural population to clean drinking water and sanitation. Another priority for the government is to enhance the energy efficiency of production, and to develop the energy sector, based on renewable energy sources such as hydropower. The country is working toward creating intersectoral synergies through the nexus approach.

## Current status of SDG 6 on clean water and sanitation

A study on implementation of SDGs within the SPECA region (Alzhanova 2017) identified the most important goals for the Aral Sea Basin countries. The results of this analysis in an aggregated form are presented in the figure below (Figure 13.1). *SDG 6 on Clean Water and Sanitation* was referred to as one of the most critical for the region, while *SDG 13 on Climate Action* was also mentioned as a priority across all states.

The framework of SDGs provides not only individual goals and targets but also relevant indicators. Through vigorous discussions, these indicators were agreed upon by the Member States to ensure fair monitoring of the progress in sustainable development by each country, both developing and developed. In the case of Goal 6, 11 indicators correspond to the eight targets in total. Table 13.1 below aggregates information on the status of progress in achieving each indicator of SDG 6 by country within the Aral Sea Basin (by numerical values and the color code). These indicators are discussed one by one in the following sections of this chapter.

The statistical information that is used in the table below and in the charts introduced later in this chapter were collected primarily from the UN Statistics Division's SDG Global Database, an online data hub that was opened in 2018 (UNSD 2019). In situations when no information on particular indicators was available through the Global Database or when

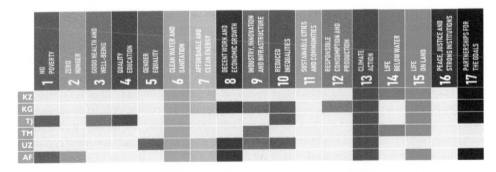

*Figure 13.1* Main national priorities in terms of SDGs implementation. The UN/LOCODE codes are used for the countries' names: KZ – Kazakhstan; KG – Kyrgyzstan; TJ – Tajikistan; TM – Turkmenistan; UZ - Uzbekistan; AF – Afghanistan

Source: Constructed by authors based on data from Alzhanova (2017)

data from this source were clearly insufficient, data from other international organizations were used, particularly: 'Proportion of population using at least basic drinking water services' and 'Proportion of population using at least basic sanitation services', from WHO/UNICEF Joint Monitoring Programme (JMP) for Water Supply, Sanitation and Hygiene (WHO/UNICEF, 2019); *UN-Water Global Analysis and Assessment of Sanitation and Drinking-Water* (GLAAS) report (UN-Water 2017); *Progress on Water Use Efficiency – Global Baseline for SDG 6 Indicator 6.4.1* (FAO 2018b); the AQUASTAT Main Database (FAO 2016); and progress reports published by UN Environment (UN Environment 2018b; UN Environment 2018c). Information that was retrieved from articles and national sources is discussed in relation to particular targets in the following sections and is not included in Table 13.1.

Color-coding represents the status of progress in terms of how close each country is to achieving the goal in the current situation (target achieved or almost achieved; challenges remain; significant challenges remain; major challenges remain; or NA—information unavailable). It does not represent the likelihood of the target being achieved within the timeframe (in most cases 2030). The chances of doing so are discussed in more detail in the sections on each indicator below. The color-code classes were adopted from the *SDG Index and Dashboards Report 2018* (Sach *et al.* 2018).

Data availability and the status of progress in achieving each target can vary greatly even within the Basin, as can be noticed from the table above. However, this kind of disparity is also common for other regions of the world. In the case of the Aral Sea Basin, the lack of information on *Target 6.3 Improvement of water quality* is most noticeable. It is worth mentioning, though, that the data related to this target and corresponding indicators are currently available mainly for high- and middle-income developed countries, with little to no data existing for developing countries in Africa or Central and Southeast Asia (WHO & UN-HABITAT 2018; UN Environment 2018a). Even though the methodology is already developed, it remains challenging for many countries to collect this kind of information since it relies heavily on existing national monitoring systems. Another issue is related to the continuous long-term data collection that is important in some cases, especially for *Indicator 6.4.1 Change in water-use efficiency over time*. However, this indicator was first introduced only in SDGs and its baseline value was only recently calculated on a global scale (FAO 2018). The actual indicator is based on the level of change over time, which is very country- and region-specific and for which not enough data are yet available. For these reasons Indicator 6.4.1, as well as Indicator 6.a.1, are not color-coded.

## Progress toward SDG 6 targets and indicators

Nonetheless, available targets and indicators can be used to assess the situation and develop plans for the future, bearing in mind that the goal must not be restricted to solely achieving these artificial targets, but in getting closer to attaining sustainable growth, improving the livelihood of people, and achieving environmental, social, and economic prosperity.

### Target 6.1. *Indicator 6.1.1: Proportion of population using safely managed drinking water services*

The issue with one of the most basic needs, the access to safe drinking water, still is not fully resolved in the Aral Sea Basin countries. The region can be characterized by limited availability of safe drinking water, which becomes an even more serious issue during dry

Table 13.1 Status of progress in achieving SDG 6 in the Aral Sea Basin

| Targets and indicators | Kazakhstan | Kyrgyzstan | Tajikistan | Turkmenistan | Uzbekistan | Afghanistan |
|---|---|---|---|---|---|---|
| *Target 6.1 By 2030, achieve universal and equitable access to safe and affordable drinking water for all* | | | | | | |
| Indicator 6.1.1 Proportion of population using at least basic drinking water services (used instead of 'safely managed drinking water services' due to lack of data and to ensure comparison between countries) (%) | 91.2 | 87.3 | 74.1 | 94.5 | 91.5 | 63.0 |
| *Target 6.2 By 2030, achieve access to adequate and equitable sanitation and hygiene for all and end open defecation, paying special attention to the needs of women and girls and those in vulnerable situations* | | | | | | |
| Indicator 6.2.1 Proportion of population using (a) safely managed sanitation services and (b) a hand-washing facility with soap and water | | | | | | |
| Sub-indicator: proportion of population practicing open defecation (%) | 0.0 | 0.0 | 0.1 | 0.1 | 0.0 | 13.6 |
| Sub-indicator: proportion of population using at least basic sanitation services (used instead of 'safely managed sanitation services' due to lack of data) (%) | 97.8 | 96.6 | 95.5 | 96.6 | 100 | 39.2 |
| Sub-indicator: proportion of population with basic handwashing facilities on premises (%) | 96.4 | 89.2 | 72.5 | 97.7 | NA | 38.3 |
| *Target 6.3 By 2030, improve water quality by reducing pollution, eliminating dumping and minimizing release of hazardous chemicals and materials, halving the proportion of untreated wastewater and substantially increasing recycling and safe reuse globally* | | | | | | |
| Indicator 6.3.1 Proportion of wastewater safely treated (%) | NA | NA | NA | NA | NA | NA |
| Indicator 6.3.2 Proportion of bodies of water with good ambient water quality (%) | NA | NA | NA | NA | NA | NA |

*Target 6.4 By 2030, substantially increase water-use efficiency across all sectors and ensure sustainable withdrawals and supply of freshwater to address water scarcity and substantially reduce the number of people suffering from water scarcity*

| | | | | | | |
|---|---|---|---|---|---|---|
| Indicator 6.4.1 Change in water-use efficiency over time (USD/m3) | 6.9 | 0.5 | 0.4 | 0.4 | 0.6 | 0.3 |
| Indicator 6.4.2 Level of water stress: freshwater withdrawal as a proportion of available freshwater resources (%) | 28.1 | 44.0 | 71.4 | 162.8 | 138.8 | 43.7 |

*Target 6.5 By 2030, implement integrated water resources management at all levels, including through transboundary cooperation as appropriate*

| | | | | | | |
|---|---|---|---|---|---|---|
| Indicator 6.5.1 Degree of integrated water resources management implementation (0–100) | 30 | NA | NA | NA | 45 | 12 |
| Indicator 6.5.2 Proportion of transboundary basin area with an operational arrangement for water cooperation (%) | 72.4 | NA | NA | NA | 59.3 (rivers, lakes) | 51.7 (rivers, lakes) |

*Target 6.6 By 2020, protect and restore water-related ecosystems, including mountains, forests, wetlands, rivers, aquifers, and lakes*

| | | | | | | |
|---|---|---|---|---|---|---|
| Indicator 6.6.1 Change in the extent of water-related ecosystems over time (loss, % of the baseline value) | 7.5 | 0.7 | 1.2 | 0.6 | 44.3 | 24.0 |

*Target 6.a By 2030, expand international cooperation and capacity-building support to developing countries in water and sanitation-related activities and programs, including water harvesting, desalination, water efficiency, and wastewater treatment, recycling, and reuse technologies*

| | | | | | | |
|---|---|---|---|---|---|---|
| Indicator 6.a.1 Amount of water- and sanitation-related official development assistance that is part of a government-coordinated spending plan Sub-indicator: total official development assistance (gross disbursement) for water supply and sanitation, by recipient countries (millions of constant 2016 USD) | 0.2 | 39.0 | 56.6 | 0.2 | 108.3 | 152.6 |

*Target 6.b Support and strengthen the participation of local communities in improving water and sanitation management*

Indicator 6.b.1 Proportion of local administrative units with established and operational policies and procedures for participation of local communities in water and sanitation management

*continued . . .*

Table 13.1 Continued

| Targets and indicators | Kazakhstan | Kyrgyzstan | Tajikistan | Turkmenistan | Uzbekistan | Afghanistan |
|---|---|---|---|---|---|---|
| Sub-indicator: countries with procedures in law or policy for participation by service users/communities in planning program in rural drinking–water supply, by level of definition in procedures | Clearly defined | Not clearly defined | Clearly defined | NA | NA | Clearly defined |
| Sub-indicator: countries with users/communities participating in planning programs in rural drinking-water supply, by level of participation | High | NA | Moderate | NA | NA | Moderate |
| Sub-indicator: countries with procedures in law or policy for participation by service users/communities in planning program in water resources planning and management, by level of definition in procedures | NA | Not clearly defined | Clearly defined | NA | Clearly defined | Clearly defined |
| Sub-indicator: countries with users/communities participating in planning programs in water resources planning and management, by level of participation | NA | NA | Moderate | NA | Moderate | Low |

Target achieved or almost achieved

Challenges remain

Significant challenges remain

Major challenges remain

NA—information unavailable

*Data sources*: UNSD (2019); WHO/ UNICEF (2019); FAO (2016); UN Environment (2018b); UN Environment (2018c).

years (Gupta & Gupta 2016). Freshwater is unevenly distributed in the region. For instance, the areas to the south of the Aral Sea (Karakalpakstan, Khorezm, and Bukhara provinces of Uzbekistan) are experiencing severe water deficit and contamination, which corresponds to a much higher number of water-related diseases compared to other areas within the basin (Gupta & Gupta 2016). The rural population is especially vulnerable to drinking water shortages. Wherever there is piped drinking water in rural areas, it is available for only two to six hours per day in some places; even though this problem has long been recognized and various actions were taken to improve access to safe drinking water, it remains a problem, especially for the rural population (Bekturganov *et al.* 2016).

Unfortunately, the data for *Indicator 6.1.1* are currently missing for two of the analyzed countries (Kazakhstan and Afghanistan). Figure 13.2 illustrates countries' progress using the data available. As shown, Turkmenistan is most likely to reach the target, while Tajikistan is falling significantly behind and Uzbekistan is stagnating.

In order to compare all countries within the Basin, a similar indicator was used—proportion of population using at least basic drinking water services (Figure 13.3). Similar to the previous

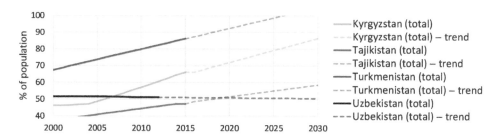

*Figure 13.2* Proportion of population using safely managed drinking water services

Source: Constructed by authors based on data from UNSD (2019)

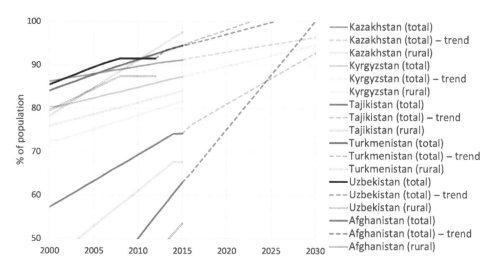

*Figure 13.3* Proportion of population using at least basic drinking water services

Source: Constructed by authors based on data from WHO/ UNICEF (2019)

graph, Turkmenistan is the leader among Basin countries. This situation can be explained by the success in implementing the 'General Programme for Clean Drinking Water in Residential Areas', which was adopted by the state in 2011 and which is going to end in 2020. In the case of every other basin country, there is a significant gap between the total population and rural population (very limited statistical data were available for urban areas). However, if the current progress trend keeps its pace, it can be expected that each country will ensure almost universal access to basic drinking services by 2030.

## Target 6.2. Indicator 6.2.1: Proportion of population using safely managed sanitation services, including a handwashing facility with soap and water

Based on the existing data, most of the region is relatively close to reaching Target 6.2. However, many areas, especially rural areas, can still be characterized by limited access to sanitation and handwashing facilities. Proper water management and an acceptable approach to sanitation play a critical role in ensuring better health conditions among the local population. High rates of illnesses in the whole region are on a major part caused by inadequate water supply (Reimov & Fayzieva 2014). While many aspects of the Aral Sea disaster are well documented and studied, health impacts of this crisis are still a much-less-explored area of research. Water quality is a very important component of this system, especially considering the effects of climate change in the Aral Sea Basin, which could result in a growing number of outbreaks of waterborne disease (Reimov & Fayzieva 2014). The area on the south of the Aral Sea, the Karakalpak Republic (Uzbekistan), is among the least developed areas in the region and characterized by the worst health condition in the entire basin (Micklin *et al.* 2014).

At the same time, the gender aspect of the problem is not that well explored in the existing literature. Women are proven to be more affected by the overall environmental degradation in the Aral Sea Basin, which is particularly evident through the analysis of their health—statistically, females in the region are more predisposed to anemia and some types of cancer (Lioubimtseva 2014). Many Central Asian countries can still be characterized by partially traditional societies, which are even more common in remote rural areas. Moreover, in some of the basin countries there is a trend to revive traditional gender stereotypes, making women more vulnerable, changing their roles in society and economy, resulting in corresponding regress to disparities (Usha 2016).

To properly monitor the progress, Indicator 6.2.1 is divided into three sub-indicators: proportion of population practicing open defecation; per centage of population using at least basic sanitation services; and per centage of population with basic handwashing facilities on premises. Since almost no data are available on safely managed sanitation services, an approximate indicator was used for the analysis—access to at least basic sanitation services. In the case of this indicator and the indicator on open defecation, every country within the Aral Sea Basin had reached the target, with the exception of Afghanistan. On Figure 13.4, Afghanistan's current and expected progress in these fields is shown. We can anticipate that the country will be able to eliminate the open defecation practice by 2030; however, it still will be far from reaching the goal in the case of access to even basic sanitation services, especially taking into account the huge gap between urban and rural population.

The third sub-indicator on access to handwashing facilities remains more challenging for the region as a whole. It can be noticed from the Figure 13.5 that stagnation is typical for most countries in the Basin, except Turkmenistan, most likely because of the country's clearly

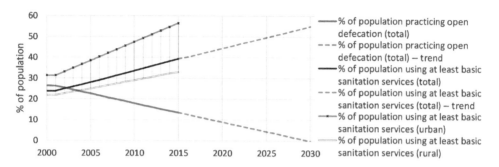

*Figure 13.4* Afghanistan's progress

Source: Constructed by authors based on data from UNSD (2019); WHO/ UNICEF (2019)

defined priorities in ensuring universal access to clean water and sanitation for its citizens. Based on the current trends, Afghanistan will remain far from reaching the goal by 2030, with a particularly difficult situation for the rural population. Unfortunately, not enough data are available for Uzbekistan to monitor this country's progress. Based on the national reports, only around 38 per cent of its population had access to a sanitation system in 2013 (ranging from 65 per cent in cities to 1 per cent in remote rural areas) (Abduraimov 2016). However, Uzbekistan adopted an 'Integrated development and modernization program on drinking water supply and sewage system for 2017–2021', which is focusing on the improvement of people's access to safe water and sanitation, with special attention to rural areas.

## Target 6.3. *Indicator 6.3.1: Proportion of wastewater safely treated*

The sewerage system in the Basin remains quite unevenly distributed and generally can be described as inadequate. In the state of chronic water shortages and Central Asia's hot and dry climate, this becomes an urgent issue (Micklin *et al.* 2014). The lack of appropriate wastewater

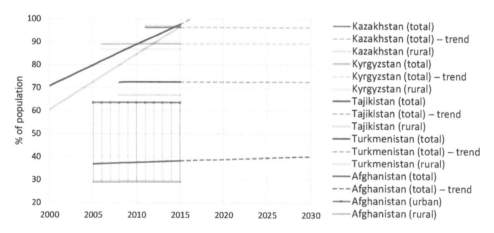

*Figure 13.5* Proportion of population with basic handwashing facilities on premises

Source: Constructed by authors based on data from UNSD (2019)

treatment of both municipalities and industries can have a serious effect on other components of the system—the quality of water bodies, ecosystems, people's health, access to clean drinking water (Bekturganov *et al.* 2016). Currently, more than half of all sewage water in the region is discharged into waterways untreated and thus only aggravates the existing large-scale contamination of the Basin environment (Reimov & Fayzieva 2014).

Methodologically, Indicator 6.3.1 is subdivided into two parameters: domestic wastewater and industrial wastewater, while Indicator 6.3.2 assesses the overall impact of untreated sewage, including agricultural run-off. Unfortunately, not enough reliable data are currently available to monitor the progress of these indicators. Analyzed national reports include in most cases statistics mainly on urban wastewater treatment, as the sewage system in rural areas is very often underdeveloped and the industrial sector not strongly controlled.

Kazakhstan seems to have the most detailed and open statistics available among the Basin countries. According to them, the proportion of industrial wastewater treated in the two regions of the country that fall within the Aral Sea Basin (Kyzylorda and Turkistan regions) was around 95 per cent in 2016 (ME of KZ 2017), while the proportion of treated urban wastewater was 99 per cent in 2017 (MNE of KZ 2019). Such impressive figures can be explained by the country's strong policies on wastewater treatment. However, it is hard to say whether these numbers would fully represent the country's SDG progress. Currently, most of Kazakhstan's wastewater is discharged without secondary-level treatment, and a major part of the country's infrastructure is aged and outdated, leading to significant water losses. However, other Basin countries are in a more complicated situation with respect to sewage water treatment, particularly due to the lack of data and overall unaccountability for water use and discharge.

In Kyrgyzstan, only one-fifth of the population is provided with a central drainage system (NCSD of KG 2018), while in the industrial sector there is often almost zero control over the wastewater discharge. Taking this into account, a national report stated that in 2014 around 94.5 per cent of wastewater was treated (STEPF of KG 2016). In Tajikistan, many small cities and urban settlements do not have access to a sewage system, while around 70 per cent of the existing treatment plants are old and require replacement (UNECE 2017). However, over the last years, there was no significant rehabilitation of infrastructure. At the same time, since 2010, Tajikistan's Agency of Statistics stopped collecting data on water use and pollution (UNECE 2017). A similar situation exists in Turkmenistan—a considerable part of the urban population still does not have access to the sewage network, while the condition of the existing treatment plants remains poor (UNECE 2012). In Uzbekistan, only around 60 per cent of the population in big cities has access to wastewater systems, with almost no access to sanitation in rural areas. Industrial wastewater is often discharged without proper treatment or is not treated at all. However, Afghanistan is in a much more difficult situation compared to other Basin countries. Wastewater management is practically not developed in Afghanistan, due to the lack of finances, expertise, and overall limited water resources. At the same time, it is important to keep in mind that almost 99 per cent of the water withdrawal is in the agricultural sector, which is not covered by Indicator 6.3.1.

## Target 6.3. Indicator 6.3.2: Proportion of bodies of water with good ambient water quality

This indicator represents combined impact of untreated wastewater and other pollutants on inland water (open water bodies, rivers, and groundwater). It includes the agricultural sector, which was not covered by the previous indicator. From the historical perspective, water pollution, as well as overall environmental degradation, became a serious issue in the Basin since the beginning of the extensive use of irrigation along the Amu Darya and Syr Darya—rivers supplying the Aral Sea (Zhuang 2016). Fertilizers and pesticides from the agricultural areas polluted waterways as well as the Aral Sea itself, and as the process of drying out was progressing, the chemicals were accumulating in the lake. Not only did the quality of the Sea's water drastically decline, but the groundwater was also contaminated by salts and other minerals (Lioubimtseva 2015). However, limited availability of clean surface water and growing competition for water resources in the region can force some countries to consider a large-scale use of groundwater resources in the near future (Rakhmatullaev *et al.* 2010).

At the same time, there is a rare example of a significant improvement in water quality in the Basin—the case of the Small (Northern) Aral in Kazakhstan (Lioubimtseva 2015). A new dam separated the northern part of the lake and, as a result, not only did the water level in this section of the Sea increase and the salinity level drop, but even the biodiversity in the Small Aral slowly started to be restored (Aladin *et al.* 2016).

Existing water quality monitoring activities in the region do not allow the riparian countries to provide enough information and properly assess the status of indicator 6.3.2. However, based on national data, an overview of the situation can be made. According to Kazakhstan's national report, the Syr Darya, as well as all transboundary rivers between Kazakhstan and Kyrgyzstan, were assessed as 'moderately polluted' (ME of KZ 2017). It is noticeable from the Kyrgyz report that water quality has slightly declined in recent years (STEPF of KG 2016). Turkmenistan's and Uzbekistan's reports assess the Amu Darya's waters as 'moderately polluted', with more than 5 billion $m^3$ of agricultural runoff entering the river flow every year (UNECE 2012; SCNP of UZ 2013). In the case of Tajikistan, comprehensive data on pollution are not available since the Agency of Statistics has stopped collecting water-related information (UNECE 2017).

At the moment, there are not enough data to monitor the progress of both Indicators 6.3.1 (wastewater) and 6.3.2 (water quality) on the Basin level. However, based on the national reports we can assume that the situation is far from perfect. Thus, Target 6.3 remains one of the most pressing for the region, particularly due to the absence of accurate and reliable information. Availability of information on freshwater is a recognized issue, as most countries in the world cannot guaranty quality and consistency of monitoring data (UN Environment 2018a). In the case of the Aral Sea Basin, typically too many governmental authorities are involved in the monitoring activities, while too many standards inherited from the Soviet system still exist in the legislation and must be followed. The work of responsible agencies is often uncoordinated and insufficiently funded, while the scattered data which are actually collected are inaccessible for the public or even for other state organizations.

## Target 6.4. Indicator 6.4.1: Change in water use efficiency over time

The water infrastructure (for municipal use and irrigation) of the Aral Sea Basin countries is a legacy mostly from the Soviet Union time and has been deteriorating ever since. Even more developed countries of the region, like Kazakhstan, still rely on a technically depreciated water distribution system that is resulting in around 40-per cent water losses in transportation (ME of KZ 2017). Other countries are in a similar situation. One of the reasons concerns undeveloped principles of pricing and tariffs on water supply that cannot cover operating and investment costs. The problem can be addressed through the valuing of water principles, as discussed in Chapter 12, and covering the costs associated with operation and management of irrigation infrastructure by raising the currently low service fees, which cover only a small portion of required expenses. It is also important to considerably increase local and international invest-ments in maintenance and replacement of irrigation infrastructure and the introduction of new technologies. In order to reduce the socio-economic pressure on water resources, modern water-saving technologies and recycling water supply systems should be introduced, ensuring a reduction in losses and in unaccounted water for agricultural, industrial and municipal purposes.

The baseline data for Indicator 6.4.1 were only recently calculated by the FAO and UN-Water (FAO 2018b) since it was introduced in the SDGs for the first time and an entirely new methodology had to be developed. Water use efficiency is defined as the 'value added per unit of water used, expressed in USD/m³', and covers three major economic sectors: agricul-ture, industry, and services (FAO 2018b). While the indicator itself is formulated to address the trend in water use efficiency over time, since it is such a new parameter, at the moment, available data are not sufficient to analyze the progress, but only to identify the baseline. Figure 13.6 comprises available data for both *Indicator 6.4.1 on water use efficiency* (yellow) and *Indicator 6.4.2 on level of water stress* (blue). From that figure, it is clear that Kazakhstan is the leader in the region while all other Basin countries are on the same level. This indicator should be regarded as country- and region-specific; however, on a global scale, countries of the Aral Sea Basin are quite obviously falling behind—on average, water use efficiency in the world is around 15 USD/m³, in Central and Southern Asia, 2 USD/m³ (FAO 2018b), while for the Basin coun-tries (apart from Kazakhstan) this parameter equals 0.44 USD/m³ (Figure 13.6).

### Target 6.4. Indicator 6.4.2: Level of water stress: freshwater withdrawal as a proportion of available freshwater resources

In Central Asia, human activities have significantly altered the Aral Sea Basin hydrological cycle, while growing population density in some regions causes higher pressure on limited water resources (Der Beek *et al.* 2011). Considering water scarcity of the region and the lack of agree-ment between nations, ensuring efficient and fair use of this shared resource is both essential for the region and quite difficult to achieve. While the ongoing competition between agricultural and energy sectors is causing the main conflict in the Basin's freshwater use (Rakhmatullaev *et al.* 2017), severe water stress that is already present in riparian countries is very likely to increase in the future (Lioubimtseva 2015). A related issue is the fact that current projections and water use models are quite uncertain about future development trends (Malsy *et al.* 2015).

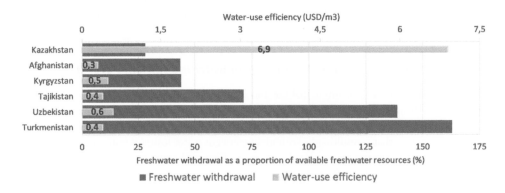

*Figure 13.6* Water-use efficiency and level of water stress (freshwater withdrawal)

Source: Constructed by authors based on data from UNSD (2019); FAO (2018)

The most detailed data on this indicator can be accessed through the FAO database, AQUASTAT (FAO 2016), though they are often scattered and incomplete. Relatively outdated information is available for Basin countries for only one to three separate years (since 2000), which does not permit assumptions on future trends. However, data inconsistency, data being outdated, weak monitoring, coordination and reporting are issues that are recognized on a global scale (FAO 2018a). Figure 13.6 shows the current level of water stress in the Basin countries (freshwater withdrawal as a proportion of available freshwater resources). It can be noticed that, similar to the previous indicator, Kazakhstan is leading in the region. At the same time, Turkmenistan and Uzbekistan are experiencing extreme levels of water stress, close to the levels experienced by countries in Northern Africa.

Water balance in the region could be achieved in case all Basin countries would shift to water saving practices and improve water accounting and auditing and through this reduce their water consumption. Experience in implementing integrated water resources management (IWRM) in the Ferghana Valley, which is shared by Kyrgyzstan, Tajikistan, and Uzbekistan, has shown that it is possible to actually reduce water withdrawal from transboundary rivers by 20 per cent (Sokolov 2006). Water conservation is a laborious activity; however, it is possible to reach a 1-per cent rate of water saving in the region. Taking into account the transborder nature of the issue and high competition over water resources in the Basin, effective communication and cooperation between all involved stakeholders is needed, supported by educational programs that would make next generations more aware and conscious about the water crisis, and prepare them for the future challenges. Through this example, it is clear how closely *Target 6.4 on water-use efficiency* is connected to *Target 6.5 on the promotion of IWRM and transboundary cooperation* in the Basin conditions.

### Target 6.5. Indicator 6.5.1: Degree of integrated water resources management implementation (0–100)

The first steps toward implementation of integrated water resources management (IWRM) at the Basin level in Central Asia started in the 1980s when the need for a single water management organization for the whole Basin was recognized in response to the analysis of water shortages (Dukhovny & Sokolov 2003). In 1993 the International Fund for Saving the Aral Sea (IFAS) was established, whose mission is to coordinate regional and national activities and ensure water security in the Aral Sea Basin. More specifically, IFAS is responsible for the implementation of the Aral Sea Basin Program (ASBP)—the core long-term action program focused on sustainable development. The last Program (ASBP-3), which finished in 2015, included a special activity area for the proposed projects—Direction 1 on Integrated Use of Water Resources (ASBP-3 2012). It had some relevant priorities that were formulated to define the actions and development within the program's time frame. According to the recently proposed concept of the new ASBP-4 development (project to be presented in late 2019), IWRM implementation is listed as a number-one priority.

Despite the fact that there is a good foundation for the introduction of IWRM into the water strategy of the Aral Sea Basin, the level of IWRM implementation in the riparian states, at this time, is low. As an example, Kazakhstan's National IWRM program was developed together with UNDP back in 2006, but has still not been adopted by the government. However, Kazakhstan has included the IWRM concept in its national water strategy. In 2015, Kyrgyzstan introduced a Water Code based on IWRM. However, the development of water strategy on a regional level, which was proposed by the World Bank, is still under consideration by some countries of the Basin. There are a number of obstacles in the way of IWRM implementation in the region. Different interpretations of this concept, which is believed by some experts to be equal to the 'complex schemes of water use and protection of water resources' term used in the Soviet Union, lead to disputes over realizing IWRM. Another issue is the difficulty of changing the institutional system of post-Soviet countries as they have centralized and hierarchical government structures, and lack initiative and the formal requirement of public participation in decision-making, as well as having limited financial and human resources (Uvarov *et al.* 2015).

The progress report published by UN Environment in 2018 included a detailed assessment of the current degree of IWRM implementation for only three of the Basin countries—Kazakhstan, Uzbekistan, and Afghanistan (Tajikistan provided incomplete data, while there was no information from Kyrgyzstan and Turkmenistan) (UN Environment 2018b). Indicator 6.5.1 itself is measured on a scale from 0 to 100 and is evaluated through the country's self-assessment questionnaire, which covers four dimensions (Enabling environment; Institutions and participation; Management instruments; Financing). Figure 13.7 represents the status of the three countries that had submitted their answers. While Uzbekistan has the highest score, it is important to mention that the UN report expresses concern that countries whose scores are less than 50 are unlikely to meet the global target (UN Environment 2018b).

### Target 6.5. Indicator 6.5.2: Proportion of transboundary basin area with an operational arrangement for water cooperation

The earliest formal agreement on the collaborative use and protection of transboundary water resources of the Aral Sea Basin was accepted in 1992 and included five out of six countries that share this Basin (with the exception of Afghanistan). Later cooperation agreements and organizations traditionally involved five initial riparian countries, which is also true for the recent major agreement that was signed in 2008—the Statute of the Interstate Commission for Water Coordination of Central Asia (ICWC). Starting from 2018, work on the establishment of the new Aral Sea Basin Program (ASBP-4) is ongoing. The conceptual framework of ASBP-4, developed in close cooperation with GIZ, was recently approved by the International Fund for Saving the Aral Sea (IFAS).

Considering high competition for water resources in the region, intergovernmental coordination of organization and comprehensive programs is essential to ensure transition to

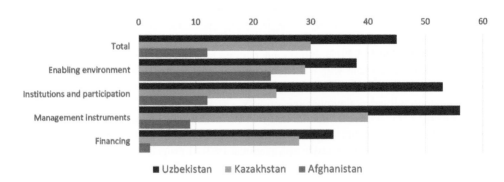

*Figure 13.7* Degree of integrated water resources management implementation (0–100)

Source: Constructed by authors based on data from UN Environment (2018b)

more sustainable resource management. A core conflict in the region exists between the use of water for energy by the upstream countries and for irrigation by downstream countries (Rakhmatullaev *et al.* 2017). It seems that involved countries in the Basin are not yet capable of reaching consensus. Each nation promotes its own policies related to the water–food–energy nexus focusing on economic growth while overlooking the water security issue on a regional scale (Lee & Jung 2018).

Indicator 6.5.2 includes two sub-indicators: on the transboundary river and lake basins, and on transboundary aquifers. While some countries of the Basin provided data on the state of agreements on rivers and lakes, there is almost no information on the groundwater component (UNECE/UNESCO 2018). While in Kazakhstan, 100 per cent of the country's rivers and lakes are covered by an operational arrangement for water cooperation, it was reported that 0-per cent of its underground aquifers are covered by such agreements, resulting in the final score of 72.4 per cent. It is worth noticing that Kazakhstan was the only country in the Basin that provided information on the state of its transboundary underground water. Most probably, other countries have also not yet established working agreements on their shared groundwater sources, as this topic remains very complicated and sensitive for the region. In Uzbekistan only 59.3 per cent of river and lakes are covered by the agreement, while in Afghanistan this parameter equals 51.7 per cent (not participating in the Interstate Commission for Water Coordination of Central Asia). Information submitted by Turkmenistan still needs clarification, while there was no response from Kyrgyzstan and Tajikistan (UNECE/UNESCO 2018).

In the current situation, it is possible to say that the existing regional agreement on transboundary water resources of the Aral Sea Basin formally fulfils the conditions of one of the sub-indicators. However, without the agreements on groundwater resources, *Target 6.5* cannot be reached in the Basin. Competition for water and related conflicts of interest, lack of coordination, and inefficient work of the ICWC led to a serious deterioration of the situation in the Aral Sea Basin years after the first comprehensive agreement was signed (McCracken & Meyer 2018). The current situation indicates that the cooperation activities in the region at some point have become simply formalistic (Xu 2017). The focus should be on the quality of cooperation among Basin countries, on facilitating the collaborative activities and improving the water management system (McCracken & Meyer 2018).

## Target 6.6. Indicator 6.6.1: Change in the extent of water-related ecosystems over time

The most devastating environmental crisis that has happened in the Basin is, of course, the desiccation of the Aral Sea, which caused massive transformations not only in the vicinity of the lake, but also in a wider area, resulting in a change of ecosystems, local climate, land cover, and water quality (Lioubimtseva 2015). Reduction of river inflow resulted in the drastic decrease of many species of birds and fish, leading not only to the loss of biodiversity in the region but also undermining livelihoods of the local population. However, there are cases that have helped to mitigate the consequences of the Aral Sea disaster for the surrounding ecosystems. One example is the creation of a system of artificial lakes, which has helped preserve a number of migratory bird species, especially in the Karadzhar, Sudochinsk, and Mezhdurechensk areas of Uzbekistan (Duhovniy 2017).

In 2006, Central Asian countries signed the 'Framework Convention on Environmental Protection for Sustainable Development in Central Asia', which specifically highlights the importance of the Aral Sea Basin. The aim of the Convention is 'to ensure the effective protection of the environment for sustainable development in Central Asia, including the improvement of the ecological environment, rational use of natural resources, as well as reduce and prevent transboundary environmental damage through the harmonization and coordination of environmental policies and actions of the Contracting Parties and by establishing mutual rights and responsibilities'.

Indicator 6.6.1 methodologically incorporated five categories of water-related ecosystems (vegetated wetlands, rivers and estuaries, lakes, aquifers, and artificial water bodies), while there are three extent components that must be monitored (spatial extent, quality, quantity) (UN Environment 2018c). One of the easiest and most efficient ways to assess the changes in spatial extent is using earth observations (remote sensing) that can be done without the involvement of national institutions, and which form the main data available for the progress assessment of the Basin countries. Since 2005, all countries have experienced loss in the extent of open-water bodies, as can be seen on Figure 13.8. Each state is affected by shrinkage of water bodies when comparing statistics for two time periods (2006–2010 and 2011–2015). The worst situation is in Uzbekistan, as it has lost almost 45 per cent of open-water bodies since 2005.

One of the important technical constraints in reaching Target 6.6 is related to the fact that the deadline for this target is 2020, not 2030. Considering the drastic changes in the Aral Sea Basin ecosystems that have happened over the last decades as well as the lack of success in restoring the environment around the Aral Sea, this target will be quite impossible to achieve. However, countries should focus on reaching this and other water-related targets, learning from success stories (for instance, the Small Aral restoration), to ensure restoration of ecosystems and people's livelihoods, which often rely on the surrounding environment.

### Target 6.a. *Indicator 6.a.1: Amount of water- and sanitation-related official development assistance that is part of a government-coordinated spending plan*

Right after the fall of the Soviet Union, international donors started providing financial and other types of aid to Aral Sea Basin countries. In total, 630.8 million USD of grants and loans were offered to the region in the 1995–2005 period. The main contributors were the World Bank, the Asian Development Bank, Germany, Japan, Kuwait, the USA, the Global Infrastructure Facility, the European Union, France, and Switzerland (UNDP 2008). Over the period 2005–2010, the Government of Kyrgyzstan's water-related investments accounted for 32.71 million USD on average per year, which is only 0.1 per cent of total government

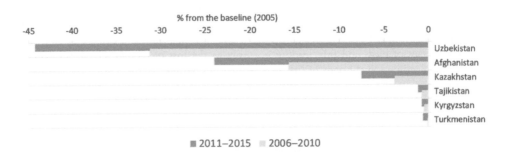

*Figure 13.8* Change in the extent of open-water bodies over time

Source: Constructed by authors based on data from UN Environment (2018c)

expenditures (UN-Water 2013). However, the government's investments are increasing, and with the support from development partners, estimated funding for improving public access to water resources and sanitation services for the 2018–2040 period will be at least 330 million USD (NCSD of KG 2018).

In the period 2010–2016, more than 63 water-related projects were implemented in Uzbekistan with the support of international donors, whereas 68 per cent of them were financed through grants and the rest through the loans and credits (Djumaboev *et al.* 2017). However, the grant budgets with an average of 1.8 million USD are really small in comparison to the 160-million-USD average of loan/credit-financed projects. The Uzbek government recently adopted a state program for 2017–2021 intended to 'improve the living conditions and quality of life of the region's population providing socio-economic development aid for the Aral Sea Basin' through the implementation of 67 projects with an estimated budget of 2.58 billion USD.

Last year Ashgabat hosted an international conference, 'Partnership for Financing Development in the Heart of the Great Silk Road', which looked at global and regional trends and prospects for financing the SDGs in Central Asia. The same year, leaders of the Central Asian countries met at the first Aral Sea summit since 2009, where they discussed the current situation in the light of the Sustainable Development Agenda and the needs for better transboundary cooperation, a review of the IFAS's role and operations, and the raising of awareness and funds.

Statistical representation of Indicator 6.a.1 is through total official development assistance for water supply and sanitation, a parameter that measures international aid flow. However, it does not directly demonstrate success in international cooperation activities. Figure 13.9 shows data for Basin countries. While Afghanistan is leading, two other countries can be characterized by having received minimal assistance—Kazakhstan (which was receiving substantial funds in 2008/2009) and Turkmenistan (for which data were available only until 2011).

### Target 6.b. Indicator 6.b.1: Proportion of local administrative units with established and operational policies and procedures for participation of local communities in water and sanitation management

The involvement of all relevant stakeholders, including local communities and private industry, is key for effective water resource management, especially in such a complex basin as the Aral Sea. The successful engagement of all parties concerned can be facilitated by corresponding supportive national-level policies (Qadir *et al.* 2018). Traditional watershed management practices often

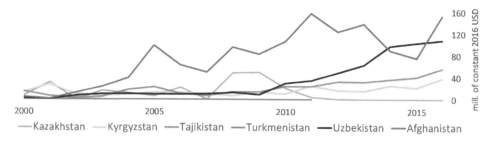

*Figure 13.9* Total official development assistance (gross disbursement) for water supply and sanitation
Source: Constructed by authors based on data from UNSD (2019)

overlook the importance of public participation. Local communities can help shape the most appropriate objectives and define goals, as they are the ones benefiting from proposed projects. Additional resources and time must be devoted to ensuring an inclusive decision-making process; however, reluctance in shifting to such an approach can only lead to failure in the future (Wang *et al.* 2016).

Historically, civil society was not much involved in the decisions around projects' implementation in the region. The lack of success in the past should be a sign for state governments that municipalities and local communities must play a bigger role in recovery activities and sustainable development. It is possible to say that in the current situation local administrations and civil society do not take any significant part in the water resources management in the Aral Sea region. Local cooperative societies are often still in the process of formation and do not have much in the way of rights or recognition from the government. At the same time, non-governmental organizations (NGOs), which are more common, might not have strong connections on the local level as they are predominantly found in the major cities.

## Other water-related SDGs

As was mentioned many times in the previous sections of the chapter, in case of the Aral Sea Basin, many water-related issues must be addressed together, otherwise none can be fully solved. This is true for both individual Sustainable Development Goals, as they were meant to be interconnected, and for some particular groups of targets and indicators, for example, the first three targets of SDG 6 (*Target 6.1 on safe drinking water*, *Target 6.2 on sanitation*, and *Target 6.3 on water quality*) as well as for *Target 6.4 on efficient water use* and *Target 6.5 on IWRM and transboundary cooperation*. Considering the situation in the Basin, where conflicts over scarce water resources are frequent, adequate and operational cooperation mechanisms between countries are desperately needed. Each one of the SDGs represents a critical issue that cannot be resolved on its own, separately from others. Water is one of the most crucial factors in the region, connecting SDGs from different dimensions of sustainability (environment, society, economy, governance). Besides SDG 6, a number of other goals include targets and indicators that are related to water and sanitation issues, even though they are not explicitly mentioned. The current status of some of these indicators in the Basin countries are presented in Table 13.2.

*Table 13.2* Status of other water-related indicators.

| Goal and indicator | KZ | KG | TJ | TM | UZ | AF |
|---|---|---|---|---|---|---|
| **Goal 1: End poverty in all its forms everywhere** | | | | | | |
| Indicator 1.2.1 Proportion of population living below the national poverty line (%) | | 2.7 | 25.4 | 31.3 | no data | 14.1 (2013) | 35.8 (2011) |
| **Goal 2: End hunger, achieve food security and improved nutrition, and promote sustainable agriculture** | | | | | | |
| Indicator 2.1.1 Prevalence of undernourishment (%) | <2.5 | 6.4 | 30.1 | 5.5 | 6.3 | 23 |
| **Goal 3: Ensure healthy lives and promote well-being for all at all ages** | | | | | | |
| Indicator 3.9.2 Mortality rate attributed to unsafe water, unsafe sanitation, and lack of hygiene (deaths per 100,000 population) | 0.4 | 0.8 | 2.7 | 4.0 | 0.4 | 13.9 |

**Goal 7: Ensure access to affordable, reliable, sustainable and modern energy for all**

| | | | | | | |
|---|---|---|---|---|---|---|
| Indicator 7.2.1 Renewable energy share in the total final energy consumption (%) | 1.6 | 23.3 | 44.7 | 0.0 | 3.0 | 18.4 |

**Goal 11: Make cities and human settlements inclusive, safe, resilient, and sustainable**

| | | | | | | |
|---|---|---|---|---|---|---|
| Indicator 11.5.1, Series: number of deaths and missing persons attributed to disasters per 100,000 population (number) (same as for SDG 13) | 0.2 | 0.0 | no data | no data | 0.0 | no data |
| Indicator 11.5.1, Series: number of directly affected persons attributed to disasters per 100,000 population (number) (same as for SDG 13) | 68.0 | 1.6 | no data | no data | 0.0 | no data |
| Indicator 11.5.2 Direct economic loss attributed to disasters (millions of current USD) | 12.3 | 0.1 | no data | no data | 0.0 | no data |

**Goal 15: Protect, restore, and promote sustainable use of terrestrial ecosystems, sustainably manage forests, combat desertification, and halt and reverse land degradation and halt biodiversity loss**

| | | | | | | |
|---|---|---|---|---|---|---|
| Indicator 15.1.2 Average proportion of Freshwater Key Biodiversity Areas (KBAs) covered by protected areas (%) | 17.4 | 31.1 | 34.6 | 13.1 | 10.4 | 0.1 |

*Data source:* UNSD (2019). For the countries' names the UN/LOCODE codes are used: KZ – Kazakhstan; KG – Kyrgyzstan; TJ – Tajikistan; TM – Turkmenistan; UZ – Uzbekistan; AF – Afghanistan.

People living in poverty are among the most vulnerable to water shortages as they have limited access to such basic services as safe drinking water and adequate sanitation. Eradication of poverty addressed in SDG 1 would help to ensure the basic human right for water supply, adequate sanitation, and hygiene (WASH), as the elite are not affected by water scarcity. Poor WASH, in return, is directly connected to health security (SDG 3), threatening the well-being of people. Efficient water management can provide food and jobs, secure the economy, contribute to SDG 2 (ending hunger), and support productive employment and work for all (SDG 8). A major part of developing countries' freshwater resources is used for agricultural purposes. The need for improved nutrition is linked to healthy lifestyles of vulnerable groups of population, especially for children (SDG 3). People living in poverty, women, children, and elderly are the most vulnerable groups of the population with respect to disasters. Water-related natural disasters are the most common disasters and they are particularly addressed by SDG 11 (sustainable cities and communities) and SDG 13 (climate action).

Nations have to ensure energy security (SDG 7) and encourage the shift to sustainable and clean energy production, which implies the use of renewable energy sources such as sun, wind, and water. This, in turn, requires addressing the issues of industrialization, innovation, and infrastructure (SDG 9). Hydropower plays an important part in the energy balance of all countries in the Aral Sea Basin, being the main source of electricity in Tajikistan (more than 98 per cent) and Kyrgyzstan. Thus, these states could focus on the modernization and expansion of hydrological and transmission facilities.

Failure to implement IWRM on the basin level can lead to irreversible consequences for the environment. On the other hand, human needs must be met without major disruptions of the ecosystems (SDG 15). Drastic changes of the hydrological regime in the Aral Sea Basin due to the construction of dams along the main tributaries (the Syr Darya and Amur Darya Rivers) can change the local microclimate and ecosystems upstream and downstream, potentially intensifying extreme weather conditions, causing droughts and floods. Improvement of the resilience and robustness of the basin system under climate change conditions is an important part of IWRM (SDG 13). Transboundary cooperation and proper water management of international river basins have been recognized as a crucial part of sustainable development and are explicitly mentioned not only in SDG 6, but also are essential parts of SDG 16 (on the promotion of peace) and SDG 17 (on global partnership). IWRM provides a framework for consensus between food, ecosystem, and energy production, as water can and must play a unifying role between conflicting parties.

## Conclusions

At the current stage, efforts in the Aral Sea Basin to achieve Sustainable Development Goal 6 'Clean Water and Sanitation' face a number of challenges. Even some of the most basic human rights (like access to safe water and adequate sanitation) are still not provided universally among riparian states. The situation in Afghanistan stands out as the most severe, with less than 63 per cent of the population using basic drinking water services, 13 per cent still practicing open defecation (eradicated in other Basin countries), and only around 39 per cent having access to basic sanitation services and handwashing facilities on premises. In this regard, there is also still a lot to be done in Tajikistan and Kyrgyzstan.

SDG 6 progress is uneven. Afghanistan is lagging far behind the other countries, but there are disparities between these as well, depending on the indicator. Some disparities between states will remain, but the existing drastic contrast between the urban and rural population needs to be addressed. Nonetheless, taking into account existing trends, by 2030 all countries will be able to reach Target 6.1 (drinking water), while Afghanistan will be able to significantly improve some of its indicators. In the case of another common issue in the region, access to handwashing facilities, development can be expected to be less remarkable. This fact, as well as the generally poor water quality, can have serious impacts on the local population's health.

The twenty-first century offers innovations and improved technologies (more efficient irrigation equipment, water saving methods, etc.), which could support more rational water use. The Basin is currently lagging behind in terms of water-use efficiency, and is at the same time subject to high levels of water stress across all countries. Maximum levels of more than 100 per cent are attained in Turkmenistan and Uzbekistan. Promotion of water-efficient practices and cost-effective alternative energy production technologies may reduce the dependence of upstream countries on hydropower and improve the availability of water resources for the downstream areas that heavily rely on agriculture. As introduction of new technologies would require significant investments, it is crucial to find different sources of financing for water-related projects, such as increasing state funding, attracting international donors, and raising water fees.

Operational transboundary water cooperation agreements, functioning regional organizations, and active dialogue between stakeholders are urgently needed to ensure the decent livelihood of the Aral Sea Basin's population. The tension between upstream and downstream countries in the region, which has built up over the last three decades, may be released by enhanced cooperation between industries and national governments. Applying the concepts of the water–energy–food nexus as well as a 'benefits sharing' approach within the integrated water resources management framework on the Basin level should help riparian countries reach SDGs. These issues will be further aggravated by changes in water balance due to climate change.

Another important dimension of SDG 6 progress in the Basin is data availability and, correspondingly, national monitoring systems. While the lack of comprehensive statistical data is a common issue for SDG indicators in general (particularly for Target 6.3 on water quality and wastewater), this situation requires special attention in the case of the Aral Sea Basin, where water plays such a crucial role. After the fall of the Soviet Union, a large number of hydrological monitoring stations in Central Asia were closed. Thus, data availability and accuracy of such vital parameters as water discharge and water quality are quite limited now. Considering the complexity of the river network within the Basin and the importance of adequate monitoring of such indicators not only for SDGs but, for instance, for IWRM and efficient water-use, national monitoring systems must be reconstructed and enhanced.

# References

Abduraimov, M. (2016). Uzbekistan – drinking water supply and sanitation, existing problems and potential solutions in the light of achieving strategic development goals [Узбекистан – питьевое водоснабжение и санитария, существующие проблемы и пути их решения в свете достижения стратегических целей развития]. Regional Information System on Water and Land Resources in the Aral Sea Basin (CAWater-IS). Retrieved 10 March 2019 from www.cawater-info.net/pdf/abduraimov16.pdf.

Aladin, N. V., Micklin, P. T., Plotnikov, I. S., Ermakhanov, Z. K., & Aladin, I. N. (2016). The partial restoration of the Aral Sea and the biological, socio-economic and health conditions in the region. *Человеческий Капитал и Профессиональное Образование*, *3*, 62–73.

Alzhanova, A. (2017). *Implementation of the Sustainable Development Goals in the SPECA Region*. Study commissioned by UNECE. United Nations Special Programme for the Economies of Central Asia (SPECA). Twelfth session of the SPECA Governing Council, Dushanbe, Tajikistan, 5–6 December 2017. Retrieved 10 March 2019 from www.unece.org/fileadmin/DAM/SPECA/documents/gc/session12/UNECE_Study_GC_English.pdf.

ASBP-3 (2012). Program of actions on providing assistance to the countries of the Aral Sea Basin for the period of 2011–2015 (ASBP-3). International Fund for Saving the Aral Sea. Retrieved 8 March 2019 from www.cawater-info.net/library/eng/asbp3_e.pdf.

Bekturganov, Z., Tussupova, K., Berndtsson, R., Sharapatova, N., Aryngazin, K., & Zhanasova, M. (2016). Water related health problems in central asia: a review. *Water*, *8*(6), 219.

Crighton, E. J., Elliott, S. J., Upshur, R., van der Meer, J., & Small, I. (2003). The Aral Sea disaster and self-rated health. *Health & Place*, *9*(2), 73–82.

Der Beek, T. A., Voß, F., & Flörke, M. (2011). Modelling the impact of global change on the hydrological system of the Aral Sea basin. *Physics and Chemistry of the Earth, Parts A/B/C*, *36*(13), 684–695.

Djumaboev, K., Anarbekov, O., Holmatov, B., & Hamidov, A. (2017). Overview of water-related programs in Uzbekistan. Project report of the Sustainable Management of Water Resources in Rural Areas in Uzbekistan. Component 1: National policy framework for water governance and integrated water resources management and supply part. Colombo, Sri Lanka: International Water Management Institute (IWMI).

Duhovniy, V. (ed.) (2017). *Aral Sea and Aral Region* [Аральское море и Приаралье]. Tashkent, Uzbekistan: Baktria Press.

Dukhovny, V. & Sokolov, V. (2003). Lessons on cooperation building to manage water conflict in the Aral Sea basin. In *Technical Documents in Hydrology, 11*. Paris: UNESCO.

FAO (2016). AQUASTAT Main Database, Food and Agriculture Organization of the United Nations (FAO). Retrieved from www.fao.org/nr/water/aquastat/data/query/.

FAO (2018a). *Progress on Level of Water Stress – Global Baseline for SDG 6 Indicator 6.4.2*. Rome: FAO/UN-Water.

FAO (2018b). *Progress on Water Use Efficiency – Global Baseline for SDG 6 Indicator 6.4.1*. Rome: FAO/UN-Water.

Gupta, A. & Gupta, A. (2016). Environmental challenges in Aral Sea basin: impact on human health. *International Research Journal of Social Sciences*, 6, 419–440.

Lee, S. O. & Jung, Y. (2018). Efficiency of water use and its implications for a water-food nexus in the Aral Sea Basin. *Agricultural Water Management*, 207, 80–90.

Lioubimtseva, E. (2014). Impact of climate change on the Aral Sea and its Basin. In Micklin, P., Aladin, N. V. & Plotnikov, I. (eds), *The Aral Sea: The Devastation and Partial Rehabilitation of a Great Lake*. Berlin: Springer, pp 405–427.

Lioubimtseva, E. (2015). A multi-scale assessment of human vulnerability to climate change in the Aral Sea Basin. *Environmental Earth Sciences*, 73(2), 719–729.

Malsy, M., aus der Beek, T., & Flörke, M. (2015). Evaluation of large-scale precipitation data sets for water resources modelling in Central Asia. *Environmental Earth Sciences*, 73(2), 787–799.

McCracken, M. & Meyer, C. (2018). Monitoring of transboundary water cooperation: Review of Sustainable Development Goal Indicator 6.5.2 methodology. *Journal of Hydrology*, 563, 1–12.

ME of AF (2017). SDGs' progress report – Afghanistan. Voluntary national review at the high level political forum. Ministry of Economy, General Directorate of Policy & RBM Kabul, Afghanistan. Retrieved 10 March 2019 from https://sustainabledevelopment.un.org/content/documents/16277Afghanistan.pdf.

ME of KZ (2017). National report on the state of the environment and natural resource management of the Republic of Kazakhstan in 2016 [Национальный доклад о состоянии окружающей среды и об использовании природных ресурсов Республики Казахстан за 2016 год]. Ministry of Energy of the Republic of Kazakhstan, Astana.

Micklin, P., Aladin, N. V. & Plotnikov, I. (eds) (2014). *Aral Sea: The Devastation and Partial Rehabilitation of a Great Lake*. Berlin: Springer.

MNE of KZ (2019). Committee on statistics - Ecological indicators of environmental monitoring and assessment. Ministry of National Economy of Republic of Kazakhstan. Retrieved 10 March 2019 from http://stat.gov.kz/faces/homePage.

NCSD of KG (2018). Draft of the National Sustainable Development Strategy for the Kyrgyz Republic for the period of 2018–2040 [Проект Национальной Стратегии Устойчивого Развития Кыргызской Республики на 2018–2040]. National Council for Sustainable Development (NCSD) of the Kyrgyz Republic. Retrieved 10 March 2019 from www.president.kg/files/docs/Files/proekt_strategii_final_russ.pdf.

Qadir, M., Schubert, S., Oster, J. D., Sposito, G., Minhas, P. S., Cheraghi, S. A., Murtaza, G., Mirzabaev, A., & Saqib, M. (2018). High-magnesium waters and soils: Emerging environmental and food security constraints. *Science of The Total Environment*, 642, 1108–1117.

Rakhmatullaev, S., Abdullaev, I., & Kazbekov, J. (2017). Water-energy-food-environmental nexus in Central Asia: from transition to transformation. In S.S. Zhiltsov, I.S. Zonn, A.G. Kostianoy, & A.V. Semenov (eds), *The Handbook of Environmental Chemistry n 55: Water Resources in Central Asia: International Context*. Berlin; Heidelberg: Springer, p 1–18.

Rakhmatullaev, S., Huneau, F., Kazbekov, J., Le Coustumer, P., Jumanov, J., El Oifi, B., Motelica-Heino, M., & Hrkal, Z. (2010). Groundwater resources use and management in the Amu Darya river basin (Central Asia). *Environmental Earth Sciences*, 59(6), 1183–1193.

Reimov, P. & Fayzieva, D. (2014). The present state of the south Aral Sea area. In Micklin, P., Aladin, N. V. & Plotnikov, I. (eds), *The Aral Sea: The Devastation and Partial Rehabilitation of a Great Lake*. Berlin: Springer, pp 171–206.

Sachs, J., Schmidt-Traub, G., Kroll, C., Lafortune, G., & Fuller, G. (2018). *SDG Index and Dashboards Report 2018*. New York: Bertelsmann Stiftung and Sustainable Development Solutions Network (SDSN).

SCNP of UZ (2013). National report on the state of the environment and natural resource management of the Republic of Uzbekistan (2008–2011) [Национальный доклад о состоянии окружающей среды и использовании природных ресурсов в Республике Узбекистан (2008–2011 гг.)]. State Committee for Nature Protection of the Republic of Uzbekistan.

Sokolov, V. (2006). Experiences with IWRM in the Central Asia and Caucasus Regions, *Water International*, 31(1), 59–70.

STEPF of KG (2016). National report on the state of the environment of the Kyrgyz Republic in 2011–2014 [Национальный доклад о состоянии окружающей среды Кыргызской Республики за 2011–2014 годы]. State Agency on Environment Protection and Forestry of the Kyrgyz Republic, Bishkek, Kyrgyzstan.

Törnqvist, R., Jarsjö, J., & Karimov, B. (2011). Health risks from large-scale water pollution: trends in Central Asia. *Environment International*, 37(2), 435–442.

UN (2015). Transforming Our World: The 2030 Agenda for Sustainable Development. Resolution adopted by the General Assembly on 25 September 2015 (A/RES/70/1). New York, USA.

UN (2018). Global Indicator Framework for the Sustainable Development Goals and Targets of the 2030 Agenda for Sustainable Development (A/RES/71/313). Resolution adopted by the General Assembly on Work of the Statistical Commission pertaining to the 2030 Agenda for Sustainable Development. Annex II. Annual refinements (E/CN.3/2018/2). New York, USA.

UN Environment (2018a). *Progress on Ambient Water Quality: Piloting the Monitoring Methodology and Initial Findings for SDG Indicator 6.3.2.* Published by UN Environment on behalf of UN-Water. New York, USA.

UN Environment (2018b). *Progress on Integrated Water Resources Management. Global Baseline for SDG 6 Indicator 6.5.1: Degree of IWRM Implementation.* Published by UN Environment on behalf of UN-Water. New York, USA.

UN Environment (2018c). *Progress on Water-Related Ecosystems: Piloting the Monitoring Methodology and Initial Findings for SDG Indicator 6.6.1.* Published by UN Environment on behalf of UN-Water. New York, USA.

UNDP (2008). *Review of Donor Assistance in the Aral Sea Region (1995–2005).* UNDP, Tashkent, Uzbekistan.

UNECE (2012). *Turkmenistan. Environmental Performance Reviews. First Review.* United Nations Economic Commission for Europe. New York; Geneva: UN.

UNECE (2017). *Tajikistan. Environmental Performance Reviews. Third Review.* United Nations Economic Commission for Europe. New York; Geneva: UN.

UNECE/UNESCO (2018). *Progress on Transboundary Water Cooperation - Global Baseline for SDG Indicator 6.5.2.* Published by United Nations and UNESCO on behalf of UN-Water. Paris, France.

UNSD (2019). Global SDG Indicators Database. *UN Statistics Division.* Retrieved 10 March 2019 from https://unstats.un.org/sdgs/indicators/database/.

UN-Water (2013). UN-Water country brief – Kyrgyzstan. AQUASTAT Programme of the FAO on behalf of UN-Water. Retrieved 10 March 2019 from https://www.zaragoza.es/contenidos/medioambiente/onu/1026_eng_res6_unwater_briefs_kyrgyzstan.pdf.

UN-Water (2017). *Global Analysis and Assessment of Sanitation and Drinking Water (GLAAS) 2017 Report: Financing Universal Water, Sanitation and Hygiene under the Sustainable Development Goals.* Geneva: World Health Organization. Retrieved from www.who.int/water_sanitation_health/monitoring/investments/glaas-2016-2017-cycle/en/.

Usha, K. B. (2016). The Aral Sea crisis in Central Asia: environment, human security and gender concerns. *IUP Journal of International Relations, 10*(2), 7–29.

Uvarov, D., Groll, M., & Opp, C. (2015). Integrated water resources management in Kazakhstan: status quo and challenges. *Geographica Augustana, 17,* 35–43.

Wang, G., Mang, S., Cai, H., Liu, S., Zhang, Z., Wang, L., & Innes, J. L. (2016). Integrated watershed management: evolution, development and emerging trends. *Journal of Forestry Research, 27*(5), 967–994.

WHO & UN-HABITAT (2018). *Progress on Safe Treatment and Use of Wastewater: Piloting the Monitoring Methodology and Initial Findings for SDG Indicator 6.3.1.* Geneva: World Health Organization and UN-HABITAT.

WHO (2012). *Number of Deaths by Cause.* Global Health Observatory Data Repository. Geneva: World Health Organization.

WHO/UNICEF (2019). Joint Monitoring Programme (JMP) for Water Supply, Sanitation and Hygiene. Retrieved 10 March 2019 from https://data.worldbank.org/indicator/.

Word Bank (2006). *Water for Growth and Development.* A Theme Document of the 4th World Water Forum. Retrieved from http://siteresources.worldbank.org/INTWRD/Resources/FINAL_0601_SUBMITTED_Water_for_Growth_and_Development.pdf.

Xu, H. (2017). The study on eco-environmental issue of Aral Sea from the perspective of sustainable development of Silk Road economic belt. *IOP Conference Series: Earth and Environmental Science, 57*(1), 1–9.

Zhuang, W. (2016). Eco-environmental impact of inter-basin water transfer projects: a review. *Environmental Science and Pollution Research, 23*(13), 12867–12879.

# INDEX

Printed and bound by CPI Group (UK) Ltd, Croydon, CR0 4YY

24/10/2024

01778292-0001